Grasses occupy a greater area of the world's land surface than any other plant family, occurring in almost every terrestrial environment and providing a vital source of food for humans and animals. This volume presents the most recent information on their population biology, bringing together contributions from researchers studying both applied and fundamental aspects of this important group of plants. Demographic, physiological, ecological and molecular approaches to understanding grass populations are considered in relation to reproduction and to aspects of life history patterns such as dispersal, germination, seedling establishment, population dynamics and reproduction. Other areas covered include the role of genetic variation and phenotypic plasticity in shaping life history traits, the impact of biotic factors, and the ecology of specific species in major grass-dominated ecosystems in Africa, Australia and Japan. Given the importance of grasses in many ecosystems, this book contains much that will be of interest to botanists, ecologists and conservationists throughout the world.

T0275782

POPULATION BIOLOGY OF GRASSES

POPULATION BIOLOGY OF GRASSES

Edited by

G. P. CHEPLICK PhD
College of Staten Island, City University of New York

CAMBRIDGE
UNIVERSITY PRESS

PUBLISHED BY THE PRESS SYNDICATE OF THE UNIVERSITY OF CAMBRIDGE
The Pitt Building, Trumpington Street, Cambridge CB2 1RP, United Kingdom

CAMBRIDGE UNIVERSITY PRESS
The Edinburgh Building, Cambridge CB2 2RU, United Kingdom
40 West 20th Street, New York, NY 10011–4211, USA
10 Stamford Road, Oakleigh, Melbourne 3166, Australia

© Cambridge University Press 1998

This book is in copyright. Subject to statutory exception
and to the provisions of relevant collective licensing agreements,
no reproduction of any part may take place without
the written permission of Cambridge University Press.

First published 1998

Typeset by 10/13 Times New Roman [SE]

A catalogue record for this book is available from the British Library

Library of Congress Cataloguing in Publication data

Population biology of grasses / edited by G. P. Cheplick.
   p. cm.
Based in part on a symposium held on 8 August 1995 at the annual
meeting of the Botanical Society of America in San Diego,
California.
Includes index.
ISBN 0-521-57205-3 (hardcover)
1. Grasses–Congresses. 2. Plant populations–Congresses.
I. Cheplick, G. P. (Gregory Paul), 1957–
QK495.G74P755 1998
584′.91788–dc21 97–26257 CIP

ISBN 0 521 57205 3 hardback

Transferred to digital printing 2002

Contents

Contributors

M. O. Aguilera, Instituto Nacional de Tecnología Agropecuaria, San Luis RA-5730, V. Mercedes, San Luis, Argentina

C. C. Baskin, School of Biological Sciences, University of Kentucky, Lexington KY 40506-0225, USA

J. M. Baskin, School of Biological Sciences, University of Kentucky, Lexington KY 40506-0225, USA

A. D. Bradshaw, School of Biological Sciences, University of Liverpool, Liverpool L69 3BX, UK. Present address: 58 Knowsley Road, Liverpool L19 0PG, UK

D. D. Briske, Rangeland Ecology and Management, Texas A & M University, College Station TX 77843-2126, USA

G. P. Cheplick, Department of Biology, The College of Staten Island of the City University of New York, Staten Island, NY 10314, USA

K. Clay, Department of Biology, Indiana University, Bloomington IN 47405, USA

J. D. Derner, Rangeland Ecology and Management, Texas A & M University, College Station TX 77843-2126, USA

T. M. Everson, Department of Range and Forest Resources, University of Natal, Private Bag X01, Scottsville 3209, South Africa

E. Garnier, Centre d'Ecologie Fonctionelle et Evolutive (CNRS), 1919 Route de Mende, 34293 Montpellier Cedex 5, France

M. J. W. Godt, Department of Botany, University of Georgia, Athens GA 30602, USA

J. L. Hamrick, Departments of Botany and Genetics, University of Georgia, Athens GA 30602, USA

K. H. Keeler, School of Biological Sciences, University of Nebraska–Lincoln, Lincoln NE 68588-0118, USA

W. K. Lauenroth, Rangeland Ecosystem Department, Colorado State University, Fort Collins CO 80523, USA

A. Matika, Monuments and Sites Division, Agency for Cultural Affairs, Kasumigaseki 3-2-2, Chiyoda-ku, Tokyo 100, Japan

K. K. Newsham, Institute of Terrestrial Ecology, Monks Wood, Abbots Ripton, Huntingdon, Cambridgeshire PE17 2LS, UK. Present address: British Antarctic Survey, High Cross, Madingley Road, Cambridge CB3 0ET, UK

T. G. O'Connor, Department of Range and Forest Resources, University of Natal, Private Bag X01, Scottsvile 3209, South Africa

D. M. Orr, Queensland Department of Primary Industries, Tropical Beef Centre, PO Box 5545, Rockhampton Mail Centre, Queensland 4702, Australia

J. A. Quinn, Department of Ecology, Evolution and Natural Resources, Rutgers University, New Brunswick NJ 08903-0231, USA

A. R. Watkinson, Schools of Biological and Environmental Sciences, University of East Anglia, Norwich NR4 7TJ, UK

S. D. Wilson, Department of Biology, University of Regina, Regina, Saskatchewan S4S 0A2, Canada

Preface

While there has been considerable coverage of the evolution, systematics, and reproductive biology of the grasses and numerous studies of grassland communities and ecosystems, information on the population biology of this economically important plant family has not been recently synthesized. This volume arose from the recognition of this contention and is, in part, based upon a symposium on the 'Population Biology of Grasses' held on 8 August, 1995 at the annual meeting of the Botanical Society of America in San Diego, California. Most participants agreed that a volume on this topic would be timely and of interest to a variety of botanists and ecologists throughout the world. A number of the original participants have contributed chapters to the present volume.

In this book, demographic, physiological, ecological, and molecular approaches to understanding grass populations are considered in relation to life history patterns and reproductive biology. Relevant aspects of grass life histories included are dispersal, germination, seedling establishment, population dynamics, and reproduction. Critical to understanding the microevolution of grass populations is an analysis of the role of genetic variation and phenotypic plasticity in shaping life history traits. From an ecological perspective, the widespread nature of the interactions of other plants and fungi with grasses underscores the importance of considering the impact of biotic factors on the population biology of grasses. All of these aspects of population biology are explored in various chapters in the present volume. In addition, other chapters employ a population perspective to focus on the ecology of specific species in major grass-dominated ecosystems in Africa, Australia, and Japan.

This book would not have been possible without the tremendous commitment of the various contributors. I thank them all for sending their manuscripts to me in a timely manner and for taking the time to carefully revise and edit their original submissions. The chapters have benefited greatly from the many comments and suggestions offered by the following: L. W. Aarssen, E. B. Allen, C. C. Baskin, J. M. Baskin, R. E. J. Boerner, A. D. Bradshaw, J. M. Bullock, M. L. Cain, J. R. Estes, R. C. Jackson, W. K. Lauenroth, G. C. Lewis, S. J. Novak, M. A. Parker, E. G. Reekie, B. Schmid, A. F. Schnabel, B. Shipley, G. M. Simpson, R. J. Soreng, O. W. van Auken, J. White, and S. D. Wilson. I deeply appreciate the time and effort these scientists put into reviewing early versions of the manuscripts, some of which were quite lengthy. I am also grateful to James White, University College, Dublin, for his suggestions and comments on the original book proposal.

It is hoped that this volume will help bring together ideas from researchers with both basic and applied perspectives who have used a diversity of approaches in their attempt to understand the ecology and evolution of grass populations. It would appear that, although the bridge between basic and applied plant population biology may be rather narrow, it is one bridge that grass researchers should easily be able to cross!

G.P.C.
Staten Island
February 1997

Darwin revisited: approaches to the ecological study of grasses

ANTHONY BRADSHAW

At the beginning of a book on the ecological behaviour of an important and distinctive family of plants, it is not an entirely stupid thing to contemplate and analyse what we are doing – not so much *what* we are trying to do as *how* we are trying to do it. It is possible that by such an exercise we can obtain a clearer idea of the approaches most likely to yield valuable results.

This book is about the many different properties of grass populations and species, and the differences between them. But why should such a collection of studies be put together? What value can come from it? Grasses are, of course, interesting to many people for their own sake. But they also occupy a greater area of the world's land surface than any other plant family. As both food for grazing animals, and as grain crops, the grass family provides more of our food than any other family. Unlike many other families, grasses are found in almost every environment.

A common background

But at the same time grasses have a consistent, almost curious morphology and internal structure – in particular, the leaves and the division of the stem into nodes and internodes – which give both a considerable uniformity to the whole family and a series of special properties. With the nodal structure is associated an ability to produce both roots and shoots (either aboveground tillers or underground rhizomes) freely. Anyone who wishes to understand more about the uniqueness of the Gramineae should consult the remarkable book by Arber (1934).

The ability to produce shoots means that the grasses have considerable abilities to grow and propagate themselves vegetatively. Single individuals can form large clones of genetically identical individuals – ramets. These

1

offer grasses a whole alternative means of reproduction, which in many species and situations can be more important than normal sexual reproductions. Clones of *Holcus mollis* can spread over 200 m (Harberd, 1967). This production of ramets is also invaluable in experimental work, since single individuals can so easily be propagated. With their basic similarities and these special properties, grasses are, therefore, an excellent group of plants on which to test out ecological ideas and investigate subtleties of morphological and physiological characteristics and ecological behaviour.

The discipline of ecology

Ecology is the study of organisms in relation to their environment. The aim of ecological studies of particular species cannot therefore be just to describe their characteristics, but to see how these relate to the environments where they grow. The underlying assumption is that in certain ways they must show adaptation, in ways that are worth unravelling and understanding, even although it must be remembered that adaptation is a state arrived at by the evolutionary processes that went on before. Harper (1982) suggests that it might be better therefore to talk about 'abaptation', which serves to remind us that we may not find the perfect relationships to present conditions which we so often seem to expect.

What is important is the fundamental tenet of biology that the features shown by living organisms are the result of evolution, and therefore of natural selection. This suggests that we can use the actions of evolution as a tool to obtain a clearer picture of the relationships between characters of species and their environment.

The use of evolution in ecology

People tend to think that evolution is a slow process and that natural selection takes a long time to deliver new and adapted states. This is not the place to consider all the features and mechanisms of evolution. But it is worthwhile remembering that simple mathematics shows that, if there are differences in the fitnesses of genotypes within a population, evolutionary change is inevitable. If differences in fitness are as much as 50%, enormous changes can take place in 10 generations – or less. There was a time when evolutionists were very chary of assuming such differences in fitness; Haldane (1932) assumed differences of 0.1%. But the accumulated evidence of ecologists is that differences in fitness of 50% are commonplace. Indeed, we take it for granted that there are many situations where, while one species or genotype survives, another is eliminated completely.

This, however, is using theoretical arguments. What can we find in practice? Does natural selection really deliver major differences at the level of populations? It is not always easy to see what is happening unless we have an easily recognizable character. There are many to choose from. In plants, metal tolerance has perhaps been one of the most explored, particularly in grass species, not only because of its occurrence in these species but also because of the ease of handling both populations and individual genotypes. A good example of what has been possible is the work of Wu, Thurman & Bradshaw (1975) on the populations of *Agrostis stolonifera* around a copper smelter.

The sum total of these studies (Shaw, 1990) on many different species and populations growing within, and outside, metal-contaminated sites, has shown that:

1. population differences can be so great that populations cannot survive at all in the reciprocal alternative habitats;
2. these differences can occur between populations less than 10 m apart, even in outbreeding species;
3. these differences can develop in the field in less than 10 generations;
4. under experimental conditions major population differences can be brought about in only one generation;
5. the selection can operate on individual genotypes, at both the seedling and established adult stages;
6. other characteristics can also change.

This suggests that we can get a handle on what evolution can do by comparing populations of the same species coming from different environments, even when the time-scale is not known, or is quite short. Such studies, referred to as experimental taxonomy, ecological genetics or genecology, initiated almost a century ago, have been remarkably rewarding. They have demonstrated the adaptive significance of innumerable different characteristics of plants, and the way whole constellations of characters can be selected at the same time. It is fashionable nowadays to think that work carried out more than 20 years ago is out of date, yet the detailed pioneering work of Turesson (1922) and Clausen, Keck & Hiesey (1940) deserves re-examination from an ecological point of view.

Perhaps it is the work of Clausen *et al.* (1940, 1948) which is particularly interesting because it involved careful reciprocal transplants by which the real adaptive significance of the differences found between populations could be demonstrated. Suffice it to say that the differences were considerable. Yet this work was carried out under open, spaced-plant conditions without other species being able to have any influence. It remained for people such as Snaydon and his co-workers to assess populations, of a grass species *Anthoxanthum odoratum*, under fully competitive conditions, making use of its great ease of vegetative propagation (Davies & Snaydon, 1976). This work has revealed remarkable differences in fitness in reciprocal transplants of populations originating only 20 m apart in the plots of the Park Grass Experiment, in the liming treatments started less than 60 years previously.

It is impossible to do justice to the characteristics on which natural selection has been shown to have had effect in plant populations. Grasses have had a full share of attention, e.g. morphological and adaptive differences in *Agrostis capillaris* (Bradshaw, 1959), startling, localized differences in morphology and salt tolerance in *Agrostis stolonifera* (Aston & Bradshaw, 1966), differences in life history characteristics in widespread prairie grasses (McMillan, 1959), and major differences in nutrient response in *Festuca ovina* (Snaydon & Bradshaw, 1961). The early work has shown that life history characters could be influenced by selection. The detail of what can occur has been investigated by Law in *Poa annua* (Law, Bradshaw & Putwain, 1977). Completely different patterns of growth exist within different populations of this species, and can evolve over short periods.

We are now rightly becoming more interested, not in the static characteristics of species, but in their ability to respond to environmental variation. Such phenotypic plasticity, or lack of it, reveals that plants in their own different way can show just as interesting features of behaviour as animals. Indeed because plants cannot run away from environmental troubles, but must endure them, such plasticity becomes of considerable adaptive significance (Sultan, 1987). It can take many forms, from an ability, or not, to elongate plant parts with shading in different *Potentilla* species (Huber, 1996), to differences in stability of grain yield in relation to drought in cultivars of barley (Finlay & Wilkinson, 1963). Plasticity is assisted greatly in grasses by their powers of clonal growth. But this does not mean that this plasticity is uniformly manifested; in *Amphibromus scabrivalvis*, for instance, plasticity of tillering, root growth and general growth rate all show genotypic variation (Cheplick, 1995).

The reason for describing the variety of characters, and character combinations, that show evolutionary differentiation within species, is that this indicates their adaptive importance. If a character has been altered by natural selection it must, *a priori*, be of adaptive significance and deserve our attention. Unfortunately the reverse does not, however, necessarily follow. Because a character does not show evolutionary differentiation it does not have to mean that it is not of adaptive significance. Its evolution may be restricted by lack of suitable variation (Bradshaw, 1991).

Some thoughts on methodology

Evolutionary differentiation is also a particularly valuable ecological tool from a more methodological standpoint. It is difficult to make sensible progress in ecology, or indeed in almost any aspects of biology, without making comparisons – the significance of one observation can only become apparent when it can be compared with another (Bradshaw, 1987). Sometimes that 'other' is just past experience; but in critical work another treatment, or other different genetic material, must be involved. At least two observations are required to make one comparison. In statistical terms we are all familiar with the fact that the number of independent comparisons, the *degrees of freedom*, is one less than the number of observations.

Traditional ecology very often compares whole species. This is a legitimate approach, which has served us well in revealing differences and their ecological consequences. But species differences cannot always be related critically to environment. Among the differences that occur between two species, many can be ancestral and not related to fitness in the present environment. They can be an outcome of aspects of the genetic make-up of the species quite unrelated to present conditions. This is most obvious with differences between families.

Population differences are quite different. These represent what the environment, through natural selection, has done in material which in other respects has a common genetic background. They are analogous to a field experiment carried out in carefully selected uniform conditions, in which the effects of a single treatment, or a series of treatments, are compared. The differences that are found must be adaptively important otherwise they would not have accumulated. There is the possibility that some differences may have accumulated by correlated response, due to pleiotropy or linkage, but the whole relationship between the differences shown by the material and the environment is much more simple and direct.

Some investigators may feel that population differences are likely to be too small to merit investigation, and feel more comfortable comparing species. Indeed, a great deal of excellent work has been done by this means, as will be obvious from the contributions to this book. But the arguments for the study of populations seem clear. Major differences exist between populations within species, certainly within grasses. If these are investigated with the sophistication of modern experimental ecology, including placing material in its natural habitats, in competitive situations simulating those natural environments in all their tough complexity, or in physiologically precise conditions (e.g. Kik, Jongman & van Andel, 1991), what matters in the adaptation of plants to their environment can be shown with a direct simplicity.

Finally, there are individual genotypic differences. In ecological work these are often considered just a source of error, because any differences between individuals serve to blur the differences between species or populations. But genotypic differences represent the raw material on which natural selection can act, and deserve their own study. In many species such differences are difficult to work with because the individual genotype cannot be replicated and therefore cannot be represented in different treatments. In grasses, however, such replication is easy.

Grasses in ecological research

Grasses are admirable material in many respects. Do not let us forget their economic importance. But from an ecological point of view what matters is that, within their relatively circumscribed basic structure, they display a remarkable range of evolutionary adaptation, annuals to perennials, minute ephemerals to tall 'trees', aquatics to desert xerophytes. This range of variation can be found both between and within species, all the product of the sifting power of natural selection.

Whichever type of material is chosen, grasses provide opportunities for elegant and critical investigations of the relationships between plants and their environment, relationships which may sometimes appear to be simple and obvious, when in reality they are subtle and complex, as the many different contributions to this book show.

References

Arber, A. (1934). *The Gramineae*. Cambridge: Cambridge University Press.
Aston, J. & Bradshaw, A. D. (1966). Evolution in closely adjacent plant populations. II. *Agrostis stolonifera* in maritime habitats. *Heredity, London*, **21**, 649–64.
Bradshaw, A. D. (1959). Population differentiation in *Agrostic tenuis* Sibth. I. Morphological variation. *New Phytologist*, **58**, 208–27.
Bradshaw, A. D. (1987). Comparison – its scope and limits. *New Phytologist*, 106 (Suppl), 3–21.
Bradshaw, A. D. (1991). Genostasis and the limits to evolution. *Philosophical Transactions of the Royal Society, London*, B **333**, 289–305.

The ecological approach: Darwin revisited 7

Cheplick, G. P. (1995). Genotypic variation and plasticity of clonal growth in relation to nutrient availability in *Amphibromus scabrivalvis*. *Journal of Ecology*, **83**, 459–68.

Clausen, J., Keck, D. D. & Hiesey, W. M. (1940). *Experimental studies on the nature of plant species. I. The effect of varied environments on western North American plants*. Carnegie Institute of Washington Publication 520.

Clausen, J., Keck, D. D. & Hiesey, W. M. (1948). *Experimental studies on the nature of plant species. III. Experimental responses of climatic races of Achillea*. Carnegie Institute of Washington Publication 581.

Davies, M. S. & Snaydon, R. W. (1976). Rapid population differentiation in a mosaic environment III Coefficients of selection. *Heredity, London*, **36**, 56–66.

Finlay, K. W. & Wilkinson, G. N. (1963). The analysis of adaptation in a plant breeding programme. *Australian Journal of Agricultural Research*, **14**, 742–54.

Haldane, J. B. S. (1932). *Causes of Evolution*. London: Longmans.

Harberd, D. J. (1967). Observation on natural clones in *Holcus mollis*. *New Phytologist*, **66**, 401–8.

Harper, J. L. (1982). After description. In *The Plant Community as a Working Mechanism*, ed. E. I. Newman, pp. 11–25. Oxford: Blackwell Scientific Publications.

Huber, H. (1996). Plasticity of internodes and petioles in prostrate and erect *Potentilla* species. *Functional Ecology*, **10**, 401–9.

Kik, C., Jongman, M. & van Andel, J. (1991). Variation in relative growth rate and survival in ecologically contrasting populations of *Agrostis stolonifera*. *Plant Species Biology*, **6**, 47–54.

Law, R., Bradshaw, A. D. & Putwain, P. D. (1977). Life history variation in *Poa annua*. *Evolution*, **31**, 233–46.

McMillan, C. (1959). The role of ecotypic variation in the distribution of the central grassland of North America. *Ecological Monographs*, **29**, 285–308.

Shaw, A. J. (1990) *Heavy Metal Tolerance in Plants: Evolutionary Aspects*. Boca Raton, Florida: CRC Press.

Snaydon, R. W. & Bradshaw, A. D. (1961). Differential responses to calcium within the species *Festuca ovina* L. *New Phytologist*, **60**, 219–34.

Sultan, S. E. (1987). Evolutionary implications of phenotypic plasticity in plants. *Evolutionary Biology*, **21**, 127–78.

Turesson, G. (1922). The genotypic response of the plant species to the habitat. *Hereditas, Lund*, **3**, 211–350.

Wu, L., Thurman, D. A. & Bradshaw, A. D. (1975). The potential for evolution of heavy metal tolerance in plants III The rapid evolution of copper tolerance in *Agrostis stolonifera*. *Heredity, London*, **34**, 165–87.

Part one

Population variation and life history patterns

1

Allozyme diversity in the grasses

MARY JO W. GODT AND J. L. HAMRICK

Since Lewontin & Hubby (1966) and Harris (1966) independently discovered allozyme polymorphisms within populations of *Drosophila* and humans, allozymes have provided the most widely used descriptors of genetic diversity in natural plant and animal populations. Their usefulness can be ascribed to several features, including codominant expression, apparent neutrality, and ease and cost of detection. Today allozymes are utilized in studies of plant populations to determine breeding systems, describe mating systems, determine paternity, estimate gene flow via seed and pollen, examine clonal structure, and determine systematic relationships (Soltis & Soltis, 1989). In conservation biology and crop science, the description of genetic structure provides sampling guidelines for the *ex situ* and *in situ* preservation of genetic diversity (e.g. Marshall & Brown, 1975; Brown & Briggs, 1991; Godt, Hamrick & Bratton, 1995; Swenson *et al.*, 1995; Godt & Hamrick, 1996). Furthermore, allozymes have been utilized to study the evolution of crop species and to identify cultivars (Doebley, 1989; Torres, 1989).

Although allozyme diversity does not necessarily reflect variation in quantitative traits, it provides a handy yardstick for comparisons of diversity within populations and across species. Over 1000 published studies report allozyme variation in seed plants. These represent a wide cross-section of the plant world, although commercially valuable species (i.e. trees and crops) and temperate species have received the most attention. As the number of allozyme studies has grown, several reviewers have described patterns in allozyme diversity and its distribution (Gottlieb, 1977; Brown, 1979; Hamrick, Linhart & Mitton, 1979; Crawford, 1983; Hamrick & Godt, 1989). Notably, these reviews have consistently shown that the distribution of genetic diversity is strongly associated with a plant's breeding system, with most of the genetic diversity in outbreeding species found within populations, while nearly half of the genetic diversity of selfing species is found among populations. Woody species tend to maintain more genetic diversity, whereas species with restricted geographic distributions harbour the least. Moreover, although patterns emerge when plant species are considered in their entirety, the ability to predict genetic diversity and its distribution in particular species remains somewhat elusive. One can only conclude that the evolutionary and ecological history of each species uniquely determines its genetic composition and structure.

The number of published allozyme studies now permits examinations of patterns of genetic diversity within several plant groups. In particular, some of the larger plant families (i.e. Asteraceae, Fabaceae and Poaceae) can be examined in some detail. This could provide further insights into genetic diversity patterns since the effects of phylogeny can be partially controlled. In this study we examine allozyme diversity within the grasses.

Materials and methods

Studies considered

In 1989, we surveyed the vascular plant literature and constructed a data base of allozyme studies that incorporated 653 studies of seed plants (Hamrick & Godt, 1989). These data form the foundation for the current review. Because we have since updated the data base for several groups of plants, the overall data are more comprehensive for some groups than for others. Specifically, the data base was updated and reviewed in 1990 for all plants (Hamrick, Godt & Sherman-Broyles, 1992), for tropical trees in 1993 (Hamrick, 1994) and for crop species in 1996 (Hamrick & Godt, 1997). For this review we updated the data base with respect to the grasses by adding studies published between 1990 and mid-1996. Although the inclusion of more studies from particular groups might be perceived to have biased the overall data set, the summary statistics for all plant species changed very little with the inclusion of additional information.

All studies that reported allozyme variation were considered for the data base. However, studies that reported electrophoretic phenotypes without genetic interpretation of the data could not be used. Some taxa have been analysed more than once; these studies are not complete duplications because they often represent studies by different labs, or by the same lab studying populations in different portions of the species' range or assessing different loci. Such species are represented in the data base more than once.

Genetic parameters

Genetic diversity statistics were extracted from or calculated for each study. These statistics describe genetic diversity within each species, within populations, and among populations within species. Within-population genetic diversity statistics represent population means, and thus are influenced by the distribution of genetic diversity among populations. (For example, within-population values will be relatively lower for species whose genetic diversity is distributed among, rather than within, populations.) For this reason, we introduced calculations that describe species' genetic diversity (Hamrick & Godt, 1989). Efforts have been made to standardize genetic statistics throughout the data base (i.e. parameters were re-calculated if we concluded the authors had used different criteria or equations). In addition, we calculated as many genetic statistics as possible from the data reported (e.g. tables of genotypes or allele frequencies were sometimes available even though the complete array of genetic statistics was not reported). Nonetheless, data were not available in every study to permit calculation of all genetic parameters; hence the number of studies utilized varied among parameters.

For the species and population means, we calculated (as shown in Table 1.1) percentage polymorphic loci (P), mean number of alleles per locus (A), effective number of alleles per locus (A_e), and genetic diversity (H_e; expected heterozygosity given Hardy–Weinberg assumptions). Parameters that are subscripted with an s indicate species values, whereas those subscripted with a p represent population means. Total genetic diversity (H_T) was calculated at polymorphic loci and partitioned into that found within populations (H_S) and among populations (D_{ST}). The proportion of total genetic diversity found among populations (G_{ST}) was then calculated. We also estimated gene flow (as measured by Nm, the number of migrants per generation) based on genetic structure (Table 1.1).

Species traits

Each species was categorized for the following traits: regional distribution (boreal-temperate, temperate or tropical), geographic range (widespread, regional, narrow or endemic), life form (annual or perennial), mode of reproduction (sexual or a mixture of asexual and sexual), breeding system (outcrossed, mixed-mating or selfed); seed dispersal mechanism (wind, explosive, animal-ingested or animal-attached, gravity or a mixture of gravity and animal-attachment), and successional status (early, mid or late). Information on the species' traits was obtained from the original

Table 1.1. *Calculations of genetic parameters*

The subscripts s and p indicate species level and population parameters
respectively.

A. *Percentage polymorphic loci*
P_s=the number of loci that exhibit two or more alleles in the study, divided by the
number of loci examined
P_p=the proportion of loci that exhibit two or more alleles within each
population, averaged across populations

B. *Mean number of alleles per locus*
A_s=the total number of alleles observed in the study, divided by the number of
loci examined
A_p=the number of alleles observed per locus within populations, averaged over
loci within populations, and averaged across populations

C. *Effective number of alleles per locus*
$A_{es}=1/(1-H_{es})$, where H_{es} is as defined below
$A_{ep}=1/\Sigma p_i^2$ where p_i is the frequency of the ith allele at a locus. Calculated for
every locus, averaged over loci and over populations

D. *Genetic diversity*
$H_{es}=1-\Sigma \bar{p}_i^2$ where $\bar{p}_i$ is the mean frequency of the ith allele, calculated for each
locus, and then averaged over loci
$H_{ep}=1-\Sigma p_i^2$ where p_i is the frequency of the ith allele in each population

E. *Among population diversity*
$H_T=1-\Sigma \bar{p}_i^2$, a measure of total genetic diversity, where $\bar{p}_i$ is the mean frequency
of the ith allele across all populations. H_T is calculated for each polymorphic
locus and averaged across these loci
$H_S=1-\Sigma p_i^2$, a measure of within-population genetic diversity, calculated for each
population where p_i is the frequency of the ith allele
$D_{ST}=H_T-H_S$, calculated for each polymorphic locus
$G_{ST}=D_{ST}/H_T$, calculated for each polymorphic locus, and averaged over loci

F. *Gene flow (assessed as Nm, the number of migrants per generation)*
$Nm=(1-G_{ST})/4G_{ST}$

---

study or from floras. When detailed trait information was unavailable, edu-
cated guesses were made based on such characteristics as seed shape, floral
morphology, or characteristics of closely related taxa. Some categories
within traits represent arbitrary divisions of a continuum (e.g. geographic
range) and thus placement of species into particular categories has been in
part subjective. For geographic range, we considered a species to be widespread if it occurred on more than one continent, regional if it occupied a
large area within a continent (e.g. the eastern US), narrow if it was found
in a more restricted area (e.g. the southeastern US) and endemic if it was
extremely localized (e.g. found only in the Florida panhandle). Species that

propagated themselves clonally by such means as rhizomes, stolons, or by agamospermy were categorized as having a mixed mode of reproduction.

Statistical analyses

The statistical analyses used paralleled those of Hamrick & Godt (1989) and Hamrick *et al.* (1992). For each species trait, means and standard errors were calculated. The analytical procedures of SAS (SAS Institute, Inc., 1987) were employed to analyse differences between categories of trait variables. Specifically, differences between categories were analysed using the general linear model (GLM) procedures of SAS coupled with a least squares means procedure (LSM/PDIFF) to test for pair-wise differences.

Results and discussion

Many allozyme studies of the Poaceae have been published but only a fraction of these provided data that could be used to describe allozyme diversity. Of 415 studies, about 57% or 237 studies contained data appropriate for this review. Forty-three genera and 143 species were represented in this data set. Not surprisingly, crop species and their close relatives were the focus of many studies.

Genetic diversity in grasses compared with other plants

In the data set, 16 or 17 allozyme loci were analysed per study for grasses as well as for other plant species (Table 1.2). The mean number of populations studied was higher for the grasses than for other plants (21 vs 12), most likely reflecting that studies of commercially valuable Poaceae often incorporate numerous accessions.

Within species and populations, a higher proportion of loci are polymorphic within the grasses and these loci have more alleles per locus with frequencies that are similar to those found in other plants (Table 1.2). Thus, overall genetic diversity is higher for grass species (H_{es}) and within grass populations (H_{ep}). However, grasses exhibit somewhat more genetic differentiation among populations (G_{ST}) than other species. About 27% of total genetic diversity is found among grass populations compared with 22% for other plant species.

Table 1.2. *Allozyme diversity in the grasses compared with other plants*

See Table 1.1 for description and calculation of genetic parameters. N is the mean
number of studies contributing to the calculation of the genetic parameters.

A. Species

Group	N	Mean no. populations	Mean no. loci	P_s	A_s	A_{es}	H_{es}
Grasses	161	21	16	60.0	2.38	1.27	0.191
Other plants	666	12	17	50.8	1.92	1.20	0.146

B. Within population

Group	N	Mean no. populations	Mean no. loci	P_p	A_p	A_{ep}	H_{ep}
Grasses	135	21	16	40.8	1.66	1.19	0.138
Other plants	684	12	17	34.1	1.52	1.15	0.112

C. Among populations

Group	N	Mean no. populations	Mean no. loci	H_T	H_S	G_{ST}
Grasses	140	21	16	0.340	0.243	0.272
Other plants	576	12	17	0.295	0.225	0.216

Associations between life history traits and genetic diversity

Life form

The effect of generation length on genetic diversity in plants is difficult to
predict. On one hand, mutations may accumulate at a faster rate within
populations of annuals because they cycle more rapidly than perennials.
On the other hand, long-lived perennials may experience more environmental variation within their life spans and may face selective pressures by
herbivores and other organisms whose populations are frequently evolving
at a more rapid rate. These potentially frequency-dependent selective pressures could lead to the maintenance of higher levels of genetic diversity
within populations of perennials.

Within the grasses, annual species have a significantly higher proportion
of polymorphic loci (65% vs 55%) and more alleles per locus (2.65 vs 2.12)
than perennials (Table 1.3). However, annual and perennial grasses do not
differ significantly in overall genetic diversity (H_{es}). Thus, the higher

Table 1.3. *Within-species allozyme diversity for grasses with different attributes.*

See Table 1.1 for description and calculation of genetic diversity parameters. N is the mean number of values contributing to P_s, A_s, A_{es} and H_{es}. (NS indicates non-significant differences; * indicates $P \leq 0.05$; ** indicates $P \leq 0.01$ and *** indicates $P \leq 0.001$). Values within a category followed by the same letter are not significantly different. The mean numbers of populations and loci analysed are given in Table 1.5. (Note that only two categories were statistically analysed under seed dispersal because of the small sample size for wind-dispersed seeds.)

Classification	N	P_s	A_s	A_{es}	H_{es}
Life form		*	**	NS	NS
Annual	80	65.0a	2.65a	1.28a	0.193a
Perennial	80	55.2b	2.12b	1.26a	0.190a
Geographic range		*	*	NS	NS
Endemic or narrow	36	56.9ab	2.16a	1.28a	0.204a
Regional	60	54.2a	2.13a	1.21a	0.165a
Widespread	65	66.8b	2.72b	1.32a	0.209a
Regional distribution		**	***	NS	NS
Boreal or temperate	128	56.6a	2.16a	1.27a	0.187a
Tropical or tropical-temperate	33	73.4b	3.14b	1.29a	0.208a
Breeding system		**	*	NS	**
Selfing	79	51.4a	2.11a	1.24a	0.162a
Mixed-mating	14	66.7b	2.15ab	1.36a	0.256b
Outcrossing	69	69.1b	2.69b	1.29a	0.212b
Seed dispersal		**	***	NS	NS
Gravity	84	54.5a	1.99a	1.25a	0.176a
Attached or gravity-attached	73	67.1b	2.88b	1.30a	0.211a
Wind	4	42.0	1.69	1.19	0.156
Mode of reproduction		**	**	NS	NS
Sexual	104	64.5a	2.57a	1.28a	0.196a
Sexual and asexual	57	51.7b	2.03b	1.26a	0.182a
Successional status		***	*	**	**
Early	131	64.7a	2.48a	1.30a	0.207a
Mid	30	40.3b	1.96b	1.15b	0.118b

number of alleles per locus within annual species must be due to more low frequency alleles. The occurrence of these rare alleles could be due in part to the electrophoretic analysis of many accessions and landraces of annual crop species (Hamrick & Godt, 1997). It could also reflect the more complete reporting of allelic diversity in species such as annual crops where there is concern for germplasm conservation (alleles found at frequencies of less than 0.01 are often not reported). Similar trends in genetic diversity appear comparing within-population diversity of annual and perennial grasses, although the differences are not significant (Table 1.4). These results do not differ substantially from patterns found when all seed plant species are considered. Within species and populations annual herbs have a higher proportion of variable loci and more genetic diversity than perennial herbs (Hamrick & Godt, 1989).

Geographic range

Species with large geographic ranges may experience a broader range of biotic and abiotic conditions that could impose varied selective pressures on populations and lead to higher genetic diversity within species. In addition, species with widespread distributions often (but not always) consist of more individuals than species with more restricted distributions. Thus overall effective population sizes of such species may be higher, leading to predictions of more genetic diversity within widespread species.

In the data set, only four grass species were categorized as 'endemics'; we pooled these species with narrowly distributed species. Most grasses had regional or widespread distributions. Genetic diversity at the species level (H_{es}) did not differ significantly among geographic range categories, although widespread species had a significantly higher number of alleles per locus and regional species had a significantly lower proportion of polymorphic loci (Table 1.3). Within populations, however, significant differences were observed among species with different geographic ranges, with widespread species having a higher proportion of polymorphic loci that resulted in higher overall genetic diversity (Table 1.4). The lack of differences between geographic range categories at the species level and the occurrence of significantly higher levels of genetic diversity within populations of widespread species indicate that widespread species exhibit less genetic divergence among their populations.

Differences between species with different geographic ranges were more striking when all plants were considered (Hamrick & Godt, 1989). In this case, widespread species had nearly twice the genetic diversity of endemics,

Table 1.4. *Within-population allozyme diversity for grasses with different attributes*

See Table 1.1 for description and calculation of genetic diversity parameters. N is the mean number of values contributing to P_p, A_p, A_{ep} and H_{ep}. (NS indicates non-significant differences; * indicates $P \leq 0.05$; ** indicates $P \leq 0.01$ and *** indicates $P \leq 0.001$). Values within a category followed by the same letter are not significantly different. The mean numbers of populations and loci analysed are given in Table 1.5. (Note that only two categories were statistically analysed under seed dispersal because of the small sample sizes for wind-dispersed seeds.)

Classification	N	P_p	A_p	A_{ep}	H_{ep}
Life form					
		NS	NS	NS	NS
Annual	80	44.2a	1.73a	1.21a	0.147a
Perennial	56	36.1a	1.56a	1.16a	0.125a
Geographic range					
		**	***	NS	**
Endemic or narrow	23	38.3a	1.44a	1.15a	0.122a
Regional	53	31.8a	1.45a	1.14a	0.104a
Widespread	59	50.6b	1.92b	1.24a	0.174b
Regional distribution					
		NS	NS	NS	NS
Boreal or temperate	114	40.0a	1.65a	1.18a	0.134a
Tropical or tropical-temperate	22	44.8a	1.74a	1.20a	0.162a
Breeding system					
		***	**	NS	***
Selfing	80	33.4a	1.51a	1.15a	0.109a
Mixed-mating	9	54.2b	2.03b	1.30a	0.218b
Outcrossing	47	51.0b	1.83b	1.23a	0.175b
Seed dispersal					
		*	NS	NS	*
Gravity	67	36.3a	1.60a	1.15a	0.117a
Attached or gravity-attached	66	45.5b	1.72a	1.22a	0.160b
Wind	3	34.9	1.05	1.16	0.131
Mode of reproduction					
		**	NS	NS	*
Sexual	96	44.7a	1.72a	1.20a	0.150a
Sexual and asexual	40	31.5b	1.52a	1.14a	0.108b
Successional status					
		**	**	*	**
Early	115	44.1a	1.74a	1.21a	0.150a
Mid	21	22.9b	1.28a	1.07a	0.066b

and 130% to 150% the mean genetic diversity of regional and narrowly distributed species, respectively (Hamrick & Godt, 1989).

Regional distribution

Terrestrial plant and animal species diversity tends to increase as one moves towards the Equator. Thus, terrestrial tropical species may experience a more biotically diverse environment compared with temperate species. This observation leads to the question of whether an increase in biotic diversity is accompanied by more genetic diversity within tropical species or their populations.

For this analysis, we pooled the nine boreal grasses analysed with the temperate species. Only 16 grass species were classified as strictly tropical; these were pooled with species whose ranges spanned temperate and tropical regions. Although cold temperate species had a lower percentage polymorphic loci and fewer alleles per locus, this difference did not translate into significantly lower genetic diversity at the species level (Table 1.3). Within populations no trends in genetic diversity were apparent between the two groups (Table 1.4). In contrast, when all plants were considered, species in boreal-temperate regions had significantly higher genetic diversity than those in other regions (Hamrick & Godt, 1989). This observation was probably due to the large number of woody species in the north temperate data set. Woody species tend to maintain high levels of genetic diversity (Hamrick *et al.*, 1992). Because relatively few tropical trees have been analysed, the overall data set is somewhat biased with regard to species composition in temperate vs tropical regions.

Breeding system

Marked differences were found between species with different breeding systems in earlier reviews of the allozyme literature (Hamrick *et al.*, 1979; Hamrick & Godt, 1989). However, these comparisons incorporated a wide range of taxa. Inherent biases in the data could have led to these results. For example, tree species have high levels of genetic diversity, yet there are very few (if any) predominantly selfing trees. This data set may present the most balanced comparison of species with different breeding systems to date because the number of selfers and outcrossers analysed is fairly equal. More importantly, in this study we partially control for phylogeny.

Within the grasses analysed, 17 species were classified as having mixed-mating systems; most were classified as selfers or outcrossers. Selfing species were genetically depauperate relative to mixed-mating and out-crossing species, both within species and populations. For selfers, genetic diversity at the species level (H_{es}) was 76% of the mean value found for out-crossers (Table 1.3) and 62% of the within population genetic diversity (H_{ep}) of outcrossers (Table 1.4). Selfing grasses typically had fewer poly-morphic loci and fewer alleles at those loci than species with more open breeding systems. The concordance of these observations within a taxo-nomic family indicates that the general observation of less genetic diversity within all selfing plant species is robust.

One explanation for diminished levels of genetic diversity within selfing grasses is the occurrence of coadapted gene complexes associated with different environmental conditions (Clegg & Allard, 1972; Hamrick & Allard, 1972; Hamrick & Holden, 1979). The selection of such gene com-plexes could lead to diminished levels of within-population diversity but cannot account for diminished genetic diversity at the species level.

A more comprehensive explanation for the occurrence of lower levels of genetic diversity in selfing species is as follows. Novel mutations are less likely to spread among populations of selfers compared with outcrossers since the genes of outcrossers are routinely spread via pollen and seed whereas selfers must rely primarily on seed movement. Furthermore, most selfing species are annuals, whose population sizes often fluctuate. Thus, the low likelihood of gene flow and the increased possibility of population extinction may lead to less frequent incorporation of new mutations into the overall gene pool of selfing species (Hamrick & Nason, 1996). An addi-tional factor is that individuals within even highly polymorphic selfing species are largely homozygous; thus, new populations may be founded by individuals carrying little genetic diversity. An alternative explanation for low genetic diversity within selfers is that many selfing species may be derived from outcrossing relatives. Because the initial number of individu-als developing reproductive isolation from the parental outcrossers may be low (e.g. one selfing individual can self-perpetuate), selfing species may acquire only a fraction of the genetic diversity present in their progenitors.

Seed dispersal

Grasses do not have the variety of seed dispersal mechanisms found in plants at large (but see Cheplick, this volume). To our knowledge, none of the grasses included in this study is adapted for dispersal by ingestion nor do any have explosive seed dispersal mechanisms. In this data set, four species were categorized as wind-dispersed and 11 as being dispersed both by attachment to animals and by gravity. Means for the wind-dispersed species were calculated, but these species were not included in statistical analyses of seed dispersal (e.g. Tables 1.3 and 1.4). The grasses dispersed by a combination of animal attachment and gravity were arbitrarily pooled with the species that were primarily dispersed via animal attachment. The result was fairly equitable numbers for species that were dispersed by some form of attachment, and those that had no specialized means of dispersal (gravity-dispersed). Species with the potential for higher gene flow via attachment tended to maintain more genetic diversity, and this was significant for the percentage polymorphic loci (P_s and P_p), the species' mean number of alleles per locus (A_s) and mean population genetic diversity (H_{ep}; Tables 1.3 and 1.4). This trend is consistent with observations made over all plant species (Hamrick & Godt, 1989).

Mode of reproduction

Nearly all vegetatively reproducing species also reproduce sexually, although sexual reproduction may be infrequent. The question we addressed was whether species with the capacity for clonal spread differ genetically from those that spread primarily through seed production. In the most comprehensive review of the plant allozyme literature, no significant differences in overall genetic diversity were found between taxa with these traits (Hamrick & Godt, 1989). Using a carefully selected but limited data set Ellstrand & Roose (1987) also suggested that clonally spreading species maintain as much genetic diversity as sexually reproducing species. This question is probably best addressed by controlling for phylogeny, which this data set accomplishes, at least partially.

Sexually-reproducing grasses had a higher percentage polymorphic loci and mean number of alleles per locus within species but did not differ in genetic diversity (H_{es}) from species that reproduced both sexually and asexually (Table 1.3). However, sexually reproducing grasses did have higher within-population diversity (H_{ep}=0.150 vs 0.108; Table 1.4). This could be due to lower effective population sizes within clonally reproducing species. Little or no sexual reproduction or recruitment within established stands of clonally reproducing species is a common observation (Eriksson, 1993).

Successional status

All the grass taxa were classified as early or mid-successional (four late-successional species were pooled with the mid-successional species). Early successional species had a higher percentage polymorphic loci, more alleles per locus and higher effective numbers of alleles per locus within their species and populations (Tables 1.3 and 1.4). This resulted in about a two-fold difference in genetic diversity (H_{es} and H_{ep}) between early colonizing grasses and mid-successional grasses. These differences could be due to the high number of crop species (considered early successional) in the data (Hamrick & Godt, 1997).

Species vs population genetic diversity

For the species analyses, significant differences in P_s and A_s were found among all seven life history traits whereas only one trait exhibited differences for A_{es} and two for H_{es}. Thus, differences in the percentage polymorphic loci and mean number of alleles per locus frequently did not translate into greater mean effective number of alleles per locus, or more genetic diversity (H_{es}). In contrast, mean population differences in the percentage polymorphic loci and mean number of alleles per locus frequently translated into significant differences in mean population diversity (H_{ep}). These differences between species and mean population analyses reflect in part the fact that rare alleles can have a large influence on P and A at the species level without corresponding changes in genetic diversity (H_{es}). While this is true for population analyses also, within-population measures represent means, and rare alleles will have less influence on the overall P and A values. Thus, it is more generally true that differences in P and A within-populations result in differences in genetic diversity (H_{ep}).

Genetic structure

Only two traits (geographic range and breeding system) had significant influences on the partitioning of genetic diversity within and among grass populations (Table 1.5). Grasses that occur on more than one continent and endemics or narrowly distributed species had less genetic divergence among their populations than species with regional distributions. *A priori*, we expect more divergence among populations of widespread species compared with species having small distributions, unless the ability of species to exchange genes among their populations is positively correlated with their geographic ranges. Thus, the occurrence of less divergence among populations of grasses with endemic or narrow ranges compared with regionally distributed species is consistent with predictions. In contrast, the low level of divergence among widespread species is contrary to expecta-

Table 1.5. *Distribution of genetic diversity among populations for grasses with different attributes*

NS indicates non-significant differences; * indicates $P \leq 0.05$; ** indicates $P \leq 0.01$ and *** indicates $P \leq 0.001$). Values within a category followed by the same letter are not significantly different.

Classification	N	Mean no. populations	Mean no. loci	H_T	H_S	G_{ST}
Life form				NS	NS	NS
Annual	83	34	18	0.327a	0.236a	0.290a
Perennial	58	7	13	0.355a	0.249a	0.252a
Geographic range				NS	**	***
Endemic or narrow	19	8	15	0.325a	0.242ab	0.231a
Regional	52	11	17	0.345a	0.183a	0.398b
Widespread	69	35	15	0.337a	0.285b	0.190a
Regional distribution				NS	NS	NS
Boreal-temperate	112	19	15	0.348a	0.245a	0.275a
Tropical or temperate-tropical	28	32	21	0.300a	0.228a	0.270a
Breeding system				NS	***	***
Selfing	73	35	18	0.329a	0.192a	0.415a
Mixed-mating	13	7	10	0.406a	0.355b	0.156b
Outcrossing	53	10	14	0.329a	0.286b	0.112b
Seed dispersal				NS	NS	NS
Gravity	64	11	16	0.336a	0.218a	0.315
Attached or gravity-attached	73	32	15	0.341a	0.262a	0.239
Mode of reproduction				NS	NS	NS
Sexual	107	28	17	0.332a	0.248a	0.269a
Sexual and asexual	33	6	14	0.360a	0.223a	0.291a
Successional status				NS	NS	NS
Early	123	23	16	0.340a	0.247a	0.275a
Mid	17	8	15	0.331a	0.201a	0.264a

tions. The lower level of divergence among widespread grass species may reflect the fact that many grasses in this category are crops (Hamrick & Godt, 1997). The distribution of genetic diversity within such species could be highly influenced by the world-wide transfer of germplasm. Differences in genetic structure among species with different geographic ranges is reflected in nearly significant differences ($P = 0.06$) in estimated gene flow between these groups. Mean Nm values (calculated by averaging Nm values

Table 1.6. *Allozyme diversity in several grass genera*

See Table 1.1 for description and calculation of genetic
diversity parameters. Values in parentheses indicate the
number of studies contributing to the mean.

Genus	H_{es}	H_{ep}	G_{ST}
Avena	0.314 (5)	0.180 (7)	0.571 (5)
Bromus	0.109 (7)	0.099 (7)	0.268 (9)
Festuca	0.246 (6)	0.140 (3)	0.158 (2)
Hordeum	0.238 (15)	0.166 (21)	0.301 (18)
Lolium	0.201 (12)	0.168 (20)	0.286 (12)
Oryza	0.289 (4)	0.268 (2)	0.330 (4)
Panicum	0.216 (4)	0.258 (2)	0.058 (2)
Secale	0.161 (9)	0.157 (10)	0.052 (11)
Sorghum	0.112 (5)	0.036 (4)	0.725 (3)
Triticum	0.131 (9)	0.073 (11)	0.482 (11)
Zea	0.264 (16)	0.204 (14)	0.156 (15)
Zizanium	0.083 (5)	0.096 (7)	0.390 (5)

across species rather than from the overall mean G_{ST}) were 1.44 for endemic
or narrowly distributed species, 1.56 for regionally distributed species and
3.06 for widespread species.

The genetic structure of grasses differed significantly among species with
different breeding systems. For outcrossing grasses, about 11% of total
genetic diversity was found among their populations. Mixed-mating species
exhibited similar genetic structure, with 16% of the total genetic diversity
found among populations. In contrast, about 42% of the genetic diversity
within selfers was found among populations. These differences in genetic
structure among species with different mating systems is a consistent
finding of reviews of the plant allozyme literature (Loveless & Hamrick,
1984; Hamrick & Godt, 1989). Indirect estimates of gene flow were lowest
for selfing species (Nm=0.96) and highest for outcrossers (Nm=3.52) with
mixed-mating species having an intermediate value (Nm=2.43). These
differences in gene flow were highly significant ($P \leq 0.0001$).

Genetic diversity and structure within grass genera

Genetic diversity statistics were calculated for 12 grass genera (Table 1.6).
These genera reflect nearly the entire range of genetic diversity found
among all plant species. For example, *Avena* and *Oryza* have quite high
levels of genetic diversity at the species level (H_{es}=0.314 and 0.289, respec-

tively), whereas *Zizanium* has less than a third of their genetic diversity (H_{es}=0.083). For over half of these 12 genera, more than 30% of their total genetic diversity is found among populations. The most genetic structure was found in *Sorghum*, which has nearly 73% of its variation distributed among its populations. The wide range of genetic diversity and structure found among the grasses confirms that the specific evolutionary history of each species plays a dominant role in determining its genetic composition and structure.

How much variation among grass species is accounted for by their life history traits?

In our previous review of the allozyme literature, relatively small proportions of the variation in genetic diversity and genetic structure among species were accounted for by individual life history traits of the species (Hamrick & Godt 1989). We concluded that the phylogenetic and recent evolutionary history of each species must be a predominant determinant of its genetic diversity and structure. Although the effects of phylogeny in this study were partially controlled by considering species within a single plant family, substantial variation exits among grass species that is not accounted for by the life history traits examined. The proportion of variation in genetic diversity accounted for by individual life history traits was low, ranging from less than 1% to 13%. Thus, at least within the grass family, none of the life history features considered has overriding importance on the level of genetic diversity found within species. The seven life history traits considered together accounted for 22% of the variation in species' genetic diversity (H_{es}) and 30% of the variation in mean population genetic diversity (H_{ep}). As in other reviews (Hamrick & Godt, 1989; Hamrick *et al.*, 1992), more of the among-species variation in genetic structure is correlated with particular life history traits. For the grasses, 35% of the variation in G_{ST} values was accounted for by the breeding system of the species, whereas geographic range accounted for 15% of the variation in G_{ST} values among species. Forty-six per cent of the variation in G_{ST} among species was accounted for by the seven life history traits considered together.

Conclusions

Previous reviews of the plant allozyme literature have found significant associations between life history traits of plant species and their genetic diversity and structure. One shortcoming of these reviews is that the effects of phylogeny were not controlled. By pooling all species in a common data set, patterns of variation within plant species can be identified. On the other hand, if the data is biased by taxa or if correlations exist between categories, these patterns may be the consequence of such biases. For example, woody plant species have high genetic diversity, but few woody plant species are selfers. Thus, inherent biases may have produced spurious associations between levels of genetic diversity and species with different breeding systems. The analysis of genetic diversity within a family eliminates, to some degree, these biases.

In concordance with previous results, we found that selfing grasses maintain less genetic diversity within species and within populations. Furthermore, the breeding systems of grasses determine the proportion of genetic diversity found within vs among populations, with selfing species having more genetic divergence than mixed-mating or outcrossing species. Thus, this analysis supports the observation that breeding systems influence not only genetic structure, but also overall genetic diversity.

When all plant species are considered, geographic distribution has no effect on the distribution of genetic diversity within species. In contrast, within the grasses, endemics and narrowly distributed species exhibit less genetic structure than regionally distributed species, but widespread species exhibit the least structure. This latter observation may be due to a number of widely dispersed crop species in this category (Hamrick & Godt, 1997).

On average, grass species maintain somewhat higher levels of genetic diversity than many other plant groups. However, compared with other seed plant species, they exhibit somewhat higher genetic divergence among populations. This higher mean divergence is probably associated with the comparatively high percentage of selfers found within the grasses relative to other plant species (45% vs <20%; J. L. Hamrick and M. J. W. Godt, unpublished data).

Consistent with our previous reviews, this analysis of the grasses has demonstrated that knowledge of a species' life history traits can assist in understanding the genetic composition of a species. This analysis, however, reinforces earlier observations that much of the variation in genetic diversity and its distribution among species is due to the unique evolutionary history of each species.

Acknowledgements

We thank Jennifer Caffarella for assistance with a variety of tasks associated with collecting and collating data for this chapter.

References

Brown, A. H. D. (1979). Enzyme polymorphisms in plant populations. *Theoretical Population Biology*, **15**, 1–42.

Brown, A. H. D. & Briggs, J. D. (1991). Sampling strategies for genetic variation in ex situ collections of endangered plant species. In *Genetics and Conservation of Rare Plants*, ed. D. A. Falk & K. E. Holsinger, pp. 99–122. New York: Oxford.

Clegg, M. T. & Allard, R. W. (1972). Patterns of genetic differentiation in the wild oat species *Avena barbata*. *Proceedings of the National Academy of Sciences of the United States of America*, **69**, 1820–4.

Crawford, D. J. (1983). Phylogenetic and systematic inferences from electrophoretic studies. In *Isozymes in Plant Breeding*, Part A, ed. S. D. Tanksley & T. J. Orton, pp. 257–87. Amsterdam: Elsevier.

Doebley, J. (1989). Isozymic evidence and the evolution of crop plants. In *Isozymes in Plant Biology*, ed. D. E. Soltis & P. S. Soltis, pp. 165–91. Portland: Dioscorides Press.

Ellstrand, N. C. & Roose, M. L. (1987). Patterns of genotypic diversity in clonal plant species. *American Journal of Botany*, **74**, 123–31.

Eriksson, O. (1993). Dynamics of genetics in clonal plants. *Trends in Ecology and Evolution*, **8**, 313–16.

Godt, M. J. W. & Hamrick, J. L. (1996). Genetic structure of two endangered pitcher plants: *Sarracenia jonesii* and *S. oreophila*. *American Journal of Botany*, **83**, 1016–23.

Godt, M. J. W., Hamrick, J. L. & Bratton, S. (1995). Genetic diversity in a threatened wetland species, *Helonias bullata* (Liliaceae). *Conservation Biology*, **9**, 596–604.

Gottlieb, L. D. (1977). Electrophoretic evidence and plant systematics. *Annals of the Missouri Botanical Garden*, **64**, 1–46.

Hamrick, J. L. (1994). Genetic diversity and conservation in tropical trees. *Proceedings International Symposium on Genetic Conservation and Production of Tropical Forest Tree Seed*, ed. R. M. Drysdale, S. E. John, E. T. Yapa & A. C. Yapa, pp. 1–9. ASEAN Canada Forest Tree Seed Centre.

Hamrick, J. L. & Allard, R. W. (1972). Microgeographical variation in allozyme frequencies in *Avena barbata*. *Proceedings of the National Academy of Sciences of the United States of America*, **69**, 2100–4.

Hamrick, J. L. & Godt, M. J. (1989). Allozyme diversity in plant species. In *Plant Population Genetics, Breeding and Germplasm Resources*, ed. A. H. D. Brown, M. T. Clegg, A. L. Kahler, & B. S. Weir, pp. 43–63. Sunderland: Sinauer.

Hamrick, J. L. & Godt, M. J. W. (1997). Allozyme diversity in cultivated crops. *Crop Science*, **37**, 26–30.

Hamrick, J. L., Godt, M. J. W. & Sherman-Broyles, S. (1992). Factors influencing levels of genetic diversity in woody plant species. *New Forests*, **6**, 95–124.

Hamrick, J. L. & Holden, L. R. (1979). The influence of microhabitat heterogeneity on gene frequency distribution and gametic phase disequilibrium in *Avena barbata*. *Evolution*, **33**, 521–33.

Hamrick, J. L., Linhart, Y. B. & Mitton, J. B. (1979). Relationships between life history characteristics and electrophoretically detectable genetic variation in plants. *Annual Reviews of Ecology and Systematics*, **10**, 173–200.

Hamrick, J. L. & Nason, J. D. (1996). Consequences of dispersal in plants. In *Population Dynamics in Ecological Space and Time*, ed. O. E. Rhodes, R. K. Chesser & M. H. Smith, pp. 203–36. Chicago: University of Chicago.

Harris, H. (1966). Enzyme polymorphism in man. *Proceedings of the Royal Society, Series B*, **164**, 298–310.

Lewontin, R. C. & Hubby, J. L. (1966). A molecular approach to the study of genetic heterozygosity in natural populations of *Drosophila pseudoobscura*. *Genetics*, **54**, 595–609.

Loveless, M. D. & Hamrick, J. L. (1984). Ecological determinants of genetic structure in plant populations. *Annual Reviews of Ecology and Systematics*, **15**, 65–95.

Marshall, D. R. & Brown, A. H. D. (1975). Optimum sampling strategies in genetic conservation. In *Crop Genetic Resources for Today and Tomorrow*, ed. O. H. Frankel & J. G. Hawkes, pp. 53–80. Cambridge: Cambridge University Press.

SAS Institute, Inc. (1987). *SAS/STAT™ Guide for Personal Computers*. Version 6 Edition. Cary, NC: SAS Institute, Inc.

Soltis, D. E. & Soltis, P. S. (1989). *Isozymes in Plant Biology*. Portland: Dioscorides Press.

Swenson, S. M., Allan, G. J., Howe, M., Elisens, W. J., Junak, S. A. & Rieseberg, L. H. (1995). Genetic analysis of the endangered island endemic *Malacothamnus fasciculatus* (Nutt.) Greene var. *nesioticus* (Rob.) Kearn. (Malvaceae). *Conservation Biology*, **9**, 404–15.

Torres, A. (1989). Isozyme analysis of tree fruits. In *Isozymes in Plant Biology*, ed. D. E. Soltis and P. S. Soltis, 192–205. Portland: Dioscorides Press.

2

Ecology of seed dormancy and germination in grasses

CAROL C. BASKIN AND JERRY M. BASKIN

Introduction

The caryopsis (hereafter called seed) germination stage of the life cycle of grasses is of much interest to (1) agriculturalists establishing fields of cereals to be harvested for grain, (2) range and grassland managers revegetating overgrazed or otherwise disturbed lands, (3) weed specialists attempting to control weeds in crops, pastures, lawns, and even in natural plant communities and (4) ecologists investigating plant life histories and the dynamics of plant populations, communities and ecosystems. Through experience, agriculturalists, land managers, weed biologists and ecologists have come to realize that the seeds of many grasses are difficult to germinate; consequently, much research has been done on their dormancy and germination. In fact, Simpson (1990) published a book entitled *Seed Dormancy in Grasses*. However, this wonderful compendium of information was written from a physiological point of view and, therefore, does not devote much attention to the ecological perspective of seed germination.

The importance of knowing about the germination ecology of grasses is illustrated by the fact that in moist temperate regions, where precipitation is not a limiting factor *per se*, seeds of grasses germinate in (1) autumn, (2) spring, (3) autumn and spring, (4) summer, (5) spring, summer and autumn or (6) spring and summer, depending on the species (Brenchley & Warington, 1930; Roberts & Ricketts, 1979; Roberts & Potter, 1980). Further, in seasonally wet regions grasses have distinct germination phenology patterns (Mott, 1972, 1978; Silva & Ataroff, 1985; Evans, Kay & Young, 1975). Why are there so many different germination phenology patterns in grasses?

Long-term survival of a grass species means that (1) environmental conditions occasionally have to be suitable for loss of seed dormancy, germination and seedling establishment, and (2) the environment subsequently must be favourable for plant growth, flowering and seed set. Thus, the timing of seed germination plays a critical role in the survival of seed-producing species at a population site. Further, mechanisms resulting in germination at the optimum time for maximum survival and eventual seed production are important adaptations of the species to their habitat. The diversity of germination phenologies in grasses represents adaptations to a wide range of habitats.

Type of seed dormancy in grasses

The way to determine the dormancy status of seeds of any species is to test them for germination over a range of temperatures in light and darkness. Dormant seeds do not germinate at any test condition, while non-dormant ones germinate over the widest range of conditions possible for the species (Baskin & Baskin, 1985*a*). When seeds germinate over only a portion of the range of conditions possible for the species, they are in conditional dormancy (Vegis, 1964; Baskin & Baskin, 1985*a*). Based on the kind(s) of endogenous and/or exogenous factor(s) preventing germination, Nikolaeva (1969, 1977) distinguished six types of seed dormancy (physiological, morphological, morphophysiological, physical, mechanical and chemical).

To determine the type of dormancy found in grass seeds, two kinds of information are needed. (1) Is the embryo fully developed? (2) Do the seeds imbibe water? According to Martin (1946), grass seeds have basal-lateral or lateral embryos, which are fully developed but vary in size depending on the genus. He does not include the Poaceae among the plant families with underdeveloped embryos, and thus grass seeds do not have morphological dormancy. Further, if morphological dormancy is absent, seeds could not possibly have morphophysiological dormancy.

In grass seeds, neither the seed coats nor the indehiscent fruit walls adhering to them have the thick palisade layers of cells characteristic of seeds with physical dormancy (Corner, 1976). Also, seeds of grasses imbibe water when they are placed on a moist substrate (Westra & Loomis, 1966; Bansal, Bhati & Sen 1980; Hardegree & Emmerich, 1993). In fact, both dormant and non-dormant seeds of *Setaria lutescens* imbibe the same amount of water and at the same rate (Kollman, 1970). Thus, seeds of grasses do not have physical dormancy.

Since (1) grass seeds do not have morphological, morphophysiological or physical dormancy, and (2) chemical and mechanical are not dormancy types *per se* (Nikolaeva, 1969, 1977), that leaves only physiological dormancy (PD) as the explanation for their dormancy. However, there are

Table 2.1. *Examples of grasses whose seeds come out of dormancy during dry storage*

Agropyrum spicatum[1]	*Eriachne mucronata*[4]
Andropogon gayanus[2]	*Eulalia fulva*[4]
Aristida congesta[3]	*Heteropogon contortus*[17]
Astrebla lappacea[4]	*Hilaria belangeri*[18]
Bothriochloa ewartiana[4]	*Hordeum spontaneum*[19]
Bouteloua curtipendula[5]	*Melica nutans*[6]
Brachypodium sylvaticum[6]	*Monchather paradoxa*[4]
Bromus sterilis[7]	*Panicum hirsutum*[20]
Calamovilfa longifolia[8]	*Pennisetum setosum*[10]
Cenchrus ciliaris[9]	*Phragmites communis*[21]
Chloris inflata[10]	*Poa sandbergii*[22]
Cymbopogon obtectus[4]	*Puccinellia nuttalliana*[23]
Cynodon dactylon[3]	*Rhynchelytrum repens*[24]
Dactylis glomerata[11]	*Schizachyrium scoparium*[25]
Deschampsia flexuosa[12]	*Schmidtia pappophoroides*[3]
Dichanthium sericeum[4]	*Sorghastrum nutans*[25]
Digitaria ammophila[4]	*Sorghum halepense*[26]
Diplachne fusca[13]	*Stipa viridula*[14]
Echinochloa colonum[10]	*Themeda triandra*[27]
Eleusine indica[10]	*Thyridolepis mitchelliana*[4]
Elymus canadensis[14]	*Tragus berteronianus*[15]
Enneapogon cenchroides[15]	*Trisetum flavescens*[28]
Eragrostis lehmanniana[16]	*Urochloa mosambicensis*[15]

Notes:
[1] Young et al., 1981a; [2] Elberse & Breman, 1989; [3] Veenendaal & Ernst, 1991; [4] Silcock et al., 1990; [5] Shaidaee et al., 1969; [6] Grime et al., 1981; [7] Hilton, 1984a; [8] Maun, 1981; [9] Hacker, 1989; [10] van Rooden et al., 1970; [11] Probert et al., 1985; [12] Scurfield, 1954; [13] Morgan & Meyers, 1989; [14] Robocker et al., 1953; [15] Ernst et al., 1991; [16] Hardegree & Emmerich, 1993; [17] Tothill, 1977; [18] Ralowicz & Mancino, 1992; [19] Gutterman, 1992; [20] Orozco-Segovia & Vazquez-Yanes, 1980; [21] Mariko et al., 1992; [22] Evans et al., 1977; [23] Macke & Ungar, 1971; [24] Popay, 1974; [25] Coukos, 1944; [26] Ghersa et al., 1992; [27] Martin, 1975; [28] Dixon, 1995.

three levels of PD: non-deep, intermediate and deep (Nikolaeva, 1977). Thus, the next question is: what type of PD is found in grass seeds?

Various pieces of evidence indicate that grass seeds have non-deep PD.

1. Non-deep PD is the only type of PD that is broken during dry storage at room temperatures (Nikolaeva, 1969, 1977), and seeds of numerous grasses come out of dormancy during dry storage (Table 2.1).
2. Injury of embryo covers promotes germination of seeds with non-deep

Table 2.2. *Examples of grasses whose seeds germinate to
high percentages when the seed is injured and/or its
covering layers are disturbed or removed*

Aelurpous lagopoides[1]	*Panicum virgatum*[17,18]
Aristida contorta[2]	*Paspalum notatum*[19,20]
Bothriochloa intermedia[3]	*Phalaris arundinacea*[21]
Brachiaria ruziziensis[4]	*Rhynchelytrum repens*[22]
Bromus rubens[5]	*Rottboellia exaltata*[22]
Buchlöe dactyloides[6,10]	*Schizachyrum scoparium*[1]
Cenchrus ciliaris[7]	*Setaria verticillata*[23]
Chloris pilosa[8]	*Spartina alterniflora*[24]
Chrysopogon serrulatus[1]	*Sporobolus cryptandrus*[25]
Dactylis glomerata[9]	*Stipa viridula*[26,27]
Desmostachya bipinnata[1]	*Taeniatherum asperum*[28]
Dichanthium sericeum[11]	*Themeda triandra*[29,30]
Distichlis stricta[12]	*Tripsacum dactyloides*[31]
Eragrostis lehmanniana[13,14]	*Uniola paniculata*[32]
Hordeum spontaneum[15]	*Urochloa mosambicensis*[23]
Hyparrhenia rufa[1]	*Zizania aquatica*[33]
Oryzopsis hymenoides[16]	

Notes:
[1] Hussain & Ilahi, 1990; [2] Mott, 1972; [3] Ahring *et al.*, 1975;
[4] Renard & Capelle, 1976; [5] Corbineau *et al.*, 1992; [6] Thornton,
1966; [7] Hacker, 1989; [8] Elberse & Breman, 1989; [9] Canode *et
al.*, 1963; [10] Ahring & Todd, 1977; [11] Lodge & Whalley, 1981;
[12] Sabo *et al.*, 1979; [13] Wright, 1973; [14] Hardegree & Emmerich,
1991; [15] Gutterman, 1992; [16] Jones & Nielson, 1992; [17] Sautter,
1962; [18] Zhang & Maun, 1989; [19] Andersen, 1953; [20] West &
Marousky, 1989; [21] Junttila *et al.*, 1978; [22] Popay, 1974; [23] Ernst
et al., 1991; [24] Plyer & Carrick, 1993; [25] Toole, 1941; [26] Wiesner
& Kinch, 1964; [27] Frank & Larson, 1970; [28] Nelson & Wilson,
1969; [29] Martin, 1975; [30] Sindel *et al.*, 1993; [31] Anderson, 1985;
[32] Wagner, 1964; [33] Oelke & Albrecht, 1978.

PD, and mechanical injury of grass seeds and/or disturbance or removal
of their covering layers, especially the palea and lemma, promote(s) high
germination percentages in many species (Table 2.2).

3. GA_3 promotes germination of seeds with non-deep PD, but it has little
or no effect on other kinds of dormancy (Nikolaeva, 1977). GA_3 stimulates germination in seeds of many grasses, including *Andropogon gerardii* (Kucera, 1966), *Chrysopogon fallox, C. fallax* (Mott, 1978), *Hilaria
belangeri* (Ralowicz *et al.*, 1992), *Hordeum glaucum* (Popay, 1981), *H.
vulgare* (Dunwell, 1981), *Paspalum plicatulum* (Fulbright & Flenniken,
1988), *Schizachyrium scoparium, Sorghastrum nutans* (Svedarsky &

Table 2.3. *Examples of grasses whose freshly matured seeds are non-dormant*

Life-cycle: P, perennial; SA, summer annual; WA, winter annual; Temp., optimum germination temperature or test temperature that resulted in a high germination percentage; L:D, light or dark requirements, for germination.

Species	Life cycle	Temp. (°C)	L:D[a]	Reference
Agropyron latiglume	P	22/15	L>D	Acharya, 1989
A. repens	P	30/20	L>D	Williams, 1971
Arctagrostis latifolia	P	22	–[b]	Bliss, 1958
Avena fatua	WA	16–21	–	Friesen & Shebeski, 1961
Bromus ciliatus	P	24	L=D	Hoffman, 1985
B. commutatus	WA	7–21	L=D	Froud-Williams & Chancellor, 1986
Eleusine indica	SA	35/20	L>D	Toole & Toole, 1940
Elymus glaucus	P	24/10	L>D	Hoffman, 1985
Imperata cylindrica	P	30	L>D	Dickens & Moore, 1974
Milium effusum	P	16,21	–	Thompson, 1980
Pennisetum macrourum	P	30	–	Harradine, 1980

Notes:
[a] L=D, Seeds germinate equally well in light and darkness; L>D, Seeds germinate to a higher percentage in light than in darkness.
[b] –, data are not available.

Kucera, 1970), *Sorghum plumosum*, *S. stipoideum* (Mott, 1978), *Stipa columbiana* (Young, Emmerich & Patten, 1990), *Themeda triandra* (Martin, 1975), *Tripsacum dactyloides* (Anderson, 1985) and *Uniola paniculata* (Westra & Loomis, 1966). In seeds of *Themeda triandra*, GA_3 increases the growth potential of the embryo, and thus the radicle overcomes the mechanical resistance of the glumes and splits through them (Martin, 1975).

It should be noted that although seed dormancy is very common in the Poaceae (Simpson, 1990), freshly matured seeds of some species, including winter annuals and summer annuals and perennials, are non-dormant (Table 2.3).

Variation in dormancy-breaking and germination requirements

Seed dormancy in grasses is controlled by genetics, environment and genetics × environmental interactions. Much of the research on the inheritance of seed dormancy has been done on crop species and on weeds. Since grasses fall into both of these categories, some information is available on the genetics of seed dormancy of several grasses, including *Avena fatua* (Naylor & Jana, 1976), *Eleusine* spp. (Hilu & de Wet, 1980), *Digitaria milan-jiana* (Hacker *et al.*, 1984), *Hordeum vulgare* (Boyd, Gordon & LaCroix, 1971), *Lolium perenne* (Haywood & Breese, 1966), *Oryza* spp. (Chang & Li, 1991), *Sorghum vulgare* (Gritton & Atkins, 1961) and *Triticum* sp. (Gfeller & Svejda, 1960).

Not only is seed dormancy inherited in grasses, but variations in dormancy-breaking and germination characteristics are controlled by genetics. In *Avena fatua*, the degree of seed dormancy (Adkins, Loewen & Symons, 1986) and the rate of its loss (Jana, Acharya & Naylor,1979) are controlled genetically. Genetic differences also exist with respect to the amount of influence the environment during maturation has on the duration of dormancy (Sawhney & Naylor, 1979). Germination requirements of non-dormant seeds of *Avena* spp. may be genetically controlled (Whittington *et al.*, 1970). *Lolium multiflorum* responded to selection for a delay in time of germination but not to selection for increased germination rate (Nelson, 1980).

Variations in seed dormancy may occur in response to the environment of the mother plant during the time of seed maturation. Differences in percentages and degree of dormancy in grass seeds can occur when the environment of mother plants varies (Table 2.4) with respect to many different factors, or when seeds mature at different positions on the same plant. In a number of grasses, seeds of an individual species collected at various sites differ in their dormancy/germination characteristics (Table 2.5), suggesting that ecotypes have evolved. However, without detailed studies it is difficult to separate the effects of genetics, environment or genetics × environment.

Variation in dormancy-breaking and germination requirements of some grasses is due to seed polymorphism. That is, a species produces two or more types of seeds varying in size, shape or colour, and they differ in their dormancy-breaking and/or germination requirements (Harper, Lovell & Moore, 1970). Environmental effects, especially with respect to seed size, account for somatic polymorphism in some grasses, including *Aegilops* spp. (Datta, Gutterman & Evenari, 1972), *Cenchrus longispinus* (Twentyman, 1974), *Festuca arundinacea* (Bean, 1971), *F. pratensis* (Akpan & Bean, 1977), *Hordeum vulgare* (Giles, 1990), *Lolium* spp. (Akpan & Bean, 1977), *Pennisetum typhoides* (Mohamed, Clark & Ong, 1985), *Poa pratensis* (Maun, Canode & Teare, 1969), *Sorghum bicolor* (Hamilton *et al.*, 1982), *Triticum aestivum* (Brocklehurst, Moss & Williams, 1978) and *Zea mays*

Table 2.4. *Examples of grasses that exhibit variations in dormancy/germination characteristics when the seeds mature under different conditions*

Species	Condition	Reference
Aegilops kotschyi	day length, temperature	Wurzburger & Koller, 1976
A. neglecta	position	Marañón, 1987
A. ovata	day length, temperature, position	Datta *et al.*, 1972
Agrostis curtisii	position	González-Rabanal *et al.*, 1994
Avena fatua	day length	Richardson, 1979
	position	Raju & Ramaswamy, 1983
	soil fertility	Thurston, 1951
	soil moisture, temperature	Sexsmith, 1969
A. ludoviciana	soil fertility	Thurston, 1951
A. sterilis	temperature	Somody *et al.*, 1984
Avenula marginata	position	González-Rabanal *et al.*, 1994
Cenchrus longispinus	position	Twentyman, 1974
Dactylis glomerata	temperature	Probert *et al.*, 1985
Festuca arundinacea	temperature	Boyce *et al.*, 1976
Hordeum vulgare	temperature	Khan & Laude, 1969
Lolium multiflorum	temperature	Akpan & Bean, 1977
L. perenne	temperature	Akpan & Bean, 1977
Panicum dichotomiflorum	fungal infection	Govinthasamy & Cavers, 1995
Rottboellia exaltata	day length	Heslop-Harrison, 1959
Setaria faberi	herbicides	Fawcett & Slife, 1978
Themeda australis	temperature	Groves *et al.*, 1982
Triticum sp.	soil fertility	Fox & Albrecht, 1957
T. aestivum	temperature	Belderok, 1961
Zea mays	competition	Jordan *et al.*, 1982
	soil fertility	Roy & Everett, 1963

(Lambert, Alexander & Rodgers, 1967). However, seed polymorphism in *Agropyron intermedium* (Hunt & Miller, 1965), *Anthoxanthum odoratum* (Antonovics & Schmitt, 1986), *Avena sativa* (Murphy & Frey, 1962), *Bromus inermis* (Christie & Kalton, 1960), *Panicum antidotale* (Wright, 1976), *Sorghum vulgare* (Voight, Gardner & Webster, 1966) and *Zea mays* (Leng, 1949) can be attributed to heredity. In reality, seed polymorphism in many species is an interaction between genetics and the environment. In *Anthoxanthum odoratum*, 17% of the variance in seed weight was due to maternal genetics, 3% to paternal genetics and 80% to environmental effects (Antonovics & Schmitt, 1986).

The consequences of seed polymorphism can be important in the recruit-

Table 2.5. *Examples of grasses that exhibit variations in germination*
characteristics when seeds are collected from different sites

Species	Collection sites vary with respect to	Reference
Alopecurus myosuroides	elevation, habitat	Naylor & Abdulla, 1982
Avena barbata	latitude	Paterson *et al.*, 1976
Avena fatua	latitude	Paterson *et al.*, 1976
Dactylis glomerata	air temperature	Pannangpetch & Bean, 1984
	latitude	Junttila, 1977
Danthonia sericea	soil moisture	Lindauer & Quinn, 1972
Digitaria milanjiana	soil moisture	Hacker, 1984
Festuca pratensis	elevation	Tyler *et al.*, 1978
Poa annua	elevation	Naylor & Abdalla, 1982
	latitude	Standifer & Wilson, 1988
Poa trivialis	light quality	Hilton *et al.*, 1984
Rottboellia exaltata	associated crops	Pamplona & Mercado, 1981
Taeniatherum asperum	rainfall	McKell *et al.*, 1962
Themeda australis	rainfall	Groves *et al.*, 1982

ment of new individuals into the population. Large seeds of *Aegilops* spp. (Marañón, 1987), *Anthoxanthum odoratum* (Roach, 1987), *Lolium perenne* (Naylor, 1980) and *Panicum maximum* (Mejia, Romero & Lotero, 1978) germinated to higher percentages than small ones, but small seeds of *Pennisetum typhoides* (Mohamed *et al.*, 1985) and *Festuca pratensis* (Akpan & Bean, 1977) germinated to higher percentages than large ones. However, germination of seeds of *Oryza sativa* (Krishnasamy & Seshu, 1989) and *Zea mays* (Eagles & Hardacre, 1979) was independent of size.

Most flowers of grasses are open (chasmogamous), and the stigmas may receive pollen from other flowers. However, some grass flowers do not open (cleistogamous), and seeds are produced as a result of autogamy. A few grasses such as *Sporobolus subinclusus* and *Tetrapogon spathaceus* produce only cleistogamous flowers (Uphof, 1938). Cleistogamous flowers of grasses may be produced (1) on the upper portion of the plant (Harlan, 1945), (2) near the ground and hidden by leaves – these are called cleisto-genes (Campbell *et al.*, 1983) or (3) on subterranean stems – this condition is called amphicarpy (Connor, 1979; Cheplick & Quinn, 1982). Seeds produced by cleistogamous and chasmogamous flowers of grasses may vary in size (Clay, 1983) and in degree of dormancy (Dobrenz & Beetle, 1966; McNamara & Quinn, 1977; Bell & Quinn, 1985).

Dormancy-breaking requirements

Dormant seeds of grasses can be divided into two categories, depending on whether dormancy in nature is broken by high summer or by low winter temperatures. Regardless of which environmental conditions are required to break dormancy, however, the transition from dormancy to non-dormancy is a series (continuum) of stages, rather than an abrupt change (Baskin & Baskin, 1985a). The series of stages between dormancy and non-dormancy collectively are included in conditional dormancy (*sensu* Vegis, 1964).

High temperatures

The high temperatures associated with summer are required for loss of dormancy in seeds of most grasses behaving as winter annuals in regions with a temperate or Mediterranean climate and in those of perennial species in regions with Mediterranean or tropical (and sometimes temperate) climates (Table 2.6). In wet tropical climates, seeds may be imbibed while they are subjected to high summer temperatures, and thus they would receive warm stratification treatments. However, due to dryness during summer, especially in regions with a Mediterranean climate, dormancy loss may occur without warm stratification treatments, i.e. similar to dry storage of seeds in the laboratory. Seeds of *Avena fatua* require high summer temperatures, as well as low moisture conditions, for loss of dormancy. As the temperature increases during the dormancy-breaking period, seed moisture content must decrease for maximum loss of dormancy to occur. Thus, seeds come out of dormancy at temperatures of 20–40 °C, if seed moisture contents are 5–29% (Foley, 1994).

Information on the temperature requirements for germination when seeds are passing from dormancy (D) through conditional dormancy (CD) to non-dormancy (ND) helps to explain how the timing of germination is controlled in nature. At maturity, seeds of winter annual grasses may be dormant and thus not germinate at any temperature (Newman, 1963), or they may be in CD and germinate at low temperatures (Hulbert, 1955; Dunwell, 1981; Popay, 1981). As dormant seeds of winter annuals first enter CD, they germinate only at low temperatures (Newman, 1963). Thus, seeds in CD cannot germinate in the habitat, even if summer rains wet the soil, because temperatures are above those required for germination. During the summer dormancy-breaking period, there is an increase in the maximum temperature for germination (Table 2.7) and an increase in germination rates (Dunwell, 1981; Corbineau, Belaid & Côme, 1992) and

Table 2.6. *Examples of grasses whose seeds require high summer temperatures for loss of dormancy*

Life cycle: A, annual; P, perennial; SA, summer annual; WA, winter annual; Temp., optimum germination temperature or test temperature that resulted in a high germination percentage; L:D, light or dark requirements for germination.

Species	Life cycle	Temp. (°C)	L:D[a]	Reference
Aegilops cylindrica	WA	10–20[b]	L=D	Morrow *et al.*, 1982
Aira praecox	WA	15/6	L>D	Newman, 1963
Alopecurus myosuroides	WA	25/15	L>D	Wellington & Hitchings, 1965
Andropogon gayanus	P	30–22	L=D	Felippe *et al.*, 1983; Elberse & Breman, 1989
Apera spica-venti	WA	20/15	L>D	Wallgren & Aamisepp, 1977
Aristida contorta	SA	30	L	Mott, 1972
A. ramosa	P	30, 35	L=D	Lodge & Whalley, 1981
Avena fatua	WA	21, 16	L>D	Hsiao & Simpson, 1971
Bothriochloa macra	P	25	L>D	Lodge & Whalley, 1981
Bromus japonicus	WA	15/6	L=D	Baskin & Baskin, 1981
B. rubens	WA	20–25	L=D	Hammouda & Bakr, 1969
B. secalinus	WA	15	–[c]	Steinbauer & Grigsby, 1956
B. sterilis	WA	15	D	Hilton, 1984*a*
B. tectorum	WA	15	D>L	Hulbert, 1955
Catapodium rigidum	WA	25	L=D	Clark, 1974
Cenchrus biflorus	P	35	–	Kumar *et al.*, 1971
Chrysopogon fallax	P	24–35	L>D	Mott, 1978
C. latifolius	P	24–35	L>D	Mott, 1978
Dactyloctenium aegyptium	SA	30	L>D	Longman, 1969; Kumar *et al.*, 1971
Echinochloa turnerana	SA	30	L>D	Conover & Gieger, 1984*a*
Festuca octoflora	WA	20	L>D	Hylton & Bass, 1961
Hordeum pusillum	WA	30/20	L=D	Fischer *et al.*, 1982
H. spontaneum	WA	20	D>L	Gutterman & Nevo, 1994
Lolium multiflorum	WA	10/5	–	Young *et al.*, 1975
L. persicum	WA	10–15	L>D	Banting & Gebhardt, 1979
Oryza sativa	SA	30	–	Cohn & Hughes, 1981
Phleum arenarium	WA	5	–	Ernst, 1981
Poa annua	WA	10,15	–	Standifer & Wilson, 1988
Puccinellia festucaeformis	P	10	–	Onnis & Miceli, 1975
Sorghum intrans	SA	25–35	–	Andrew & Mott, 1983
S. plumosum	P	24–35	L=D	Mott, 1978
S. stipoideum	SA	25–35	L=D	Mott, 1978; Andrew & Mott, 1983
Sporobolus arabicus	P	35	L=D	Sheikh & Mahmood, 1986
S. elongatus	P	25	L	Lodge & Whalley, 1981
Stipa bigeniculata	P	20/10	L>D	Hagon, 1976
Taeniatherum asperum	WA	10–15	–	Young *et al.*, 1968
Themeda australis	P	24–35	D>L	Hagon, 1976; Mott, 1978
Vulpia bromoides	WA	17–25	L>D	Dillon & Forcella, 1984
V. myuros	WA	22	L>D	Dillon & Forcella, 1984

Notes:
[a] L=D, seeds germinate equally well in light and darkness.
L>D, Seeds germinate to a higher percentage in light than in darkness.
D>L, Seeds germinate to a higher percentage in darkness than in light.
D, Seeds require darkness for germination.
L, Seeds require light for germination.
[b] Range of germination temperatures.
[c] –, data are not available.

Table 2.7. *Germination percentages of seeds of the winter annual grass*
Alopecurus carolinianus *after 15 days of incubation in light (14 h daily*
photoperiod) over a range of 12 h/12 h daily thermoperiods following 0 to
16 weeks of burial in soil at natural summer–autumn temperatures in
Lexington, Kentucky, USA

Age of seeds (weeks)	Month of test	Test temperatures (°C)				
		15/6	20/10	25/15	30/15	35/20
0	June	0	0	0	0	0
4	July	2	90	3	0	0
8	August	97	100	100	1	0
12	September	100	100	100	11	0
16	October	100	100	100	100	0

Source: Baskin, Baskin & Chester, unpublished data.

percentages (Table 2.7). By the end of summer, seeds are ND, and they are
capable of germinating over a range of temperatures; the maximum temperature for germination depends on the species. Thus, when the soil
becomes moist in autumn, germination occurs, because decreasing temperatures in the habitat overlap with those required for germination.

Seeds of the winter annual *Bromus japonicus* are in a latter phase of CD
when they mature in early summer, and all dormancy is lost by mid-
summer. However, seeds do not germinate in summer, because they are
retained on the dead upright mother plants until autumn, at which time dispersal begins and seeds germinate (Baskin & Baskin, 1981).

Seeds of many perennial grasses also come out of D during summer. As
seeds of *Themeda triandra* come out of D during the dry season (summer),
the rate of germination increases; however, optimum and minimum temperatures for germination decrease (Baxter & van Staden, 1993; Baxter *et
al.*, 1993). The growth potential of the embryo in seeds of *T. triandra*
increases during the dormancy-breaking period, as indicated by an increase
in their ability to germinate at decreased water potentials created by the
osmoticum polyethylene glycol (PEG) 6000 (Martin, 1975).

Low temperatures

Seeds of most summer annual grasses in temperate regions come out of D
during winter, thus D break occurs during cold stratification (Table 2.8). As
dormant seeds first enter CD, they germinate only at high temperatures

(Baskin & Baskin, 1983, 1986, 1988; Baskin, Baskin & Chester, 1993). Thus, seeds in CD cannot germinate in late autumn because temperatures of the habitat are below those required for germination. During the winter dormancy-breaking period, there is a decrease in the minimum temperature for germination (Table 2.9). By the end of winter, seeds are ND, and they can germinate over a range of temperatures, including those that were too low for germination the preceding autumn.

Not all seeds of grasses that behave as summer annuals come out of D during cold stratification. Dormant seeds of *Leptochloa panicoides* buried in moist soil at 5 °C, 15/6 °C, 20/10 °C, 25/15 °C and 30/15 °C for 12 weeks came out of D more completely at 25/15 °C and 30/15 °C than at the other temperatures. When seeds were exposed to natural autumn-winter-spring-summer temperatures, some D loss occurred during winter, but the remainder took place as temperatures increased from late winter through early summer (Baskin *et al.* 1993).

Seeds of many perennials come out of D during cold, moist stratification (Table 2.8), and in nature they would receive such a treatment in temperate regions during winter. As seeds come out of D, they gain the ability to germinate over an increasing range of environmental conditions. As dormancy was broken, there was an increase in the maximum temperature for germination in seeds of *Diarrhena americana* (Baskin & Baskin, 1988) and a decrease in the minimum temperature for germination in those of *Muhlenbergia schreberi* (Baskin & Baskin, 1985b). As seeds of *Dactylis glomerata* came out of dormancy, the light and alternating temperature requirements for germination decreased (Probert, Smith & Birch, 1985), and in those of *Sorghum halepense* the alternating temperature requirement for germination decreased (Ghersa *et al.*, 1992). Most seeds of *Leymus cinereus* are ND in autumn, but germination is prevented due to slow germination rates and rapid drying of the soil. Cold stratification increases germination rates, thus seeds of *L. cinereus* can germinate in spring during the periods when soil moisture is non-limiting (Meyer *et al.*, 1995).

Dormancy cycles

When seeds of various species are buried in soil in the temperate region, many, but not all, of them change from one dormancy state to another in response to seasonal temperature cycles (Baskin & Baskin, 1983, 1985c, 1986). Changes in dormancy state are determined by exhuming samples of buried seeds at regular intervals throughout the year and testing them for germination over a range of temperatures in light and darkness. Thus,

Table 2.8. *Examples of grasses whose seeds come out of dormancy during cold stratification*

Life cycle: P, perennial, SA, summer annual; Temp., optimum germination temperature or test temperature that resulted in a high germination percentage; L:D, light or dark requirements for germination.

Species	Life cycle	Temp. (°C)	L:D[a]	Reference
Agropyron trachycaulon	P	17/12	–[b]	McDonough, 1970
Aristida longespica	SA	35	–	Baskin & Caudle, 1967
Bouteloua hirsuta	P	21	–	Tolstead, 1941
Buchlöe dactyloides	P	35/20	–	Pladeck, 1940; Wenger, 1941
Cenchrus longispinus	SA	30/10	L=D	Boydston, 1989
Dactylis glomerata	P	28/10	L>D	Sprague, 1940
Danthonia spicata	P	25/10	–	Toole, 1939
Diarrhena americana	P	20/10	L>D	Baskin & Baskin, 1988
Digitaria ischaemum	SA	35/20	L>D	Baskin & Baskin, 1988
D. sanguinalis	SA	35/20	L>D	Toole & Toole, 1941
Diplachne fusca	SA	27/15	–	Morgan & Meyers, 1989
Echinochloa crus-galli	SA	30/10, 30	L=D	Watanabe & Hirokawa, 1975b; Barrett & Wilson, 1983
Eragrostis spectabilis	P	30	L=D	Tolstead, 1941
E. trichodes	P	30/20	–	Ahring et al., 1963
Festuca ovina	P	17/12, 20/15	–	McDonough, 1970 Young et al., 1981b
F. pratensis	P	25	–	Linnington et al., 1979
Hordeum jubatum	P	27/15, 20/18	D>L	Banting, 1979 Badger & Ungar, 1989
Leptochloa filiformis	SA	35/20	L	van Rooden et al., 1970
Leymus arenarius	P	30/20	D>L	Greipsson & Davy, 1994
Molinia caerulea	P	22/12	L>D	Pons, 1989
Muhlenbergia schreberi	P	35/20	L	Baskin & Baskin, 1985b
Nardus stricta	P	25/20	L>D	Grime et al., 1981
Oryzopsis hymenoides	P	15/5	D>L	Clark & Bass, 1970
Panicum capillare	SA	30/15	L>D	Baskin & Baskin, 1986
P. dichotomiflorum	SA	35/20	L>D	Baskin & Baskin, 1983
P. flexile	SA	30/15	L>D	Baskin & Baskin, 1988
P. virgatum	P	30	L>D	Hsu et al., 1985; Zhang & Maun, 1989
Paspalum plicatulum	P	35/20	L>D	Fulbright, Pedente & Wilson, 1983
Pennisetum alopecuroides	P	16–20	–	Washitani & Masuda, 1990
Poa pratensis	P	10, 15	–	Sprague, 1940
Sasa tsuboiana	P[c]	spring	–	Makita et al., 1993
Setaria glauca	SA	30/15	L>D	Norris & Schoner, 1980; Baskin & Baskin, 1988
S. macrostachya	P	35/20	–	Toole, 1940
S. viridis	SA	25	–	Vanden Born, 1971
Sinarundinaria fangiana	P[c]	23	–	Taylor & Zisheng, 1988
Sorghastrum nutans	P	28/18	L>D	Emal & Conard, 1973

Table 2.8.(*cont.*)

Life cycle: P, perennial, SA, summer annual; Temp., optimum germination temperature or test temperature that resulted in a high germination percentage; L:D, light or dark requirements for germination.

Species	Life cycle	Temp. (°C)	L:D[a]	Reference
Sorghum halepense	P	40/20	–	Taylorson & McWhorter, 1969
S. vulgare	SA	30/20	–	Burnside, 1965
Spartina anglica	SA	21/11	–	Probert & Longley, 1989
Spinifex hirsutus	P	25/20	D>L	Harty & McDonald, 1972
Sporobolus airoides	P	35/20	–	Toole, 1941
S. aspera	P	35/20	–	Toole, 1941
S. contractus	P	35/20	–	Toole, 1941
S. cryptandrus	P	35/20	L>D	Toole, 1941
S. vaginiflorus	SA	20	–	Baskin & Caudle, 1967
Stipa viridula	P	30/15,	–	Rogler, 1960
		15		Young & Evans, 1980
Tripsacum dactyloides	P	30/20	–	Ahring & Frank, 1968
Uniola sessiliflora	P	40/20	D>L	Wolters, 1970
Zizania aquatica	SA	20/15	–	Muenscher, 1936
Z. palustris	SA	20/15	–	Probert & Longley, 1989;
		16		Kovach & Bradford, 1992a

Notes:
[a] See footnotes for Table 2.6.
[b] –, data are not available.
[c] Monocarpic perennial.

Table 2.9. *Germination percentages of seeds of the summer annual grass* Panicum dichotomiflorum *incubated in light (14 h daily photoperiod) over a range of 12 h/12 h daily thermoperiods for 15 days following 0–24 weeks of burial in moist soil at natural autumn to early spring temperatures in Lexington, Kentucky, USA*

Age (Weeks)	Month of test	Test thermoperiods (°C)				
		15/6	20/10	25/15	30/15	35/20
0	October	0	0	0	1	1
4	November	0	0	1	4	7
8	December	0	0	5	23	85
12	January	0	0	31	99	100
16	February	0	1	68	100	100
20	March	0	41	100	100	100
24	April	8	92	100	100	100

Note:
Data from graphs in Baskin & Baskin (1983).

changes in dormancy state are reflected by changes in germination responses. Buried seeds may cycle between D and ND (Baskin & Baskin, 1986) or between CD and ND (Baskin & Baskin, 1983). Buried seeds of some species lose their initial D or CD and remain ND, regardless of changing seasonal conditions (Baskin & Baskin, 1985c). However, we do not know of an example of such lack of dormancy cycles in buried seeds of a grass, which probably is due to lack of research. All angiosperm seeds that have been found to have dormancy cycles also have non-deep PD.

Only a few studies have been done on buried seeds of winter annual grasses. Seeds of *Alopecurus myosuroides* buried in soil in England in April germinated to *c.* 99%, 90%, 0% and 60% in light at 20/10 °C the following autumn, spring, summer and autumn, respectively (Froud-Williams, Drennan & Chancellor, 1984), indicating that they had either an annual D/ND or a CD/ND cycle. Seeds of *A. carolinianus* buried in a non-heated greenhouse in Lexington, Kentucky (USA) exhibited an annual D/ND cycle. Further, dormant seeds of *A. carolinianus* came out of D when buried in moist soil at 25/15 °C and 30/15 °C but not at 5 °C or 15/6 °C (C. C. Baskin, J. M. Baskin & E. W. Chester, unpublished). Low temperatures (near 0 °C) induced ND seeds of *Bromus japonicus* into D (Haferkamp *et al.*, 1994).

Annual dormancy cycles have been found in buried seeds of several summer annual grasses. Buried seeds of *Setaria glauca* (Baskin, Baskin & El-Moursey, 1996) and *Leptochloa panicoides* (Fig. 2.1) subjected to seasonal temperature cycles in the temperate region exhibited an annual CD/ND cycle. Freshly matured seeds are dormant when dispersed in autumn, and they become ND during winter. Seeds enter CD in summer or autumn, depending on the species, and they come out of CD the following winter. When seeds are ND in spring, they germinate over the full range of temperature possible for the species, but when they enter CD they lose the ability to germinate at low temperatures. Thus, seeds can germinate in the habitat from late spring until late summer–early autumn, if light and soil moisture are non-limiting.

Buried seeds of the summer annuals *Panicum dichotomiflorum* (Baskin & Baskin, 1983) and *P. capillare* (Fig. 2.2) exhibit an annual D/ND cycle. Freshly matured seeds are dormant when dispersed in autumn; they come out of D during late autumn and winter and are ND by spring. Seeds re-enter D by autumn and become ND again by the following spring. Thus, seeds can germinate in the field from mid-spring to mid- to late summer, if light and soil moisture are non-limiting. Seeds of *Echinochloa crus-galli* var. *praticola* buried in Japan gained the ability to germinate at 25/10 °C during

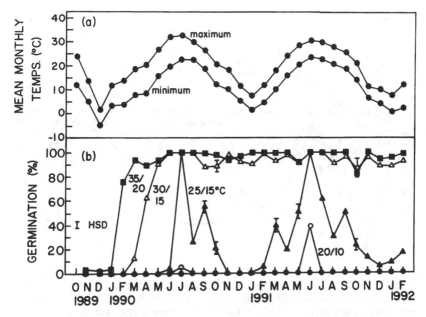

Fig. 2.1. Annual conditional dormancy/non-dormancy cycle in buried seeds of the summer annual grass *Leptochloa panicoides*. (*a*) Mean daily maximum and minimum monthly temperatures in a non-heated greenhouse in Lexington, Kentucky (USA), where seeds were buried in pots of soil. (*b*) Germination percentages (mean±SE, if ≥5%) of seeds incubated at 12 h/12 h daily thermoperiods at a 14 h daily photoperiod for 15 days following 0 to 27 months of burial in soil. (Reproduced from Baskin, Baskin & Chester (1993), *Acta Oecologica*, 14, pp. 693–704, with permission of Gauthier-Villars Publisher.)

winter but lost this ability in late August (Watanabe & Hirokawa, 1975a). Thus, seeds had either an annual D/ND or CD/ND cycle.

Studies have not been done to determine whether or not buried seeds of perennial grasses have annual dormancy cycles.

Soil seed banks

Seeds in (on) the soil are either in a transient or persistent seed bank. Seeds in the transient seed bank live until the first germination season following maturation, and those in the persistent seed bank live until the second (or some subsequent) germination season. Many grasses have transient seed banks only (Andrew & Mott, 1983; Silva & Ataroff, 1985), but others form persistent seed banks (Table 2.10).

To document the presence of a persistent seed bank, soil samples should

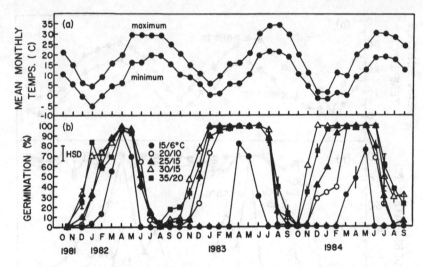

Fig. 2.2. Annual dormancy/non-dormancy cycle in buried seeds of the summer
annual grass *Panicum capillare*. (*a*) Mean daily maximum and minimum monthly
temperatures in a non-heated greenhouse in Lexington, Kentucky (USA), where
seeds were buried in pots of soil. (*b*) Germination percentages (mean±SE, if
≥5%) of seeds incubated at 12 h/12 h daily thermoperiods at a 14 h daily
photoperiod for 15 days following 0 to 35 months of burial. (Reproduced from
Baskin & Baskin (1986), *Weed Science*, 34, pp. 22–4, with permission of the Weed
Science Society of America.)

be collected in the population site of a species after the annual germination
season is completed, but before seeds of the current year's seed crop are
dispersed. After collection, samples need to be given dormancy-breaking
treatments and then subjected to temperature and light/dark conditions
appropriate for germination of the species being studied. Also, techniques
are available for separating seeds from soil samples. However, documenting
the presence of seeds in soil samples does not necessarily mean that they
would have been in the soil for longer than one germination season or that
they would have lived until the next germination season. Consequently,
seeds in soil samples may or may not be part of the persistent seed bank.
Thus, the best method for learning about persistent seed banks probably is
to put samples in a slathouse, well-ventilated shed or non-heated greenhouse, where they are subjected to near-normal seasonal temperature
changes, and then water, stir and check them regularly for newly germinated seedlings over a period of several years. This is the method used with
such great success by Brenchley & Warington (1930).

    Seed-bank studies have been conducted in a wide range of plant com-

Table 2.10. *Examples of grasses whose seeds have been found in persistent seed banks*

Only one species is listed for each genus.

Species	Habitat	Density (m⁻²)	Depth (cm)	Reference
Agrostis exarata	pastures	430	10	Hayashi & Numata, 1971
Agropyron pectinatum	grassland	252	15	Howard, 1973
Aira praecox	pasture	78	17.5	Milton, 1936
Alopecurus agrestis	field	3,061	15	Brenchley & Warington, 1930
Andropogon virginicus	forest	14	5	Naka & Yoda, 1984
Anthoxanthum odoratum	pasture	1,123	17.5	Milton, 1936
Arctagrostis latifolia	Arctic low land	71	10	Ebersole, 1989
Aristida adscencionis	grassland	42	2	Dwyer & Aguirre V., 1978
Arundinella hirta	successional	3[a]	1	Numata *et al.*, 1984
Avena barbata	grassland	8–12	2.5	Young *et al.*, 1981c
Brachiaria miliiformis	pasture	120	5	McIvor & Gardener, 1991
Brachypodium sylvaticum	successional	2	7	Donelan & Thompson, 1980
Bouteloua eriopoda	grasslands	40	2	Dwyer & Aguirre V., 1978
Bromus tectorum	semi-desert	14	5	Hassan & West, 1986
Catapodium rigidum	successional	2	7	Donelan & Thompson, 1980
Cortaderia selloana	forest	8	5	Enright & Cameron, 1988
Cynosurus cristatus	pasture	128	17.5	Milton, 1936
Dactylis glomerata	successional	11	7	Donelan & Thompson, 1980
Dactyloctenium aegyptium	pasture	10	5	McIvor & Gardener, 1991
Danthonia setacea	forest	5	3	Vlahos & Bell, 1986
Deschampsia flexuosa	forest	5	3.7	Granstrom, 1982
Digitaria ischaemum	swamp	1	10	Schneider & Sharitz, 1986
Echinochloa colonum	field	22	4.2	Kellman, 1974
Eleusine indica	field	540	10	Kellman, 1980
Enneapogon desvauxii	grassland	7	2	Dwyer & Aguirre V., 1978
Eragrostis tenuifolia	forest	2	5	Enright, 1985
Erioneuron pulchellum	grassland	41	2	Dwyer & Aguirre V., 1978
Festuca rubra	pasture	380	17.5	Milton, 1936
Heteropogon contortus	savanna	50	5	O'Connor & Pickett, 1992
Hilaria belangeri	grassland	113	5	Kinucan & Smeins, 1992
Holcus lanatus	pasture	3,050	17.5	Milton, 1936
Hordeum vulgare	successional	1	7	Donelan & Thompson, 1980
Imperata cylindrica	successional	9[a]	1	Numata *et al.*, 1964
Leersia oryzoides	marsh	190	6	Leck & Simpson, 1987
Leptochloa filiformis	field	9	4.2	Kellman, 1974
Lolium perenne	pasture	50	17.5	Milton, 1936
Miscanthus sinensis	forest	95	10	Nakagoshi, 1984a,b
Molinia caerulea	pasture	25	17.5	Milton, 1936
Muhlenbergia uniflora	lakeshore	2	4	Wisheu & Keddy, 1991
Nardus stricta	pasture	128	17.5	Milton, 1936
Oryzopsis hymenoides	semi-desert	3	5	Hassan & West, 1986
Panicum capillare	grassland	4	2	Dwyer & Aguirre V., 1978
Paspalum thunbergii	pasture	240	10	Hayashi & Numata, 1971
Phleum pratense	forest	100	5	Kramer & Johnson, 1987

Table 2.10. (cont.)

Only one species is listed for each genus.

Species	Habitat	Density (m^{-2})	Depth (cm)	Reference
Poa compressa	forest	100	15	Howard, 1973
Puccinellia maritima	salt marsh	1,664	2	Ungar & Woodell, 1993
Sacciolepis stricta	lake bed	235	10	Fiore & Putz, 1992
Scolochloa festucacea	wetland	14[b]	5	Poiani & Johnson, 1988
Setaria faberi	successional	54	5	Rothrock et al., 1993
Sieglingia decumbens	successional	40	7	Donelan & Thompson, 1980
Sitanion hystrix	semi-desert	1	5	Hassan & West, 1986
Sporobolus flexuosus	grassland	104	2	Dwyer & Aguirre V., 1978
Stipa comata	semi-desert	1	5	Hassan & West, 1986
Taeniatherum asperum	grassland	10–20	2.5	Young et al., 1981c
Themeda triandra	savanna	250	5	O'Connor & Pickett, 1992
Triodia decumbens	pasture	205	17.5	Milton, 1936
Urochloa mosambicensis	pasture	20	5	McIvor & Gardener, 1,991
Vulpia megalura	grassland	8–18	2.5	Young et al., 1981c
Zoysia japonica	pasture	9,340	10	Hayashi & Numata, 1,971

Notes:
[a] Seeds per 400 cc of soil.
[b] Seeds per 50 cc of soil.

munities, but the techniques used in some of them make it impossible to distinguish transient from persistent seed banks. In a survey of 102 published papers dealing with seed banks, only 49 were done in such a way to ensure that the transient seed bank was not included in the soil samples; 39 of the 49 publications list grasses. From these 39 publications, at least 61 genera and 99 species of grasses form persistent seed banks (Table 2.10).

In a given habitat, seeds in the persistent seed bank are not evenly distributed throughout the soil, and sometimes they are aggregated (Chauvel, Gasquez & Darmency, 1989; Dessaint, Chadoeuf & Barralis 1991). The actual location of seeds at a site depends on distribution of the seed-producing plants and on movement of seeds following dispersal, e.g. earthworms moved seeds of Capsella bursa-pastoris 17–18 cm into the soil (Hurka & Haase, 1982). Ploughing also moves seeds both horizontally (Wilson, Ingersoll & Roush, 1989) and vertically (van Esso, Ghersa & Soriano, 1986; Wilson et al., 1989). In maize and wheat fields in Argentina, one ploughing moved 80% of the Sorghum halepense seeds originally in the top 4 cm of soil to depths of 4–20 cm (van Esso et al., 1986).

Seeds of grasses have been buried in soil under natural conditions and their viability determined after various periods of time (Table 2.11). A few seeds (<1%) of Setaria glauca lived for 30 years in the Beal buried seed

Table 2.11. *Longevity of grass seeds buried in soil under natural conditions for known periods of time*

Species	Duration of study (years)	Length of viability (years)	Average % of seeds surviving	Reference
Aegilops cylindrica	2	2	<8.0	Donald, 1991
Agropyron repens	39	10	0.7	Toole & Brown, 1946
Agrostis tenuis	3	3	14.0	Rampton & Ching, 1966
	20	4[a]	17.3	Lewis, 1973
Alopecurus myosuroides	20	4[a]	28.3	Lewis, 1973
Avena fatua	20	4[a]	1.7	Lewis, 1973
	39	1	11.7	Toole & Brown, 1946
Bromus mollis	20	1[b]	5.3	Lewis, 1973
B. racemosus	39	6	24.7	Toole & Brown, 1946
B. secalinus	100	1	0	Beal, 1911
Chionochloa macra	1.1	1.1	5.5	Spence, 1990
Dactylis glomerata	6	2	2.0	Dorph-Petersen, 1925
	5	2	2.0	Kjaer, 1940
	3	2	0.3	Rampton & Ching, 1966
	20	4[a]	<1	Lewis, 1973
Dactyloctenium aegyptium	6	6	54.0	Juliano, 1940
Deschampsia flexuosa	5	5	<1.0	Granstrom, 1987
Digitaria sanguinalis	10	10	1.0	Burnside *et al.*, 1981
	5.5	5.5	<1	Egley & Chandler, 1983
Echinochloa crusgalli	15	9	3.0	Dawson & Bruns, 1975
	5.5	4.5	<1	Egley & Chandler, 1983
Eleusine indica	39	6	1.0	Toole & Brown, 1946
	5	5	1.8	Schwerzel, 1974
	5.5	5.5	4.0	Egley & Chandler, 1983
Elymus canadensis	39	1	0.7	Toole & Brown, 1946
E. triticoides	39	1	7.3	Toole & Brown, 1946
E. virginicus	39	1	14.0	Toole & Brown, 1946
Festuca arundinacea	3	3	0.3	Rampton & Ching, 1966
	20	1[b]	19.0	Lewis, 1973
F. pratensis	20	1[b]	3.7	Lewis, 1973
F. rubra	3	2	2.1	Rampton & Ching, 1966
	20	1[b]	18.0	Lewis, 1973
Holcus lanatus	20	4[a]	13.7	Lewis, 1973
Hordeum distichon	5	<1	0	Kjaer, 1940
Lolium italicum	20	4[a]	1.7	Lewis, 1973
L. multiflorum	3	3	3.3	Rampton & Ching, 1966
L. perenne	3	2	0.1	Rampton & Ching, 1966
	20	4[a]	2.0	Lewis, 1973
Panicum dichotomiflorum	10	10	2	Burnside *et al.*, 1981
P. flavidum	6	3	3.0	Juliano, 1940
P. miliaceum	3.5	3.5	0.2	Cavers *et al.*, 1992
P. texanum	5.5	5.5	5.0	Egley & Chandler, 1983

Table 2.11. (cont.)

Species	Duration of study (years)	Length of viability (years)	Average % of seeds surviving	Reference
P. virgatum	39	3	12.3	Toole & Brown, 1946
Phalaris arundinacea	39	27	0.7	Toole & Brown, 1946
Phleum nodosum	20	20	<1	Lewis, 1973
P. pratense	5	5	20.0	Kjaer, 1940
	39	21	6.3	Toole & Brown, 1946
	20	4[a]	46.7	Lewis, 1973
Poa annua	7	3	5.0	Rampton & Ching, 1970
P. pratensis	39	39	1.0	Toole & Brown, 1946
	3	3	1.3	Rampton & Ching, 1966
Rottboellia exaltata	5	5	1.6	Schwerzel, 1974
Setaria glauca	39	39	0.3	Toole & Brown, 1946
	15	9	<1	Dawson & Bruns, 1975
	100	30	<1	Kivilaan & Bandurski, 1981
S. verticillata	39	39	0.7	Toole & Brown, 1946
S. viridis	39	39	0.3	Toole & Brown, 1946
	15	9	<1	Dawson & Bruns, 1975
	10	10	3.0	Burnside, et al., 1981
Sorghum halepense	4	4	2.0	Goss, 1939
	5.5	5.5	48.0	Egley & Chandler, 1983
Sporobolus airoides	39	21	2.0	Toole & Brown, 1946
S. cryptandrus	39	39	0.3	Toole & Brown, 1946
Triticum sativum	5	<1	0	Kjaer, 1940

Notes:
[a] Seeds lived for more than 4 years, but less than 20.
[b] Seeds lived for more than 1 year, but less than 4.

study (Kivilaan & Bandurski, 1981) and 0.3–2.0% of the *Poa pratensis*, *Setaria glauca*, *S. verticillata*, *S. viridis* and *Sporobolus cryptandrus* seeds were alive when the Duvel buried seed experiment was ended after 39 years (Toole & Brown, 1946). However, in the majority of grasses few seeds are alive after 5 years of burial, and seeds of some grasses live for less than 1 year. In studies of buried seeds from archaeologically dated sites in Denmark and Sweden, no viable seeds of grasses were reported to be ≥20 years old (Ødum, 1974).

Buried seeds in the transient seed bank of *Alopecurus myosuroides* (Thurston, 1966; Moss, 1985), *Digitaria sanguinalis*, and *Setaria viridis* (Burnside et al., 1981) exhibit a Deevey Type III curve, while those in the persistent seed bank of these species exhibit a Deevey Type II or negative

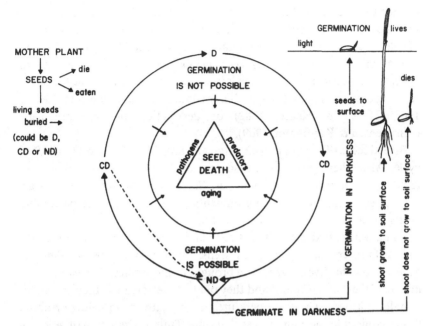

Fig. 2.3. Factors affecting persistence and depletion of buried seeds of grasses and the possible annual changes in dormancy state. This model combines the ideas of Schafer & Chilcote (1969), H. A. Roberts (1970), E. H. Roberts (1972) and Buchele *et al.* (1991).

exponential curve. Buried seeds of *Panicum dichotomiflorum* (Burnside *et al.*, 1981) and *Poa annua* (Roberts & Dawkins, 1967) have only a Deevey Type II curve.

Schafer & Chilcote (1969) developed a model showing what could happen to seeds in the soil, which was modified by H. A. Roberts (1970) and E. H. Roberts (1972). Unfortunately, the dormancy terminology used in this model is not an appropriate way to describe the dormancy cycle in seeds with non-deep PD. Thus, the ideas of Schafer & Chilcote (1969), Roberts (1970) and Roberts (1972) have been superimposed onto a dormancy cycle model (Buchele, Baskin & Baskin, 1991) to show what could happen to seeds of grasses or other species with non-deep PD when they are buried in the soil (Fig. 2.3).

Many seeds of grasses are eaten by animals within a few days following dispersal (Janzen, 1976), and thus they are in the transient seed bank for only a short period of time. Seeds are lost from the persistent seed bank in various ways, but *in situ* germination is a major cause of depletion. Seedlings from some seeds germinating near the soil surface have a better chance to become established than those from deeply buried seeds, which may die before the shoot reaches the surface. An increase in depth of burial (Toole & Brown, 1946; Bridgemohan, Brathwaite & McDavid, 1991) and lack of soil disturbance (Roberts & Dawkins, 1967) tend to decrease *in situ* germination for many species. Seeds in the seed bank may die due to natural senescence or ageing (Villiers, 1971), predation by soil organisms (Newell, 1967; Hurka & Haase, 1982) or destruction by microorganisms (Christensen & Kaufmann, 1969).

Buried seeds of grasses may undergo an annual D/ND cycle or a CD/ND cycle. Also, there probably are some grasses whose seeds remain ND, regardless of how long they are buried. Seeds can germinate when they are ND (or sometimes CD), if environmental conditions become favourable, i.e. temperature, moisture and light are non-limiting. If light is required for germination, buried seeds do not germinate until they are brought to the soil surface. In short-lived grass seeds, many may die before they are exposed to light. Even if seeds are brought to the surface, they cannot germinate if they are dormant, and they may not germinate if they are conditionally dormant, i.e. if temperatures in the habitat are below or above those required for germination at that time. Thus, germination of seeds on the soil surface may be delayed until D or CD is broken.

Germination requirements of nondormant seeds

Temperature

For grasses whose seeds require high summer temperatures for loss of dormancy, the mean (± SE) optimum temperature for germination is 21.6 ± 1.3 °C; 16.2 ± 1.1 °C for winter annuals, 27.2 ± 2.0 °C for perennials and 29.9 ± 0.1 °C for summer annuals (Table 2.6). The relatively low optimum germination temperature for winter annuals explains why seeds germinate in autumn, when habitat temperatures are decreasing, whereas the high optimum temperatures for perennials and summer annuals explain why seeds germinate in mid- to late spring or summer, when habitat temperatures are increasing. However, the optimum germination temperature for seeds of the perennial halophyte *Puccinellia festucaeformis* is 10 °C (Onnis & Miceli, 1975); consequently, seeds germinate at late winter or early spring temperatures, when soil moisture is high and salinity is low.

The mean optimum germination temperature is 22.4 ± 0.9 °C in grasses whose seeds come out of dormancy at low winter temperatures; 24.8 ± 1.1 °C for summer annuals and 23.5 ± 1.1 °C for perennials (Table 2.8).

Table 2.12. *Summary of light (L): dark (D) requirements for germination of seeds of 30 species of grasses that become non-dormant during exposure to summer temperatures and 24 that become non-dormant during exposure to winter temperatures*

Number (percentage) of seeds in each L:D requirement category is given.

Natural period of dormancy loss	Light : dark requirements				
	L^a	D	L>D	D>L	L=D
Summer	2 (7)	1 (3)	14 (47)	3 (10)	10 (33)
Winter	2 (8)	1 (4)	14 (58)	4 (17)	3 (13)

^{a} See footnotes for Table 2.6.

Thus, the optimum germination temperatures for summer annuals and perennials are very similar, and seeds should germinate in mid- to late spring in temperature regions. However, seeds of a few perennials and summer annuals germinate at average temperatures of 15 °C or less, indicating that they could germinate in late winter or early spring in temperate regions.

Light:dark

Data on the light:dark requirements for germination are available for 30 of the species whose seeds come out of dormancy in summer (Table 2.6) and for 24 of those whose seeds come out of dormancy in winter (Table 2.8). Regardless of the season when dormancy loss occurs, seeds of 28 of the 54 species germinate to higher percentages in light than in darkness, but those of 13 species are indifferent to light (Table 2.12). Seeds of *Aristida contorta* (Mott, 1972), *Leptochloa filiformis* (van Rooden, Akkermans & van der Veen, 1970) and *Muhlenbergia schreberi* (Baskin & Baskin, 1985*b*) require light for germination, but those of *Leymus arenarius* (Greipsson & Davy, 1994) and *Bromus sterilis* (Hilton, 1984*b*) require darkness. Light inhibits germination of *Avena fatua* seeds when they first enter conditional dormancy, but as dormancy loss continues during the summer-dormancy breaking period they also gain the ability to germinate in light (Hsiao & Simpson, 1971).

Germination responses of grass seeds to the light:dark environment are controlled by the phytochrome system, usually with red light promoting

54   C. C. Baskin and J. M. Baskin

and far-red light inhibiting germination (van Rooden *et al.*, 1970; Williams, 1983*a*; Hou & Simpson, 1990). In seeds of *Bromus sterilis* (Hilton, 1984*a*) and *Aristida murina* (Ginzo, 1978), however, red light inhibits germination. White light (Hilton, 1982), prolonged exposure to red light (Hilton, 1982; Hou & Simpson, 1991) and continuous exposure to light with wavelengths of 700–740 nm [High Energy Reaction, which overrides the phytochrome system (Borthwick *et al.*, 1969)] inhibits germination of some grass seeds (Ellis, Hong & Roberts, 1986*a*,*b*).

Seeds of grasses vary in their response to light filtered through green leaves (i.e. light with a low red/far-red ratio). Light filtered through green leaves reduced germination in seeds of *Agropyron repens, Apera spica-venti, Avena fatua, Festuca sulcata* (Górski, Górska & Nowicki, 1977), *Agrostis alba, Festuca pratensis, Poa palustris, P. trivialis, Phalaris arundinacea* (Górski, Górska & Rybicki, 1978), *Chloris pycnothrix* and *Setaria verticillata* (Fenner, 1980). However, filtered light did not significantly reduce germination of *Avena sativa, Hordeum jubatum, Poa alpina, Secale cereale, S. silvestre, Triticum vulgare, Zea mays* (Górski *et al.*, 1977), *Alopecurus pratensis, Festuca duriuscula, Lolium perenne* (Górski *et al.*, 1978) or *Aristida adscensionis* (Fenner, 1980). Responses of seeds to a low red/far-red ratio may play a role in preventing some grasses native to open habitats from becoming established in shady habitats. Hilton, Froud-Williams & Dixon (1984) found that seeds of *Poa trivialis* from open, sunny, arable habitats were more inhibited by light with a low red/far-red ratio than those from closed, shady grassland habitats.

Alternating temperatures

Seeds of some grasses require alternating (vs constant) temperatures for germination (Morinaga, 1926; Matumura, Takese & Hirayoshi, 1960; Thompson, Grime & Mason, 1977; Thompson & Grime, 1983; Goedert & Roberts, 1986; Williams, 1983*a*), but the stimulatory effects may interact with the presence or absence of light. Seeds of *Agrostis tenuis, Bromus erectus, Dactylis glomerata, Festuca ovina, F. rubra, Koeleria cristata, Lolium perenne* and *Phleum pratense* germinate at alternating and constant temperatures in light and darkness, and those of *Agropyron repens, Deschampsia caespitosa, Phalaris pratensis* and *Sieglingia decumbens* germinate in light at alternating temperatures (Thompson & Grime, 1983). Seeds of *Avena fatua* germinate in darkness at constant or alternating temperatures, and those of *Agropyron repens, Deschampsia caespitosa* and *Poa pratensis* germinate in darkness at alternating temperatures (Thompson &

Grime, 1983). Seeds of *Panicum turgidum* require darkness for germination at constant temperatures, but they will germinate in light or darkness at alternating temperatures (Koller & Roth, 1963).

In many species, the amplitude of the daily temperature fluctuation required for maximum germination in darkness is 8–12 °C (Thompson *et al.*, 1977; Williams 1983*a*). In *Dactylis glomerata*, maximum germination occurred when seeds were exposed to the low (11 °C) temperature for 16 hours and then the high (21 °C) one for 8 hours (Probert *et al.*, 1986). Alternating and constant temperatures can interact with light and nitrate in controlling germination (Morinaga, 1926; Williams, 1983*b*; Hilton, 1985; Goedert & Roberts, 1986).

The ecological implications of temperature fluctuations promoting germination of seeds in darkness are twofold.

1. Since the amplitude of daily temperature fluctuations decreases with an increase in soil depth (Ghersa *et al.*, 1992), germination of seeds that require alternating temperatures to germinate in darkness decreases with increases in depth of burial. Thus, response to fluctuating temperatures is a depth-detecting mechanism (Thompson & Grime, 1983; Ghersa *et al.*, 1992). However, as seeds of *Sorghum halepense* aged they became less sensitive to temperature fluctuations and germination increased at constant temperatures (Ghersa *et al.*, 1992).
2. Since the amplitude of daily temperature fluctuations is lower under a canopy of green foliage than in the open (Ghersa *et al.*, 1992), germination of seeds that require alternating temperatures to germinate in either light or darkness decreases with an increase in shading of the soil surface. Thus, response to fluctuating temperatures is a gap-detecting mechanism (Thompson & Grime, 1983). Germination of buried seeds of *Sorghum halepense* decreased under a vegetation cover and under artificial shade. On the other hand, when soil under a leaf canopy was heated, thereby causing an increase in daily temperature fluctuations, seeds germinated to the same percentage as those buried in plots where the vegetation cover was removed (Benech Arnold *et al.*, 1988).

Increases in soil temperatures appear to be the explanation for an increase in germination of grass seeds in monsoonal climates when fires remove plant canopies and/or litter (Tothill, 1969, 1977). In areas that have been burned, many grass seedlings appear shortly after the beginning of the wet season (Shaw, 1957). Heat *per se* from fires does not cause an increase in germination, because when plant material is removed mechanically germination also increases (Shaw, 1957).

Table 2.13. *Level of moisture stress (−MPa) that reduces germination of grass seeds to 50% or less*

Controls germinated to 80–100%, depending on the species.

Species	Moisture stress (−MPa)	Species	Moisture stress (−MPa)
Agropyron desertorum[1]	0.51	*Elymus scabrus*[3]	0.50
A. inerme[1]	0.51	*Enteropogon ramosus*[3]	0.75
A. intermedium[1]	1.02	*Eragrostis curvula*[3]	0.75
A. smithii[2]	0.7	*E. lehmanniana*[8]	1.11
A. trichophorum[1]	1.02	*Festuca altaica*[9]	0.41
Andropogon inerme[1]	0.51	*Hilaria jamesii*[2]	1.30
Bothriochloa macra[3]	0.25	*H. mutica*[7]	1.11
Bouteloua curtipendula[2]	>1.60	*Muhlenbergia porteri*[7]	1.52
B. eriopoda[7]	2.03	*M. wrightii*[2]	1.30
B. gracilis[2]	>1.60	*Panicum miliare*[10]	0.5
Bromus inermis[1]	1.02	*Sorghum halepense*[10]	0.50
Buchlöe dactyloides[5]	0.10	*S. plumosum*[4]	0.80
Chloris truncata[3]	0.25	*S. stipoideum*[4]	0.40
Chrysopogon fallax[4]	1.00	*Sporobolus airoides*[11]	0.81
C. latifolius[4]	1.00	*S. contractus*[2]	1.0
Distichlis stricta[2]	1.60	*S. cryptandrus*[2]	1.0
Echinochloa turnerana[6]	1.50	*Themeda australis*[4]	0.90
Ehrharta junceus[1]	0.04		

[1] McGinnies, 1960; [2] Sabo *et al.*, 1979; [3] Maze *et al.*, 1993; [4] Mott, 1978; [5] Bokhari *et al.*, 1975; [6] Conover & Geiger, 1984b; [7] Knipe & Herbel, 1960; [8] Martin & Cox, 1984; [9] Grilz *et al.*, 1994; [10] Sinha & Gupta, 1982; [11] Toole, 1941.

Soil moisture

Grass seeds of different species vary in germination response to a particular soil moisture stress level (Table 2.13). The mean (± SE) level of moisture stress required to reduce germination percentages of grasses to 50% is −0.78± −0.09 MPa (−7.8± −0.9 bars); controls germinated to 80–100%, depending on the species.

Temperature can interact with soil moisture in controlling germination of grass seeds (Itabari, Gregory & Jones, 1993). When temperatures were increased from 10 °C to 25 °C at a rate of 0.5 °C per day, seeds of *Festuca altaica* (Romo *et al.*, 1991; Grilz *et al.*, 1994) and *Bromus inermis* (Grilz, Romo & Young, 1994) germinated to a higher percentage and at a higher rate over a broader range of osmotic potentials than they did when temperatures were decreased from 25 °C to 10 °C at 0.5 °C per day. Thus, water stress in spring, when temperatures are increasing, would be less likely to inhibit germination than it would in autumn, when temperatures are decreasing.

Substrate moisture also interacts with light in controlling germination. White, blue, red and far-red light inhibited germination (compared with germination in darkness) of *Avena fatua* seeds subjected to moisture stress, but they promoted germination when moisture was non-limiting (Hsiao & Simpson, 1971). In seeds of *Bromus sterilis*, the inhibitory effects of red light (P_{fr} inhibits germination in this species) increase with an increase in moisture stress (Hilton, 1984b). Thus, low soil moisture and light could be very important in preventing germination of *A. fatua* and *B. sterilis* seeds on the soil surface during summer, when the soil is only wet for brief periods of time following showers.

Successful establishment of grass seedlings under moisture stress in the field may depend on their ability to penetrate the crust on the soil surface. Using simulated (wax) soil crusts, Frelich, Jensen & Gifford, (1973) found that increasing the moisture stress from 0 to -0.84 MPa did not reduce the ability of seedlings of *Agropyron trichophorum*, *A. elongatum*, *Bromus inermis*, *Elymus cinereus*, *E. junceus* or *Festuca arundinacea* to emerge through a soft crust. With a medium or hard crust, however, an increase in moisture stress up to -0.84 MPa reduced emergence in all species. In the absence of moisture stress (0 MPa), increasing the strength of the crust reduced emergence of *A. trichophorum*, *B. inermis*, *E. cinereus* and *E. junceus*, but not of *A. elongatum* or *F. arundinacea* seedlings.

Moisture stress resulting from the presence of salts in the soil can reduce germination of grass seeds (e.g. Sautter, 1962). However, there are species-specific variations in tolerances of osmotically induced moisture stress and of presence of different kinds of salts (Ryan, Miyamoto & Stroehlein, 1975; Ries & Hofmann, 1983). The NaCl concentration at which germination of grasses of salt marshes and salt deserts is reduced to about 10% is variable: *Distichlis spicata*, 0.09 M (Cluff, Evans & Young, 1983); *Hordeum jubatum*, 0.17 M, (Ungar, 1974), 0.31 M (Badger & Ungar, 1989); *Phragmites communis*, 0.34 M (Chapman, 1960); *Polypogon monspeliensis*, 0.34 M (Partridge & Wilson, 1987); *Puccinellia distans*, 0.45 M (Harivandi, Butler & Soltanpour, 1982); *P. festucaeformis*, 0.75 M (Onnis & Miceli, 1975); *P. lemmoni*, 0.45 M (Harivandi *et al.*, 1982); *Spartina alterniflora*, 0.68 M (Mooring, Cooper & Seneca, 1971) and *Sporobolus virginicus*, 0.26 M (Breen, Everson & Rodgers, 1977), with a mean ($\pm$ SE) of 0.38 $\pm$ 0.06 M. In the field, germination occurs after precipitation has decreased the salinity of the soil solution by dilution and/or leaching of salts (Flowers, Hajibagheri & Clipson, 1986). Further, germination of *Hordeum jubatum*

(Badger & Ungar, 1989) and *Puccinellia festucaeformis* (Onnis, Pelosini & Stefani, 1981) increases with a decrease in temperature. Thus, rains during the cool season in temperate regions decrease salinity at a time when temperatures are most favourable for germination.

In addition to salt marsh and salt desert habitats, the arrival of the wet season, when the soil becomes soaked by rain water, plays a major role in controlling the timing of germination of non-dormant seeds of many grasses. In arid regions of Western Australia (Mott, 1972), savanna woodlands in the Northern Territory of Australia (Mott, 1978; Torssell & McKeon, 1976), tropical savannas of western Venezuela (Silva & Ataroff, 1985), annual grasslands of California, USA (Evans *et al.*, 1975) and the Mojave Desert in south-western USA (Went, 1948), grass seeds germinate following rains that provide a sufficient amount of water to keep the soil moist for several days. The length of time the soil must be wet to promote germination varies. Seeds of *Aristida contorta* germinate to 50% within 24 hours (Mott, 1972), while 5 to 6 days are required for the first ones of *Rhynchelytrum repens* and *Rottboellia exaltata* to germinate (Popay, 1976).

Microsites

If soil moisture is non-limiting, germination depends on whether or not seeds are in species-specific 'safe sites' (*sensu* Fowler, 1986). Seeds of *Aristida contorta* must be on the soil surface where their light requirement for germination can be fulfilled after they are imbibed (Mott, 1972), and those of *Panicum turgidum* germinate to higher percentages if the flat side of the seed touches the wet substrate than if the convex side touches it (Koller & Roth, 1963). Seeds of *Bouteloua curtipendula* and *Aristida longiseta* germinate to maximum percentages in microsites (e.g. under litter), where desiccation is decreased (Fowler, 1986).

Wind and water (Mott & McComb, 1974), as well as hygroscopic awns (Peart, 1979; Sindel *et al.*, 1993), cause seeds of some grasses, including *Danthonia, Dichelachne, Heteropogon, Schizachyrium, Stipa* (Peart, 1979) and *Themeda* (Lock & Milburn, 1971) to move along the soil surface until they lodge in a microsite and/or become buried in the soil. Burial is especially important for the survival of grass seeds in fire-prone habitats. Seeds of *Themeda triandra* reach a mean soil depth of 11 mm *via* action of their hygroscopic awn (Lock & Milburn, 1971). At this depth, temperatures would be about 100 °C for only a few minutes during a fire (Bentley & Fenner, 1958), and this temperature is tolerated by dry seeds of many species (Crosier, 1956).

Wetting and drying

Seeds on, or near, the soil surface may be imbibed and dehydrated (wetted and dried) many times before they become non-dormant and/or environmental conditions become favourable for germination. Wetting and drying increased germination rates of seeds of *Avena sativa* (Berrie & Drennan, 1971), *Echinochloa turnerana* (Conover & Geiger, 1984*a*), *Hordeum leporinum* (Cocks & Donald, 1973), *Lolium rigidum* (Cocks & Donald, 1973; Lush & Groves, 1981) and *Triticum aestivum* (Lush, Groves & Kaye, 1981), but they had no effect on those of *Chrysopogon fallax*, *C. latifolius*, *Sorghum plumosum*, *S. stipoideum* or *Themeda australis* (Mott, 1978). Wetting and drying increased protease activity in *Avena sativa* seeds, even after they were dried (Berrie & Drennan, 1971), and it caused an enhancement of protein and RNA synthesis and an early induction of DNA replication in seeds of *Secale cereale*, when they subsequently were placed under continuously wet conditions for germination (Sen & Osborne, 1974).

Priming treatments

Pretreatments with osmotic water stress also potentially have an influence on germination. Seeds of *Bouteloua curtipendula*, *Cenchrus ciliaris*, *Eragrostis lehmanniana* and *Panicum coloratum* were placed over a range of water potentials from -1.5 to -7.7 MPa in PEG solutions for up to 14 days (priming treatment) and then moved to deionized water. Seeds exposed to low water potentials for long periods of time germinated to lower percentages than the controls, whereas those exposed to the high water potentials for short periods of time germinated to higher percentages than the controls (Hardegree & Emmerich, 1992). A 7-day period at 25 °C in PEG, ranging from -2.5 MPa for *Pseudoroegneria spicata* and *Leymus cinereus* to -1.0 MPa for *Poa sandbergii* and *P. canbyi* seeds, enhanced germination rates by 4–8 days (Hardegree, 1994). Germination of *Setaria faberi* seeds increased from 55% to 90% after a 24 hour pretreatment in PEG (-0.3 MPa) at 20 °C. Pretreatment at 35 °C for 4 hours increased germination from 55% to 80%, but after 24 hours of pretreatment germination declined to 25% (Taylorson, 1986). Thus, water stress prior to the time that soil moisture becomes non-limiting for germination in the field may, or may not, promote germination, depending on the species, temperature and the severity and duration of the stress pretreatment.

Hydropedesis

At water potentials of −0.5 to −0.12 MPa, seeds of some grasses, including *Dicanthium sericeum*, *Setaria porphyrantha*, *Danthonia linkii* and *Cenchrus ciliaris* (Watt, 1974, 1978, 1982; Lambert *et al*., 1990), imbibe water and the emerging plant parts break through the seed coat, but further growth does not occur; this is called hydropedesis (Watt, 1974). Some hydropedetic grass seeds can be stored dry for many weeks without loss of viability, and they may germinate more rapidly than non-treated ones when placed under non-limiting soil moisture conditions (Watt, 1978). Hydropedesis is an adaptive mechanism for seedling establishment in habitats where the soil surface dries rapidly following each rain (Watt, 1978).

Drying intolerance

Extreme cases of intolerance to water stress are found in seeds of *Zizania palustris* and *Spartina anglica*, which are reported to be recalcitrant (Probert & Longley, 1989). 'Recalcitrant' means that the seeds die if their moisture content drops below 20–45%, depending on the species. Also, preliminary data indicate that seeds of *Porteresia coarctata* probably are recalcitrant (Probert & Longley, 1989). However, Kovach & Bradford (1992*b*) dehydrated seeds of *Z. palustris* to 6–8% moisture content (fresh weight basis) at >25 °C and slowly rehydrated them over a period of 3 weeks at temperatures between 10 and 25 °C without loss of viability.

Flooding

In some grasses, the presence of an excess of water, or an oversaturated substrate, prevents germination (e.g. Blank & Young, 1992). However, in emergent hydrophytes, e.g. *Zizania aquatica* (Muenscher, 1936), *Scolochloa festucacea* (Smith, 1972), *Panicum hirsutum* (Orozco-Segovia & Vazquez-Yanes, 1980) and *Leersia oryzoides* (Rosa & Corbineau, 1986), seeds are dispersed into water, where they come out of dormancy and then germinate. Soaking seeds of *P. hirsutum* (Orozco-Segovia & Vazquez-Yanes, 1980) and *S. festucacea* (Smith, 1972) in water increased germination, but this treatment was not required for loss of dormancy. Sub-atmospheric levels of oxygen (3–15% by volume) promoted germination of *L. oryzoides* seeds (Rosa & Corbineau, 1986). Anaerobic conditions and light promoted germination of the aquatic annual grass *Orcuttia californica* in vernal pools in California (USA), but these conditions were not required for germination of *Tuctoria greenei*, another annual grass in the pools (Keeley, 1988).

Seeds of 19 species of Panicoideae germinated equally well in aerated and non-aerated water, and those of seven species of Panicoideae germinated to higher percentages in aerated than in non-aerated water. On the other hand, in no species of Chloridoideae or Bambusoideae was germination inhibited by non-aerated water (Clifford, 1988). Thus, there seem to be phylogenetic differences in ability to germinate under flooded conditions.

Seeds of the grass *Echinochloa turnerana* require a period of flooding for germination in the field, but in the laboratory non-flooded seeds germinated to 50% at −1.0 MPa, after 14 days. (Conover & Geiger, 1984*b*). Seeds also germinated under 10 cm of stagnant water at 30 °C in light, but not in darkness (Conover & Geiger, 1984*b*). Thus, timing of germination in the field is controlled by an interaction of factors.

1. Seeds have a light requirement for germination; therefore, they must be on the soil surface for germination to occur.
2. Seeds have a slow rate of germination, thus they cannot germinate unless they are in a site that is continuously moist for many days.
3. Seeds on the soil surface (under non-flooded conditions) cannot germinate because moisture becomes limiting within a few days following a rain.
4. Seeds have the ability to germinate at low oxygen levels, thus they are tolerant of flooding, which supplies the long period of favourable moisture conditions necessary for germination.

pH

The range of pH over which seeds of grasses will germinate to 50% or more has been determined for a few species, including *Iseilema anthephoroides*, 3.0 to 8.5; *Apluda mutica*, 6.0 to 11.0; *Dactyloctenium aegyptium*, 3.5 to 11.0 and *Sehima nervosum*, 3.5 to 9.0 (Singh, Mall & Billore, 1975). The optimum pH for germination of *I. anthephoroides* and *D. aegyptium* seeds was 5.0 and 9.0, respectively, (Singh *et al.*, 1975). Seeds of other species have been tested at pH of 3.5 to 6.5 (Justice & Reece, 1954) and 3.0 to 7.5 (Hackett, 1964). The pH at which they germinate to 50% or more are: *Phleum pratense*, 3.5 to 7.5; *Avena sativa*, 3.5 to 7.5 (Justice & Reece, 1954); *Alopecurus pratensis*, 3.5 to 6.5; *Festuca pratensis*, 3.5 to 6.5 and *Lolium perenne*, 3.0 to 6.5 (Hackett, 1964). Although based on a small number of species, it can be seen that grass seeds germinate over a fairly wide range of pH. Interestingly, calcifuge ('acid-loving') species such as *A. pratensis*, *F. pratensis* and *L. perenne* germinate to a high percentage at a pH of 6.5

(Hackett, 1964). Germination at this pH, however, does not mean that it is the maximum pH at which seeds of these calcifuge grasses will germinate; 6.5 is the highest pH that Hackett tested.

Little is known about the effects of pH on germination of grass seeds in nature. In sites with high levels of aluminium or other heavy metals, however, pH may play a role in controlling germination. High levels of aluminium and low pH reduced germination rates of *Deschampsia flexuosa* seeds less than they did those of *A. pratensis*, *F. pratensis* and *L. perenne* (Hackett, 1964).

Heavy metals

Heavy metals such as mercury, zinc, lead and cadmium, which are potentially very toxic to plants, can occur naturally in the environment. However, human activities such as mining, waste disposal and manufacturing have resulted in increased levels of these metals in various habitats (Pathak *et al.*, 1987). At low concentrations (0.1–10 ppm) mercury, cadmium (Mrozek, 1980), lead and zinc (Mrozek & Funicelli, 1982) caused an increase in rates of germination of *Spartina alterniflora* seeds in fresh water and at low (10 parts per thousand) salinity. However, with increased metal concentrations and increased salinity germination rates and percentages decreased. Seeds of *Miscanthus* spp. exhibited decreases in germination rates and percentages as concentrations of lead, mercury, copper and cadmium increased from 200 ppm to 1000 ppm. However, depending on the *Miscanthus* species some germination occurred in concentrations of 1000 ppm of each metal (Hsu & Chou, 1992). Mercury solutions of 1–100 mg l^{-1} reduced germination of *Hordeum vulgare* seeds, but addition of 1–100 mg l^{-1} of manganese to the solutions overcame most of the mercury inhibition (Pathak, Mukhiya & Singh, 1987).

Herbicides

Grass seeds may be exposed to herbicides in agricultural and/or natural ecosystems, and these compounds potentially could influence germination. Depending on the kind and concentration of the herbicide, germination percentages of grasses may be enhanced (Shaukat, 1974; Huffman & Jacoby, 1984; Morash & Freedman, 1989), reduced (Allard, DeRose & Swanson, 1946; Mitchell and Marth, 1946; Morash & Freedman, 1989) or unaffected (Hamner, Moulton & Tukey, 1946; Shaukat, 1974; Benjamini, 1986; Huffman & Jacoby, 1984). In some cases, low concentrations of a particular herbicide stimulate germination, but high concentrations inhibit it (Shaukat, 1974; Huffman & Jacoby, 1984; Morash & Freedman, 1989). Although herbicides affect germination of grass seeds under laboratory/greenhouse conditions, they may be insignificant in the field (Morash & Freedman, 1989).

Concluding thoughts

As pointed out in the Introduction of this chapter, the timing of germination can be an important adaptation of a species to its habitat. Further, the genetic variation in seed dormancy and germination characteristics of grasses and the differences in germination characteristics of seeds of some species collected in diverse habitats indicate that the timing of germination in a species can be changed to correspond closely to local conditions. When these facts are considered in view of the geographical distribution patterns and the types of plant communities and specialized habitats in which grasses grow, there is clearly a great diversity of germination ecologies in the Poaceae. How can so much diversity be accounted for in the Poaceae? The explanation for the diversity is a combination of factors.

1. Depending on the species, dormancy break occurs in response to either high summer or low winter temperatures.
2. Seeds of many grasses undergo annual dormancy cycles, and there are two kinds of dormancy cycles, CD/ND and D/ND.
3. Both annual CD/ND and D/ND cycles occur in seeds that come out of dormancy in response to high as well as low temperatures.
4. Seeds of many grasses form persistent seed banks.
5. Some seeds are ND at maturity, and there may be others that have CD or D at maturity which become ND and remain ND.
6. The combination of environmental conditions required for germination varies with the species.
7. Seeds of some species are very tolerant of unique environmental stresses such as high salinity or flooding.

References

Acharya, S. N. (1989). Germination response of two alpine grasses from the Rocky Mountains of Alberta. *Canadian Journal of Plant Science*, **69**, 1165–77.
Adkins, S. W., Loewen, M. & Symons, S. J. (1986). Variation within pure lines of wild oats (*Avena fatua*) in relation to degree of primary dormancy. *Weed Science*, **34**, 859–64.

Ahring, R. M., Dunn, N. L., Jr. & Harlan, J. R. (1963). Effect of various treatments in breaking seed dormancy in sand lovegrass, *Eragrostis trichodes* (Nutt.) Wood. *Crop Science*, **3**, 131–3.

Ahring, R. M., Eastin, J. D. & Garrison, C. S. (1975). Seed appendages and germination of two Asiatic bluestems. *Agronomy Journal*, **67**, 321–5.

Ahring, R. M. & Frank, H. (1968). Establishment of eastern gamagrass from seed and vegetative propagation. *Journal of Range Management*, **21**, 27–30.

Ahring, R. M. & Todd, G. W. (1977). The bur enclosure of the caryopses of buffalograss as a factor affecting germination. *Agronomy Journal*, **69**, 15–17.

Akpan, E. E. J. & Bean, E. W. (1977). The effects of temperature upon seed development in three species of forage grasses. *Annals of Botany*, **41**, 689–95.

Allard, R. W., DeRose, H. R. & Swanson, C. P. (1946). Some effects of plant growth-regulators on seed germination and seedling development. *Botanical Gazette*, **107**, 575–83.

Andersen, A. M. (1953). The effect of the glumes of *Paspalum notatum* Flugge on germination. *Proceedings of the Association of Official Seed Analysts*, **43**, 93–100.

Anderson, R. C. (1985). Aspects of the germination ecology and biomass production of eastern gamagrass (*Tripsacum dactyloides* L.). *Botanical Gazette*, **146**, 353–64.

Andrew, M. H. & Mott, J. J. (1983). Annuals with transient seed banks: The population biology of indigenous *Sorghum* species of tropical north-west Australia. *Australian Journal of Ecology*, **8**, 265–76.

Antonovics, J. & Schmitt, J. (1986). Paternal and maternal effects of propagule size in *Anthoxanthum odoratum*. *Oecologia*, **69**, 277–82.

Badger, K. S. & Ungar, I. A. (1989). The effects of salinity and temperature on the germination of the inland halophyte *Hordeum jubatum*. *Canadian Journal of Botany*, **68**, 1420–5.

Bansal, R. P., Bhati, P. R. & Sen, D. N. (1980). Differential specificity in water imbibition of Indian arid zone seeds. *Biologia Plantarum*, **22**, 327–31.

Banting, J. D. (1979). Germination, emergence and persistence of foxtail barley. *Canadian Journal of Plant Science*, **59**, 35–41.

Banting, J. D. & Gebhardt, J. P. (1979). Germination, afterripening, emergence, persistence and control of Persian darnel. *Canadian Journal of Plant Science*, **59**, 1037–45.

Barrett, S. C. H. & Wilson, B. F. (1983). Colonizing ability in the *Echinochloa crus-galli* complex (barnyard grass). II. Seed biology. *Canadian Journal of Botany*, **61**, 556–62.

Baskin, C. C. & Baskin, J. M. (1988). Germination ecophysiology of herbaceous plant species in a temperate region. *American Journal of Botany*, **75**, 286–305.

Baskin, C. C., Baskin, J. M. & Chester, E. W. (1993). Germination ecology of *Leptochloa panicoides*, a summer annual grass of seasonally dewatered mudflats. *Acta Oecologia*, **14**, 693–704.

Baskin, C. C., Baskin, J. M. & El-Moursey, S. A. (1996). Seasonal changes in germination responses of buried seeds of the weedy summer annual grass *Setaria glauca*. *Weed Research*, **36**, 319–24.

Baskin, J. M. & Baskin, C. C. (1981). Ecology of germination and flowering in the weedy winter annual grass *Bromus japonicus*. *Journal of Range Management*, **34**, 369–72.

Baskin, J. M. & Baskin, C. C. (1983). Seasonal changes in the germination responses of fall panicum to temperature and light. *Canadian Journal of Plant Science*, **63**, 973–9.

Baskin, J. M. & Baskin, C. C. (1985*a*). The annual dormancy cycle in buried weed seeds: A continuum. *BioScience*, **35**, 492–8.
Baskin, J. M. & Baskin, C. C. (1985*b*). Dormancy breaking and germination requirements of nimble will (*Muhlenbergia schreberi* Gmel.) seeds. *Journal of Range Management*, **38**, 513–5.
Baskin, J. M. & Baskin, C. C. (1985*c*). Does seed dormancy play a role in the germination ecology of *Rumex crispus*? *Weed Science*, **33**, 340–3.
Baskin, J. M. & Baskin, C. C. (1986). Seasonal changes in the germination responses of buried witchgrass (*Panicum capillare*) seeds. *Weed Science*, **34**, 22–4.
Baskin, J. M. & Caudle, C. (1967). Germination and dormancy in cedar glade plants. I. *Aristida longespica* and *Sporobolus vaginiflorus*. *Journal of the Tennessee Academy of Science*, **42**, 132–3.
Baxter, B. J. M. & van Staden, J. (1993). Coat imposed and embryo dormancy in *Themeda triandra* Forsk. In *Fourth International Workshop on Seeds. Volume 2*, pp. 677–82. Paris: Université Pierre et Marie Curie. Angers, France, 20–24 July 1992, ed. D. Côme & F. Corbineau.
Baxter, B. J. M., van Staden, J & Granger, J. E. (1993). Seed germination response to temperature, in two altitudinally separate populations of the perennial grass *Themeda triandra*. *South African Journal of Science*, **89**, 141–4.
Beal, W. J. (1911). The vitality of seeds buried in the soil. *Proceedings of the Society for the Promotion of Agricultural Science*, 31st annual meeting, **32**, 21–3.
Bean, E. W. (1971). Temperature effects upon inflorescence and seed development in tall fescue (*Festuca arundinacea* Schreb.). *Annals of Botany*, **35**, 891–7.
Belderok, B. (1961). Studies on dormancy in wheat. *Proceedings of the International Seed Testing Association*, **26**, 697–760.
Bell, T. J. & Quinn, J. A. (1985). Relative importance of chasmogamously and cleistogamously derived seeds of *Dichanthelium clandestinum* (L.) Gould. *Botanical Gazette*, **146**, 252–8.
Benech Arnold, R. L., Ghersa, C. M., Sanchez, R. A. & Garcia Fernandez, A. E. (1988). The role of fluctuating temperatures in the germination and establishment of *Sorghum halepense* (L.) Pers. Regulation of germination under leaf canopies. *Functional Ecology*, **2**, 311–8.
Benjamini, L. (1986). Effect of carbofuran on seed germination and initial development of seven crops. *Phytoparasitica*, **14**, 219–30.
Bentley, J. R. & Fenner, R. L. (1958). Soil temperatures during burning related to postfire seedbeds on woodland range. *Journal of Forestry*, **56**, 737–40.
Berrie, A. M. M. & Drennan, D. S. H. (1971). The effect of hydration-dehydration on seed germination. *New Phytologist*, **70**, 135–42.
Blank, R. R. & Young, J. A. (1992). Influence of matric potential and substrate characteristics on germination of Nezpar Indian ricegrass. *Journal of Range Management*, **45**, 205–9.
Bliss, L. C. (1958). Seed germination in arctic and alpine species. *Arctic*, **11**, 180–8.
Bokhari, U. G., Singh, J. S & Smith, F. M. (1975). Influence of temperature regimes and water stress on the germination of three range grasses and its possible ecological significance to a shortgrass prairie. *Journal of Applied Ecology*, **12**, 153–63.
Borthwick, H. A., Hendricks, S. B., Schneider, M. J. Taylorson, R. B. & Toole, V. K. (1969). The high energy light action controlling plant responses and development. *Proceedings of the National Academy of Sciences, USA*, **64**, 479–86.

Boyce, K. G., Cole, D. F & Chilcote, D. O. (1976). Effect of temperature and dormancy on germination of tall fescue. *Crop Science*, **16**, 15–8.

Boyd, W. J. R., Gordon, A. G. & LaCroix, L. J. (1971). Seed size, germination resistance and seedling vigor in barley. *Canadian Journal of Plant Science*, **51**, 93–9.

Boydston, R. A. (1989). Germination and emergence of longspine sandbur (*Cenchrus longispinus*). *Weed Science*, **37**, 63–7.

Breen, C. M., Everson, C & Rodgers, K. (1977). Ecological studies on *Sporobolus virginicus* (L.) Kunth, with particular reference to salinity and inundation. *Hydrobiologia*, **54**, 135–40.

Brenchley, W. E. & Warington, K. (1930). The weed seed population of arable soil. I. Numerical estimation of viable seeds and observations on their natural dormancy. *Journal of Ecology*, **18**, 235–72.

Bridgemohan, P., Brathwaite, R. A. I. & McDavid, C. R. (1991). Seed survival and patterns of seedling emergence studies of *Rottboellia cochinchinensis* (Lour.) W. D. Clayton in cultivated soils. *Weed Research*, **31**, 265–72.

Brocklehurst, P. A., Moss, J. P. & Williams, W. (1978). Effects of irradiance and water supply on grain development in wheat. *Annals of Applied Biology*, **90**, 265–76.

Buchele, D. E., Baskin, J. M. & Baskin, C. C. (1991). Ecology of the endangered species *Solidago shortii*. III. Seed germination ecology. *Bulletin of the Torrey Botanical Club*, **118**, 288–91.

Burnside, O. C. (1965). Seed and phenological studies with shattercane. *University of Nebraska College of Agriculture Experiment Station Research Bulletin*, **220**.

Burnside, O. C., Fenster, C. R., Evetts, L. L. & Mumm, R. F. (1981). Germination of exhumed weed seed in Nebraska. *Weed Science*, **29**, 577–86.

Campbell, C. S., Quinn, J. A., Cheplick G. P. & Bell, T. J. (1983). Cleistogamy in grasses. *Annual Review of Ecology and Systematics*, **14**, 411–41.

Canode, C. L., Horning, E. V. & Maguire, J. D. (1963). Seed dormancy in *Dactylis glomerata* L. *Crop Science*, **3**, 17–19.

Cavers, P. B., Kane, M. & O'Toole, J. J. (1992). Importance of seedbanks for establishment of newly introduced weeds – a case study of proso millet (*Panicum miliaceum*). *Weed Science*, **40**, 630–5.

Chang, T.-T. & Li, C.-C. (1991). Genetics and breeding. In: *Rice. Volume I. Production,* ed. B. S. Luh, pp. 23–101. Van Nostrand Reinhold, New York.

Chapman, V. J. (1960). *Salt Marshes and Salt Deserts of the World*. London: Leonard Hill [Books] Limited.

Chauvel, B., Gasquez, J. & Darmency, H. (1989). Changes of weed seed bank parameters according to species, time and environment. *Weed Research*, **29**, 213–9.

Cheplick, G. P. & Quinn, J. A. (1982). *Amphicarpum purshii* and the 'pessimistic strategy' in amphicarpic annuals with subterranean fruit. *Oecologia*, **52**, 327–32.

Christensen, C. M. & Kaufmann, H. H. (1969). *Grain Storage: The Role of Fungi in Quality Loss*. Minneapolis: University of Minnesota Press.

Christie, B. R & Kalton, R. R. (1960). Recurrent selection for seed weight in bromegrass, *Bromus inermis* Leyss. *Agronomy Journal*, **52**, 575–8.

Clark, D. C. & Bass, L. N. (1970). Germination experiments with seeds of Indian ricegrass, *Oryzopsis hymenoides* (Roem. and Schult.) Ricker. *Proceedings of the Association of Official Seeds Analysts*, **60**, 226–39.

Clark, S. C. (1974). Biological flora of the British Isles. *Catapodium rigidum* (L.) D. C. Hubbard. *Journal of Ecology*, **62**, 937–58.

Clay, K. (1983). The differential establishment of seedlings from chasmogamous and cleistogamous flowers in natural populations of the grass *Danthonia spicata* (L.) Beauv. *Oecologia*, **57**, 183–8.

Clifford, H. T. (1988). The taxonomic significance of the ability of grass ʻseedʼ to germinate under waterlogged conditions. *Kew Bulletin*, **43**, 327–8

Cluff, G. J., Evans, R. A. & Young, J. A. (1983). Desert saltgrass seed germination and seedbed ecology. *Journal of Range Management*, **36**, 419–22.

Cocks, P. S. & Donald, C. M. (1973). The germination and establishment of two annual pasture grasses (*Hordeum leporinum* Link and *Lolium rigidum* Gaud). *Australian Journal of Agricultural Research*, **24**, 1–10.

Cohn, M. A. & Hughes, J. A. (1981). Seed dormancy in red rice (*Oryza sativa*). I. Effect of temperature on dry-afterripening. *Weed Science*, **29**, 402–4.

Connor, H. E. (1979). Breeding systems in the grasses: A survey. *New Zealand Journal of Botany*, **17**, 547–74.

Conover, D. G. & Geiger, D. R. (1984*a*). Germination of Australian channel millet [*Echinochloa turnerana* (Domin) J. M. Black] seeds. I. Dormancy in relation to light and water. *Australian Journal of Plant Physiology*, **11**, 395–408.

Conover, D. G. & Geiger, D. R. (1984*b*). Germination of Australian channel millet [*Echinochloa turnerana* (Domin) J. M. Black] seeds. II. Effects of anaerobic conditions, continuous flooding, and low water potential. *Australian Journal of Plant Physiology*, **11**, 409–17.

Corbineau, F., Belaid, D. & Côme, C. (1992). Dormancy of *Bromus rubens* L. seeds in relation to temperature, light and oxygen effects. *Weed Research*, **32**, 303–10.

Corner, E. J. H. (1976). *The Seeds of Dicotyledons*. 2 volumes. Cambridge: Cambridge University Press.

Coukos, C. J. (1944). Seed dormancy and germination in some native grasses. *Agronomy Journal*, **36**, 337–45.

Crosier, W. (1956). Longevity of seeds exposed to dry heat. *Proceedings of the Association of Official Seed Analysts*, **46**, 72–4.

Datta, S. C., Evenari, M. & Gutterman, Y. (1970). The heteroblasty of *Aegilops ovata* L. *Israel Journal of Botany*, **19**, 463–83.

Datta, S. C., Gutterman, Y. & Evenari, M. (1972). The influence of the origin of the mother plant on yield and germination of their caryopses in *Aegilops ovata*. *Planta*, **105**, 155–64.

Dawson, J. H. & Bruns, V. F. (1975). Longevity of barnyardgrass, green foxtail, and yellow foxtail seeds in soil. *Weed Science*, **23**, 437–40.

Dessaint, F., Chadoeuf, R. & Barralis, G. (1991). Spatial pattern analysis of weed seeds in the cultivated soil seed bank. *Journal of Applied Ecology*, **28**, 721–30.

Dickens, R. & Moore, G. M. (1974). Effects of light, temperature, KNO_3, and storage on germination of cogongrass. *Agronomy Journal*, **66**, 187–8.

Dillon, S. P. & Forcella, F. (1984). Germination, emergence, vegetative growth and flowering of two silvergrasses, *Vulpia bromoides* (L.) S. F. Gray and *V. myuros* (L.) C. C. Gmel. *Australian Journal of Botany*, **32**, 165–75.

Dixon, J. M. (1995). Biological flora of the British Isles. *Trisetum flavescens* (L.) Beauv. (*T. pratense* Pers., *Avena flavescens* L.). *Journal of Ecology*, **83**, 895–909.

Dobrenz, A. K. & Beetle, A. A. (1966). Cleistogenes in *Danthonia*. *Journal of Range Management*, **19**, 292–6.

Donald, W. W. (1991). Seed survival, germination ability, and emergence of jointed goatgrass (*Aegilops cylindrica*). *Weed Science*, **39**, 210–16.

Donelan, M. & Thompson, K. (1980). Distribution of buried viable seeds along a successional series. *Biological Conservation*, **17**, 297–311.

Dorph-Petersen, K. (1925). Examinations of the occurrence and vitality of various weed seed species under different conditions, made at the Danish State Seed Testing Station during the years 1896–1923. *International Seed Testing Congress (Cambridge) Report*, 4, 124–38.

Dunwell, J. M. (1981). Dormancy and germination in embryos of *Hordeum vulgare* L. – Effect of dissection, incubation temperature and hormone application. *Annals of Botany*, 48, 203–13.

Dwyer, D. D. & Aguirre V., E. (1978). Plants emerging from soil under three range condition classes of desert grassland. *Journal of Range Management*, 31, 209–12.

Eagles, H. A. & Hardacre, A. K. (1979). Genetic variation in maize (*Zea mays* L.) for germination and emergence at 10 °C. *Euphytica*, 28, 287–95.

Ebersole, J. J. (1989). Role of the seed bank in providing colonizers on a tundra disturbance in Alaska. *Canadian Journal of Botany*, 67, 466–71.

Egley, G. H. & Chandler, J. M. (1983). Longevity of weed seeds after 5.5 years in the Stoneville 50-year buried-seed study. *Weed Science*, 31, 264–70.

Elberse, W. Th. & Breman, H. (1989). Germination and establishment of Sahelian rangeland species. I. Seed properties. *Oecologia*, 80, 477–84.

Ellis, R. H., Hong, T. D. & Roberts, E. H. (1986a). Quantal response of seed germination of *Brachiaria humidicola*, *Echinochloa turnerana*, *Eragrostis tef* and *Panicum maximum* to photon dose for the low energy reaction and the high irradiance reaction. *Journal of Experimental Botany*, 37, 742–53.

Ellis, R. H., Hong, T. D. & Roberts, E. H. (1986b). The response of seeds of *Bromus sterilis* L. and *Bromus mollis* L. to white light of varying photon flux density and photoperiod. *New Phytologist*, 104, 485–96.

Emal, J. G. & Conard, E. C. (1973). Seed dormancy and germination in Indiangrass as affected by light, chilling, and certain chemical treatments. *Agronomy Journal*, 65, 383–5.

Enright, N. J. (1985). Existence of a soil seed bank under rainforest in New Guinea. *Australian Journal of Ecology*, 10, 67–71.

Enright, N. J. & Cameron, E. K. (1988). The soil seed bank of a kauri (*Agathis australis*) forest remnant near Auckland, New Zealand. *New Zealand Journal of Botany*, 26, 223–36.

Ernst, W. H. O. (1981). Ecological implication of fruit variability in *Phleum arenarium* L., an annual dune grass. *Flora*, 171, 387–98.

Ernst, W. H. O., Kuiters, A. T. & Tolsma, D. J. (1991). Dormancy of annual and perennial grasses from a savanna of southeastern Botswana. *Acta Oecologia*, 12, 727–39.

Evans, R. A., Kay, B. L. & Young, J. A. (1975). Microenvironment of a dynamic annual community in relation to range improvement. *Hilgardia*, 43, 79–102.

Evans, R. A., Young, J. A. & Roundy, B. A. (1977). Seedbed requirements for germination of Sandberg bluegrass. *Agronomy Journal*, 69, 817–20.

Fawcett, R. S. & Slife, F. W. (1978). Effects of 2,4-D and dalapon on weed seed production and dormancy. *Weed Science*, 26, 543–7.

Felippe, G. M., Silva, J. C. S. & Cardoso, V. J. M. (1983). Germination studies in *Andropogon gayanus* Kunth. *Revista Brasileira de Botanica*, 6, 41–8.

Fenner, M. (1980). Germination tests on thirty-two East African weed species. *Weed Research*, 20, 135–8.

Fischer, M. L., Stritzke, J. F. & Ahring, R. M. (1982). Germination and emergence of little barley (*Hordeum pusillum*). *Weed Science*, 30, 624–8.

Fiore, E. B. & Putz, F. E. (1992). Buried dormant seeds in bottom sediments of Newmans Lake. *Florida Scientist*, 55, 157–9.

Flowers, T. J., Hajibagheri, M. A. & Clipson, N. J. W. (1986). Halophytes. *Quarterly Review of Biology*, **61**, 313–37.

Foley, M. E. (1994). Temperature and water status of seed affect afterripening in wild oat (*Avena fatua*). *Weed Science*, **42**, 200–4.

Fowler, N. L. (1986). Microsite requirements for germination and establishment of three grass species. *The American Midland Naturalist*, **115**, 131–45.

Fox, R. L. & Albrecht, W. A. (1957). Soil fertility and the quality of seeds. *Missouri Agriculture Experiment Station Research Bulletin*, **619**.

Frank, A. B. & Larson, K. L. (1970). Influence of oxygen, sodium hypochlorite, and dehulling on germination of green needlegrass seed (*Stipa viridula* Trin.). *Crop Science*, **10**, 679–83.

Frelich, J. R., Jensen, E. H. & Gifford, R. O. (1973). Effect of crust rigidity and osmotic potential on emergence of six grass species. *Agronomy Journal*, **65**, 26–9.

Friesen, G. & Shebeski, L. H. (1961). The influence of temperature on the germination of wild oat seeds. *Weeds*, **9**, 634–8.

Froud-Williams, R. J. & Chancellor, R. J. (1986). Dormancy and seed germination of *Bromus catharticus* and *Bromus commutatus*. *Seed Science and Technology*, **14**, 439–50.

Froud-Williams, R. J., Drennan, D. S. H. & Chancellor, R. J. (1984). The influence of burial and dry-storage upon cyclic changes in dormancy, germination and response to light in seeds of various arable weeds. *New Phytologist*, **96**, 473–81.

Fulbright, T. E. & Flenniken, K. S. (1988). Causes of dormancy in *Paspalum plicatulum* (Poaceae) seeds. *Southwestern Naturalist*, **33**, 35–9.

Fulbright, T. E., Pedente, E. F. & Wilson, A. M. (1983). Germination requirements of green needlegrass (*Stipa viridula* Trin.). *Journal of Range Management*, **36**, 390–4.

Gfeller, F. & Svejda, F. (1960). Inheritance of post-harvest seed dormancy and kernel colour in spring wheat lines. *Canadian Journal of Plant Science*, **40**, 1–6.

Ghersa, C. M., Benech Arnold, R. L. & Martinez-Ghersa, M. A. (1992). The role of fluctuating temperatures in germination and establishment of *Sorghum halepense*. Regulation of germination at increasing depths. *Functional Ecology*, **6**, 460–8.

Giles, B. E. (1990). The effects of variation in seed size on growth and reproduction in the wild barley *Hordeum vulgare* ssp. *spontaneum*. *Heredity*, **64**, 239–50.

Ginzo, H. D. (1978). Red and far red inhibition of germination in *Aristida murina* Cav. *Zeitschrift fur Pflanzenphysiologie*, **90**, 303–7.

Goedert, C. O. & Roberts, E. H. (1986). Characterization of alternating-temperature regimes that remove seed dormancy in seeds of *Brachiaria humidicola* (Rendle) Schweickerdt. *Plant, Cell and Environment*, **9**, 521–5.

González-Rabanal, F., Casal, M. & Trabaud, L. (1994). Effects of high temperatures, ash and seed position in the inflorescence on the germination of three Spanish grasses. *Journal of Vegetation Science*, **5**, 289–94.

Górski, T., Górska, K. & Nowicki, J. (1977). Germination of seeds of various herbaceous species under leaf canopy. *Flora*, **166**, 249–59.

Górski, T., Górska, K. & Rybicki, J. (1978). Studies on the germination of seeds under leaf canopy. *Flora*, **167**, 289–99.

Goss, W. L. (1939). Germination of buried weed seeds. *California Department of Agriculture Bulletin*, **28**, 132–5.

Govinthasamy, T. & Cavers, P. B. (1995). The effects of smut (*Ustilago destruens*) on seed production, dormancy, and viability in fall panicum (*Panicum dichotomiflorum*). *Canadian Journal of Botany*, 73, 1628–34.

Granstrom, A. (1982). Seed banks in five boreal forest stands originating between 1810 and 1963. *Canadian Journal of Botany*, 60, 1815–21.

Granstrom, A. (1987). Seed viability of fourteen species during five years of storage in a forest soil. *Journal of Ecology*, 75, 321–31.

Greipsson, S. & Davy, A. J. (1994). Germination of *Leymus arenarius* and its significance for land reclamation in Iceland. *Annals of Botany*, 73, 393–401.

Grilz, P. L., Romo, J. T. & Young, J. A. (1994). Comparative germination of smooth brome and plains rough fescue. *Prairie Naturalist*, 26, 157–70.

Grime, J. P., Mason, G., Curtis, A. V, Rodman, J., Band, S. R, Mowforth, M. A. G., Neal, A. M. & Shaw, S. (1981). A comparative study of germination characteristics in a local flora. *Journal of Ecology*, 69, 1017–59.

Gritton, E. T. & Atkins, R. E. (1961). Germination of sorghum seed as affected by dormancy. *Agronomy Journal*, 55, 169–74.

Groves, R. H., Hagon, M. W. & Ramakrishnan, P. S. (1982). Dormancy and germination of seed of eight populations of *Themeda australis*. *Australian Journal of Botany*, 30, 373–86.

Gutterman, Y. (1992). Ecophysiology of Negev upland annual grasses. In *Desertified Grasslands: Their Biology and Management* ed. G. P. Chapman, pp. 145–62. The Linnean Society Symposium Series. London: Academic Press.

Gutterman, Y. & Nevo, E. (1994). Temperatures and ecological-genetic differentiation affecting the germination of *Hordeum spontaneum* caryopses harvested from three populations: The Negev Desert and opposing slopes on Mediterranean Mount Carmel. *Israel Journal of Plant Science*, 42, 183–95.

Hacker, J. B. (1984). Genetic variation in seed dormancy in *Digitaria milanjiana* in relation to rainfall at the collection site. *Journal of Applied Ecology*, 21, 947–59.

Hacker, J. B. (1989). The potential for buffel grass renewal from seed in 16-year-old buffel grass-siratro pastures in south-east Queensland. *Journal of Applied Ecology*, 26, 213–22.

Hacker, J. B., Andrew, M. H., McIvor, J. G. & Mott, J. J. (1984). Evaluation in contrasting climates of dormancy characteristics of seed of *Digitaria milanjiana*. *Journal of Applied Ecology*, 21, 961–9.

Hackett, C. (1964). Ecological aspects of the nutrition of *Deschampsia flexuosa* (L.) Trin. *Journal of Ecology*, 52, 159–67.

Haferkamp, M. R., Karl, M. G. & MacNeil, M. D. (1994). Influence of storage, temperature, and light on germination of Japanese brome seed. *Journal of Range Management*, 47, 140–4.

Hagon, M. W. (1976). Germination and dormancy of *Themeda australis, Danthonia* spp., *Stipa bigeniculata* and *Bothriochloa macra*. *Australian Journal of Botany*, 24, 319–27.

Hamilton, R. I., Subramanian, B., Reddy, M. N. & Rao, C. H. (1982). Compensation in grain yield components in a panicle of rainfed sorghum. *Annals of Applied Biology*, 101, 119–25.

Hammouda, M. A. & Bakr, Z. Y. (1969). Some aspects of germination of desert seeds. *Phyton (Austria)*, 13, 183–201.

Hamner, C. L., Moulton, J. E. & Tukey, H. B. (1946). Effect of treating soil and seeds with 2,4-dichlorophenoxyacetic acid on germination and development of seedlings. *Botanical Gazette*, 107, 352–61.

Hardegree, S. P. (1994). Drying and storage effects on germination of primed grass seeds. *Journal of Range Management*, **47**, 196–9.

Hardegree, S. P. & Emmerich, W. E. (1991). Variability in germination rate among seed lots of Lehmann lovegrass. *Journal of Range Management*, **44**, 323–6.

Hardegree, S. P. & Emmerich, W. E. (1992). Effect of matric-priming duration and priming water potential on germination of four grasses. *Journal of Experimental Botany*, **43**, 233–8.

Hardegree, S. P. & Emmerich, W. E. (1993). Germination response of hand-threshed Lehmann lovegrass seeds. *Journal of Range Management*, **46**, 203–7.

Harivandi, M. A., Butler, J. D. & Soltanpour, P. N. (1982). Effects of sea water concentration on germination and ion accumulation in alkaligrass (*Puccinellia* spp.). *Communications in Soil Science and Plant Analysis*, **13**, 507–17.

Harlan, J. R. (1945). Cleistogamy and chasmogamy in *Bromus carinatus* Hook. & Arn. *American Journal of Botany*, **32**, 66–72.

Harper, J. L., Lovell, P. H. & Moore, K. G. (1970). The shapes and sizes of seeds. *Annual Review of Ecology and Systematics*, **1**, 327–56.

Harradine, A. R. (1980). The biology of African feather grass (*Pennisetum macrourum* Trin.) in Tasmania, I. Seedling establishment. *Weed Research*, **20**, 165–9.

Harty, R. L. & McDonald, T. J. (1972). Germination behaviour in beach spinifex (*Spinifex hirsutus* Labill.). *Australian Journal of Botany*, **20**, 241–51.

Hassan, M. A. & West, N. E. (1986). Dynamics of soil seed pools in burned and unburned sagebrush semi-deserts. *Ecology*, **67**, 269–72.

Hayashi, I. & Numata, M. (1971). Viable buried-seed population in the *Miscanthus*- and *Zoysia*-type grasslands in Japan – Ecological studies on the buried-seed population in the soil related in plant succession VI. *Japanese Journal of Ecology.*, **20**, 243–52.

Haywood, M. D. & Breese, E. L. (1966). The genetic organization of natural populations of *Lolium perenne*. I. Seed and seedling characters. *Heredity*, **21**, 287–304.

Heslop-Harrison, J. (1959). Photoperiodic effects on sexuality, breeding system and seed germinability in *Rottboellia exaltata*. *Proceedings of the Ninth International Botanical Congress*, **2**, 162–3.

Hilton, J. R. (1982). An unusual effect of the far-red absorbing form of phytochrome: Photoinhibition of seed germination in *Bromus sterilis* L. *Planta*, **155**, 524–8.

Hilton, J. R. (1984a). The influence of dry storage temperature on the response of *Bromus sterilis* L. seeds to light. *New Phytologist*, **98**, 129–34.

Hilton, J. R. (1984b). The influence of temperature and moisture status on the photoinhibition of seed germination in *Bromus sterilis* L. by the far-red absorbing form of phytochrome. *New Phytologist*, **97**, 369–74.

Hilton, J. R. (1985). The influence of light and potassium nitrate on dormancy and germination of *Avena fatua* L. (wild oat) seed stored buried under natural conditions. *Journal of Experimental Botany*, **36**, 974–79.

Hilton, J. R., Froud-Williams, R. J. & Dixon, J. (1984). A relationship between phytochrome photoequilibrium and germination of seeds of *Poa trivialis* L. from contrasting habitats. *New Phytologist*, **97**, 375–9.

Hilu, K. W. & de Wet, J. M. J. (1980). Effect of artificial selection on grain dormancy in *Eleusine* (Gramineae). *Systematic Botany*, **5**, 54–60.

Hoffman, G. R. (1985). Germination of herbaceous plants common to aspen

72   C. C. Baskin and J. M. Baskin

forests of western Colorado. *Bulletin of the Torrey Botanical Club*, **112**, 409–13.

Hou, J. Q. & Simpson, G. M. (1990). Phytochrome action and water status in seed germination of wild oats (*Avena fatua*). *Canadian Journal of Botany*, **68**, 1722–7.

Hou, J. Q. & Simpson, G. M. (1991). Effects of prolonged light on germination of six lines of wild oat (*Avena fatua*) *Canadian Journal of Botany*, **69**, 1414–17.

Howard, T. M. (1973). *Nothofagus cunninghamii* ecotonal stages. Buried viable seed in north west Tasmania. *Proceedings of the Royal Society of Victoria*, **86**, 137–42.

Hsiao, A. I.-H. & Simpson, G. M. (1971). Dormancy studies in seed of *Avena fatua*. 7. The effects of light and variation in water regime on germination. *Canadian Journal of Botany*, **49**, 1347–57.

Hsu, F. H. & Chou, C.-H. (1992). Inhibitory effects of heavy metals on seed germination and seedling growth of *Miscanthus* species. *Botanical Bulletin of Academia Sinica*, **33**, 335–42.

Hsu, F. H., Nelson, C. J. & Matches, A. G. (1985). Temperature effects on germination of perennial warm-season forage grasses. *Crop Science*, **25**, 215–20.

Huffman, A. H. & Jacoby, P. W., Jr. (1984). Effects of herbicides on germination and seedling development of three native grasses. *Journal of Range Management*, **37**, 40–3.

Hulbert, L. C. (1955). Ecological studies of *Bromus tectorum* and other annual bromegrasses. *Ecological Monographs*, **25**, 181–213.

Hunt, O. J. & Miller, D. G. (1965). Coleoptile length, seed size, and emergence in intermediate wheatgrass (*Agropyron intermedium* (Host) Beauv.). *Agronomy Journal*, **57**, 192–5.

Hurka, H. & Haase, R. (1982). Seed ecology of *Capsella bursa-pastoris* (Cruciferae): Dispersal mechanism and the soil seed bank. *Flora*, **172**, 35–46.

Hussain, R. & Ilahi, I. (1990). Germination study on some grasses. *Science Khyber*, **3**, 209–15.

Hylton, L. O., Jr. & Bass, L. N. (1961). Germination of sixweeks fescue. *Proceedings of the Association Official Seed Analysts*, **51**, 118–22.

Itabari, J. K., Gregory, P. J. & Jones, R. K. (1993). Effects of temperature, soil water status and depth of planting on germination and emergence of maize (*Zea mays*) adapted to semi-arid eastern Kenya. *Experimental Agriculture*, **29**, 351–64.

Jana, S., Acharya, S. N. & Naylor, J. M. (1979). Dormancy studies in seed of *Avena fatua*. 10. On the inheritance of germination behaviour. *Canadian Journal of Botany*, **57**, 1663–7.

Janzen, D. H. (1976). Why bamboos wait so long to flower. *Annual Review of Ecology and Systematics*, **7**, 347–91.

Jones, T. A. & Nielson, D. C. (1992). Germination of prechilled mechanically scarified and unscarified Indian ricegrass seed. *Journal of Range Management*, **24**, 175–9.

Jordan, J. L., Staniforth, D. W. & Jordan, C. M. (1982). Parental stress and prechilling effects of Pennsylvania smartweed (*Polygonum pensylvanicum*) achenes. *Weed Science*, **30**, 243–8.

Juliano, J. B. (1940). Viability of some Philippine weed seeds. *Philippine Agriculture*, **29**, 313–26.

Junttila, O. (1977). Dormancy in dispersal units of various *Dactylis glomerata* populations. *Seed Science and Technology*, **5**, 463–71.

Junttila, O., Landgraff, A. & Nilsen, A. J. (1978). Germination of *Phalaris* seeds. *Acta Horticulture*, **83**, 163–6.

Justice, O. L. & Reece, M. H. (1954). A review of literature and investigation on the effects of hydrogen-ion concentration on the germination of seeds. *Proceedings of the Association Official Seed Analysts*, **44**, 144–9.

Keeley, J. E. (1988). Anaerobiosis as a stimulus to germination in two vernal pool grasses. *American Journal of Botany*, **75**, 1086–9.

Kellman, M. C. (1974). The viable weed seed content of some tropical agricultural soils. *Journal of Applied Ecology*, **11**, 669–78.

Kellman, M. C. (1980). Geographic patterning in tropical weed communities and early secondary successions. *Biotropica*, **12** (Supplement), 34–9.

Khan, R. A. & Laude, H. M. (1969). Influence of heat stress during seed maturation on germinability of barley seed at harvest. *Crop Science*, **9**, 55–8.

Kinucan, R. J. & Smeins, F. E. (1992). Soil seed bank of a semiarid Texas grassland under three long-term (36-years) grazing regimes. *The American Midland Naturalist*, **128**, 11–21.

Kivilaan, A. & Bandurski, R. S. (1981). The one hundred-year period for Dr. Beal's seed viability experiment. *American Journal of Botany*, **68**, 1290–1.

Kjaer, A. (1940). Germination of buried and dry stored seeds. I. 1934–1939. *Proceeding of the International Seed Testing Association*, **12**, 167–88.

Knipe, D. & Herbel, C. H. (1960). The effects of limited moisture on germination and initial growth of six grass species. *Journal of Range Management*, **13**, 297–302.

Koller, D. & Roth, N. (1963). Germination-regulating mechanisms in some desert seeds. VII. *Panicum turgidum* (Gramineae). *Israel Journal of Botany*, **12**, 64–73.

Kollman, G. (1970). Germination-dormancy and Promotor-inhibitor Relationships in *Setaria lutescens* Seeds. Ph.D. thesis. Ames: Iowa State University.

Kovach, D. A. & Bradford, K. J. (1992*a*). Temperature dependence of viability and dormancy of *Zizania palustris* var. *interior* seeds stored at high moisture contents. *Annals of Botany*, **69**, 297–301.

Kovach, D. A. & Bradford, K. J. (1992*b*). Imbibitional damage and desiccation tolerance of wild rice (*Zizania palustris*) seeds. *Journal of Experimental Botany*, **43**, 747–57.

Kramer, N. B. & Johnson, F. D. (1987). Mature forest seed banks of three habitat types in central Idaho. *Canadian Journal of Botany*, **65**, 1961–6.

Krishnasamy, V. & Seshu, D. V. (1989). Seed germination rate and associated characters in rice. *Crop Science*, **29**, 904–8.

Kucera, C. L. (1966). Some effects of gibberellic acid on grass seed germination. *Iowa State Journal of Science*, **41**, 137–43.

Kumar, A., Joshi, M. C. & Babu, V. R. (1971). Some factors influencing the germination of seeds in two desert grasses. *Tropical Ecology*, **12**, 202–8.

Lambert, F. J., Bower, M., Whalley, R. D. B., Andrews, A. C. & Bellotti, W. D. (1990). The effects of soil moisture and planting depth on emergence and seedling morphology of *Astrebla lappacea* (Lindl.) Domin. *Australian Journal of Agricultural Research*, **41**, 367–76.

Lambert, R. J., Alexander, D. E. & Rodgers, R. C. (1967). Effect of kernel position on oil content in corn (*Zea mays* L.). *Crop Science*, **7**, 143–4.

Leck, M. A. & Simpson, R. L. (1987). Seed bank of a freshwater tidal wetland: Turnover and relationship to vegetation change. *American Journal of Botany*, **74**, 360–70.

Leng, E. R. (1949). Direct effect of pollen parent on kernel size in dent corn. *Agronomy Journal*, **41**, 555–8.

Lewis, J. (1973). Longevity of crop and weed seeds: Survival after 20 years in soil. *Weed Research*, **13**, 179–91.

Lindauer, L. L. & Quinn, J. A. (1972). Germination ecology of *Danthonia sericea* populations. *American Journal of Botany*, **59**, 942–51.

Linnington, S., Bean, E. W. & Tyler, B. F. (1979). The effects of temperature upon seed germination in *Festuca pratensis* var. *apennina. Journal of Applied Ecology*, **16**, 933–8.

Lock, J. M. & Milburn, T. R. (1971). The seed biology of *Themeda triandra* Forsk. in relation to fire. In *The Scientific Management of Animal and Plant Communities for Conservation*, ed. E. Duffey and A. S. Watt, pp. 337–49. Oxford: Blackwell Scientific Publications.

Lodge, G. M. & Whalley, R. D. B. (1981). Establishment of warm- and cool-season native perennial grasses on the north-west slopes of New South Wales. I. Dormancy and germination. *Australian Journal of Botany*, **29**, 111–19.

Longman, K. A. (1969). The dormancy and survival of plants in the humid tropics. *Symposia of the Society for Experimental Biology*, **23**, 471–88.

Lush, W. M. & Groves, R. H. (1981). Germination, emergence and surface establishment of wheat and ryegrass in response to natural and artificial hydration-dehydration cycles. *Australian Journal of Agriculture Research*, **32**, 731–9.

Lush, W. M., Groves, R. H. & Kaye, P. E. (1981). Presowing hydration-dehydration treatments in relation to seed germination and early seedling growth of wheat and ryegrass. *Australian Journal of Plant Physiology*, **8**, 409–25.

Macke, A. J. & Ungar, I. A. (1971). The effects of salinity on germination and early growth of *Puccinellia nuttalliana. Canadian Journal of Botany*, **49**, 515–20.

Makita, A., Konno, Y., Fujita, N., Takada, K. & Hamabata, E. (1993). Recovery of a *Sasa tsuboiana* population after mass flowering and death. *Ecological Research*, **8**, 215–24.

Marañón, T. (1987). Ecología del polimorfismo somático de semillas y la sinaptospermia en *Aegilops neglecta* Req. ex Bertol. *Anales del Jardin Botanico de Madrid*, **44**, 97–107.

Mariko, S., Kachi, N., Ishikawa, S.-I. & Furukawa, A. (1992). Germination ecology of coastal plants in relation to salt environment. *Ecological Research*, **7**, 225–33.

Martin, A. C. (1946). The comparative internal morphology of seeds. *The American Midland Naturalist*, **36**, 513–660.

Martin, C. C. (1975). The role of glumes and gibberellic acid in dormancy of *Themeda triandra* spikelets. *Physiologia Plantarum*, **33**, 171–6.

Martin, M. H. and Cox, J. R. (1984). Germination profiles of introduced lovegrasses at six constant temperatures. *Journal of Range Management*, **37**, 507–9.

Matumura, M., N. Takase, N. & Hirayoshi, I. (1960). Physiological and ecological studies on germination of *Digitaria* seeds (1) Difference in response to germinating conditions and dormancy among individual plants. *Research Bulletin of the Faculty of Agriculture, Gifu University, Japan*, **12**, 89–96.

Maun, M. A. (1981). Seed germination and seedling establishment of *Calamovilfa* on Lake Huron sand dunes. *Canadian Journal of Botany*, **59**, 460–9.

Maun, M. A., Canode, C. L. & Teare, I. D. (1969). Influence of temperature during anthesis on seed set in *Poa pratensis* L. *Crop Science*, **9**, 210–12.

Maze, K. M., Koen, T. B. & Watt, L. A. (1993). Factors influencing the germination of six perennial grasses of central New South Wales. *Australian Journal of Botany*, **41**, 79–90.

McDonough, W. T. (1970). Germination of 21 species collected from a high-elevation rangeland in Utah. *The American Midland Naturalist*, **84**, 551–4.

McGinnies, W. J. (1960). Effects of moisture stress and temperature on germination of six range grasses. *Agronomy Journal*, **52**, 159–62.

McIvor, J. G. & Gardener, C. J. (1991). Soil seed densities and emergence patterns in pastures in the seasonally dry tropics of northeastern Australia. *Australian Journal of Ecology*, **16**, 159–69.

McKell, C. M., Robinson, J. P. & Major, J. (1962). Ecotypic variation in medusahead, an introduced annual grass. *Ecology*, **43**, 686–98.

McNamara, J. & Quinn, J. A. (1977). Resource allocation and reproduction in populations of *Amphicarpum purshii* (Gramineae). *American Journal of Botany*, **64**, 17–23.

Mejia P., V., Romero, M., C. & Lotero C., J. (1978). Factores que afectan la germinación y el vigor de la semilla del pasto guinea (*Panicum maximum* Jacq.). *Revista del Instituto de Colombiana Agropecuario*, **13**, 69–76.

Meyer, S. E., Beckstead, J., Allen, P. S. & Pullman, H. (1995). Germination ecophysiology of *Leymus cinereus* (Poaceae). *International Journal of Plant Sciences*, **156**, 206–15.

Milton, W. E. J. (1936). The buried viable seeds of enclosed and unenclosed hill land. *Bulletin of the Welsh Plant Breeding Station, Series H*, **14**, 58–73.

Mitchell, J. W. & Marth, P. C. (1946). Germination of seeds in soil containing 2,4-dichlorophenoxyacetic acid. *Botanical Gazette*, **107**, 408–24.

Mohamed, H. A., Clark, J. A. & Ong, C. K. (1985). The influence of temperature during seed development on the germination characteristics of millet seeds. *Plant, Cell and Environment*, **8**, 361–2.

Morash, R. & Freedman, B. (1989). The effects of several herbicides on the germination of seeds in the forest floor. *Canadian Journal of Forest Research*, **19**, 347–50.

Morinaga, T. (1926). Effect of alternating temperatures upon the germination of seeds. *American Journal of Botany*, **13**, 148–58.

Morgan, W. C. & Meyers, B. A. (1989). Germination of the salt-tolerant grass *Diplachne fusca*. I. Dormancy and temperature responses. *Australian Journal of Botany*, **37**, 225–37.

Mooring, M. T., Cooper, A. W. & Seneca, E. D. (1971). Seed germination response and evidence for height ecophenes in *Spartina alterniflora* from North Carolina. *American Journal of Botany*, **58**, 48–55.

Morrow, L. A., Young, F. L. & Flom, D. G. (1982). Seed germination and seedling emergence of jointed goatgrass (*Aegilops cylindrica*). *Weed Science*, **30**, 395–8.

Moss, S. R. (1985). The survival of *Alopecurus myosuroides* Huds. seeds in soil. *Weed Research*, **25**, 201–11.

Mott, J. J. (1972). Germination studies on some annual species from an arid region of Western Australia. *Journal of Ecology*, **60**, 293–304.

Mott, J. J. (1978). Dormancy and germination in five native grass species from savannah woodland communities of the Northern Territory. *Australian Journal of Botany*, **26**, 621–31.

Mott, J. J. & McComb, A. J. (1974). Patterns in annual vegetation and soil microrelief in an arid region of Western Australia. *Journal of Ecology*, **62**, 115–26.

Mrozek, E., Jr. (1980). Effect of mercury and cadmium on germination of *Spartina alterniflora* Loisel seeds at various salinities. *Environmental and Experimental Botany*, **20**, 367–77.

Mrozek, E., Jr. & Funicelli, N. A. (1982). Effect of zinc and lead on germination of *Spartina alterniflora* Louisel seeds at various salinities. *Environmental and Experimental Botany*, **22**, 23–32.

Muenscher, W. C. (1936). Storage and germination of seeds of aquatic plants. *Cornell University Agriculture Experiment Station Bulletin*, **652**.

Murphy, C. F. & Frey, K. J. (1962). Inheritance and heritability of seed weight and its components in oats. *Crop Science*, **2**, 509–12.

Naka, K. & Yoda, K. (1984). Community dynamics of evergreen broadleaf forests in southwestern Japan. II. Species composition and density of seeds buried in the soil of a climax evergreen oak forest. *Botanical Magazine Tokyo*, **97**, 61–79.

Nakagoshi, N. (1984*a*). Ecological studies on the buried viable seed population in soil of the forest communities in Miyajima Island, southwestern Japan II. *Hikobia*, **9**, 109–22.

Nakagoshi, N. (1984*b*). Buried viable seed populations in forest communities on the Hiba Mountains, southwestern Japan. *Journal of Science of Hiroshima University*, **19**, 1–56.

Naylor, J. M. & Jana, S. (1976). Genetic adaptation for seed dormancy in *Avena fatua*. *Canadian Journal of Botany*, **54**, 306–12.

Naylor, R. E. L. (1980). Effects of seed size and emergence time on subsequent growth of perennial ryegrass. *New Phytologist*, **84**, 313–8.

Naylor, R. E. L. & Abdalla, A. F. (1982). Variation in germination behaviour. *Seed Science and Technology*, **10**, 67–76

Nelson, J. R. & Wilson, A. M. (1969). Influence of age and awn removal on dormancy of medusahead seeds. *Journal of Range Management*, **22**, 289–90.

Nelson, L. R. (1980). Recurrent selection for improved rate of germination in annual ryegrass. *Crop Science*, **20**, 219–21.

Newell, P. F. (1967). Mollusca. In *Soil Biology*, ed. A. Burges & F. Raw, pp. 413–33. Academic Press, London.

Newman, E. I. (1963). Factors controlling the germination date of winter annuals. *Journal of Ecology*, **51**, 625–38.

Nikolaeva, M. G. (1969). Physiology of deep dormancy in seeds. Izdatel'stvo 'Nauka,' Leningrad. (Translation from Russian by Z. Shapiro, National Science Foundation, Washington, DC).

Nikolaeva, M. G. (1977). Factors controlling the seed dormancy pattern. In *The Physiology and Biochemistry of Seed Dormancy and Germination*, ed. A. A. Khan, pp. 51–74. Amsterdam and New York: North Holland Publishing.

Norris, R. F. & Schoner, C. A., Jr. (1980). Yellow foxtail (*Setaria lutescens*) biotype studies: Dormancy and germination. *Weed Science*, **28**, 159–63.

Numata, M., Hayashi, I., Komura, T. & Oki, K. (1964). Ecological studies on the buried-seed population in the soil as related to plant succession I. *Japanese Journal of Ecology*, **14**, 207–17.

O'Connor, T. G. & Pickett, G. A. (1992). The influence of grazing on seed production and seed banks of some African savanna grasslands. *Journal of Applied Ecology*, **29**, 247–60.

Ødum, S. (1974). Seeds in ruderal soils, their longevity and contribution to the flora of disturbed ground in Denmark. *Proceedings of the British Weed Control Conference,* **12,** 1131–44.

Oelke, E. A. & Albrecht, K. A. (1978). Mechanical scarification of dormant wild rice seed. *Agronomy Journal,* **70,** 691–4.

Onnis, A. & Miceli, P. (1975). *Puccinellia festucaeformis* (Host) Parl.: dormienza e influenza della salinita sulla germinazione. *Giornale Botanico Italiano,* **109,** 27–37.

Onnis, A., Pelosini, F. & Stefani, A. (1981). *Puccinellia festucaeformis* (Host) Parl.: Germinazione e crescita iniziale in funzione della salinità del substrato. *Giornale Botanico Italiano,* **115,** 103–16.

Orozco-Segovia, A. D. L. & Vazquez-Yanes, C. (1980). La germinación de *Panicum hirsutum* Swartz: una arvense de cultivos de zonas inundables. *Boletín de la Sociedad Botánica de México,* **39,** 91–106.

Pamplona, P. P. & Mercado, B. L. (1981). Ecotypes of *Rottboellia exaltata* L. f. in the Philippines. I. Characteristics and dormancy of seeds. *Philippine Agriculture,* **64,** 59–66.

Pannangpetch, K. & Bean, E. W. (1984). Effects of temperature on germination in populations of *Dactylis glomerata* from NW Spain and central Italy. *Annals of Botany,* **53,** 633–9.

Partridge, T. R. & Wilson, J. B. (1987). Germination in relation to salinity in some plants of salt marshes in Otago, New Zealand. *New Zealand Journal of Botany,* **25,** 255–61.

Pathak, S. M., Mukhiya, Y. K. & Singh, V. P. (1987). Mercury, manganese interaction studies on barley germination and phytotoxicity. *Indian Journal of Plant Physiology,* **30,** 13–19.

Paterson, J. G., Goodchild, N. A. & Boyd, W. J. R. (1976). Effect of storage temperature, storage duration and germination temperature on the dormancy of seed of *Avena fatua* L. and *Avena barbata* Pott ex Link. *Australian Journal of Agricultural Research,* **27,** 373–9.

Peart, M. H. (1979). Experiments on the biological significance of the morphology of seed-dispersal units in grasses. *Journal of Ecology,* **67,** 843–63.

Pladeck, M. M. (1940). The testing of buffalo grass 'seed', *Buchloe dactyloides* Engelm. *Agronomy Journal,* **32,** 486–94.

Plyer, D. B. & Carrick, K. M. (1993). Site-specific seed dormancy in *Spartina alterniflora* (Poaceae). *American Journal of Botany,* **80,** 752–6.

Poiani, K. A. & Johnson, W. C. (1988). Evaluation of the emergence method in estimating seed bank composition of prairie wetlands. *Aquatic Botany* **32,** 91–7.

Pons, T. L. (1989). Dormancy, germination and mortality of seeds in heathland and inland sand dunes. *Acta Botanica Neerlandica,* **38,** 327–35.

Popay, A. I. (1974). Investigations into the behaviour of the seeds of some tropical weeds. I. Laboratory germination tests. *East African Agriculture and Forestry Journal,* **40,** 31–43.

Popay, A. I. (1976). Investigations into the behaviour of the seeds of some tropical weeds. III. Patterns of emergence. *East African Agriculture and Forestry Journal,* **41,** 304–12.

Popay, A. I. (1981). Germination of seeds of five annual species of barley grass. *Journal of Applied Ecology,* **18,** 547–58.

Probert, R. J. & Longley, P. L. (1989). Recalcitrant seed storage physiology in three aquatic grasses (*Zizania palustris, Spartina anglica* and *Porteresia coarctata*). *Annals of Botany,* **63,** 53–63.

78   C. C. Baskin and J. M. Baskin

Probert, R. J., Smith, R. D & Birch, P. (1985). Germination responses to light and alternating temperatures in European populations of *Dactylis glomerata* L. II. The genetic and environmental components of germination. *New Phytologist*, **99**, 317–22.

Probert, R. J., Smith, R. D & Birch, P. (1986). Germination responses to light and alternating temperatures in European populations of *Dactylis glomerata* L. V. The principle components of the alternating temperature requirement. *New Phytologist*, **102**, 133–42.

Raju, M. V. S. & Ramaswamy, S. N. (1983). Studies on the inflorescence of wild oats (*Avena fatua*). *Canadian Journal of Botany*, **61**, 74–8.

Ralowicz, A. E. & Mancino, C. F. (1992). Afterripening in curly mesquite seeds. *Journal of Range Management*, **45**, 85–7.

Ralowicz, A., Mancino, C. & Kopec, D. (1992). Chemical enhancement of germination in curly mesquite seed. *Journal of Range Mangement*, **45**, 507–8.

Rampton, H. H. & Ching, T. M. (1966). Longevity and dormancy in seeds of several cool-season grasses and legumes buried in soil. *Agronomy Journal*, **58**, 220–2.

Rampton, H. H. & Ching, T. M. (1970). Persistence of crop seeds in soil. *Agronomy Journal*, **62**, 272–7.

Renard, C. & Capelle, P. (1976). Seed germination in ruzizi grass (*Brachiaria ruziziensis* Germain & Evard). *Australian Journal of Botany*, **24**, 437–46.

Richardson, S. G. (1979). Factors influencing the development of primary dormancy in wild oat seeds. *Canadian Journal of Plant Science*, **59**, 777–84.

Ries, R. E. & Hofmann, L. (1983). Effect of sodium and magnesium sulfate on forage seed germination. *Journal of Range Mangement*, **36**, 658–62.

Roach, D. A. (1987). Variation in seed and seedling size in *Anthoxanthum odoratum*. *The American Midland Naturalist*, **117**, 258–64.

Roberts, E. H. (1972). Dormancy: A factor affecting seed survival in the soil. In *Viability of Seeds*, ed. E. H. Roberts, pp. 321–357. Syracuse: Syracuse Univ. Press.

Roberts, H. A. (1970). Viable weed seeds in cultivated soils. *Report of the National Vegetable Research Station (Wellesbourne) for 1969*, pp. 25–38.

Roberts, H. A. & Dawkins, P. A. (1967). Effect of cultivation on the numbers of viable weed seeds in soil. *Weed Research*, **7**, 290–301.

Roberts, H. A. & Potter, M. E. (1980). Emergence patterns of weed seedlings in relation to cultivation and rainfall. *Weed Research*, **20**, 377–86.

Roberts, H. A. & Ricketts, M. E. (1979). Quantitative relationships between the weed flora after cultivation and the seed population in the soil. *Weed Research*, **19**, 269–75.

Robocker, W. C., Curtis, J. T. & Ahlgren, H. L. (1953). Some factors affecting emergence and establishment of native grass seedlings in Wisconsin. *Ecology*, **34**, 194–9.

Rogler, G. A. (1960). Relation of seed dormancy of green needlegrass (*Stipa viridula* Trin.) to age and treatment. *Agronomy Journal*, **52**, 467–9.

Romo, J. T, Grilz, P. L., Bubar, C. J. & Young, J. A. (1991). Influences of temperature and water stress on germination of plains rough fescue. *Journal of Range Management*, **44**, 75–81.

Rosa, M. L. & Corbineau, R. (1986). Quelques aspects de la germination des caryopses de *Leersia oryzoides* (L.) Sw. *Weed Research*, **26**, 99–104.

Rothrock, P. E., Squiers, E. R. & Sheeley, S. (1993). Heterogeneity and size of a persistent seedbank of *Ambrosia artemisiifolia* L. and *Setaria faberi* Herrm. *Bulletin of the Torrey Botanical Club*, **120**, 417–22.

Roy, N. N. & Everett, H. L. (1963). Seed production, fertility levels, and cold test germination in corn. *Crop Science,* **3**, 273–5.

Ryan, J., Miyamoto, S. & Stroehlein, J. L. (1975). Salt and specific ion effects on germination of four grasses. *Journal of Range Management,* **28**, 61–4.

Sabo, D. G., Johnson, G. V., Martin, W. C. & Aldon, E. F. (1979). Germination requirements of 19 species of arid land plants. *United States Department of Agriculture, Forest Service Research Paper MR-210.*

Sawhney, R. & Naylor, J. M. (1979). Dormancy studies in seed of *Avena fatua.* 9. Demonstration of genetic variability affecting the response to temperature during seed development. *Canadian Journal of Botany,* **57**, 59–63.

Sautter, E. H. (1962). Germination of switchgrass. *Journal of Range Management,* **15**, 108–10.

Schafer, D. E. & Chilcote, D. O. (1969). Factors influencing persistence and depletion in buried seed populations. I. A model for analysis of parameters of buried seed persistence and depletion. *Crop Science,* **9**, 417–9.

Schneider, R. L. & Sharitz, R. R. (1986). Seed bank dynamics in a southeastern riverine swamp. *American Journal Botany,* **73**, 1022–30.

Schwerzel, P. J. (1974). The effect of depth of burial in soil on the survival of some common Rhodesian weed seeds. *Rhodesian Agriculture,* **73**, 97–9.

Scurfield, G. (1954). Biological flora of the British Isles. *Deschampsia flexuosa* (L.) Trin. *Journal of Ecology,* **42**, 225–33.

Sen, S. & Osborne, D. J. (1974). Germination of rye embryos following hydration-dehydration treatments: Enhancement of protein and RNA synthesis and earlier induction of DNA replication. *Journal of Experimental Botany,* **25**, 1010–19.

Sexsmith, J. J. (1969). Dormancy of wild oat seed produced under various temperature and moisture conditions. *Weed Science,* **17**, 405–7.

Shaidaee, G., Dahl, B. E. & Hansen, R. M. (1969). Germination and emergence of different age seeds of six grasses. *Journal of Range Management,* **22**, 240–3.

Shaukat, S. S. (1974). The effects of simazine, atrazine and 2,4-D on germination and early seedling growth of *Oryza sativa* L. *Pakistan Journal of Botany,* **6**, 141–9.

Shaw, N. H. (1957). Bunch spear grass dominance in burnt pastures in south-eastern Queensland. *Australian Journal of Agricultural Research,* **8**, 325–34.

Sheikh, K. H. & Mahmood, K. (1986). Some studies on field distribution and seed germination of *Suaeda fruticosa* and *Sporobolus arabicus* with reference to salinity and sodicity of the medium. *Plant and Soil,* **94**, 333–40.

Silcock, R. G., Williams, L. M. & Smith, F. T. (1990). Quality and storage characteristics of the seeds of important native pasture species in south-west Queensland. *Australian Rangelands Journal,* **12**, 14–20.

Silva, J. F. & Ataroff, M. (1985). Phenology, seed crop and germination of coexisting grass species from a tropical savanna in western Venezuela. *Acta Oecologia,* **6**, 41–51.

Simpson, G. M. (1990). *Seed Dormancy in Grasses.* Cambridge: Cambridge University Press.

Sindel, B. M., Davidson, S.J., Kilby, M. J. & Groves, R. H. (1993). Germination and establishment of *Themeda triandra* (kangaroo grass) as affected by soil and seed characteristics. *Australian Journal of Botany,* **41**, 105–17.

Singh, V. P., Mall, S. L. & Billore, S. K. (1975). Effect of pH on germination of four common grass species of Ujjain (India). *Journal of Range Management,* **28**, 497–8.

Sinha, A. & Gupta, S. R. (1982). Effects of osmotic tension and salt stress on germination of three grass species. *Plant and Soil*, **69**, 13–19.

Smith, A. L. (1972). Factors influencing germination of *Scolochloa festucacea* caryopses. *Canadian Journal of Botany*, **50**, 2085–92.

Somody, C. N., Nalewaja, J. D & Miller, S. E. (1984). The response of wild oat (*Avena fatua*) and *Avena sterilis* accessions to photoperiod and temperature. *Weed Science*, **32**, 206–13.

Spence, J. R. (1990). A buried seed experiment using caryopses of *Chionochloa macra* Zotov (Danthonieae: Poaceae), South Island, New Zealand. *New Zealand Journal of Botany*, **28**, 471–4.

Sprague, V. G. (1940). Germination of freshly harvested seeds of several *Poa* species and of *Dactylis glomerata*. *Agronomy Journal*, **32**, 715–21.

Standifer, L. C. & Wilson, P. W. (1988). A high temperature requirement for after ripening of imbibed dormant *Poa annua* L. seeds. *Weed Research*, **28**, 365–71.

Steinbauer, G. P. & Grigsby, B. (1956). Some correlations between germination and dormancy of weed seeds in the laboratory and in the field. *Proceedings of the North Central Weed Control Conference [USA]*, **13**, 35–6.

Svedarsky, D. & Kucera, C. L. (1970). Effects of gibberellic acid and post-harvest age on germination of prairie grasses. *Iowa State Journal of Science*, **44**, 513–18.

Taylor, A. H. & Zisheng, Q. (1988). Regeneration from seed of *Sinarundinaria fangiana*, a bamboo, in the Wolong Giant Panda Reserve, Sichuan, China. *American Journal of Botany*, **75**, 1065–73.

Taylorson, R. B. (1986). Water stress-induced germination of giant foxtail (*Setaria faberi*) seeds. *Weed Science*, **34**, 871–5.

Taylorson, R. B. & McWhorter, C. C. (1969). Seed dormancy and germination in ecotypes of Johnsongrass. *Weed Science*, **17**, 359–61.

Thompson, K. & Grime, J. P. (1983). A comparative study of germination responses to diurnally-fluctuating temperatures. *Journal of Applied Ecology*, **20**, 141–56.

Thompson, K., Grime, J. P. & Mason, G. (1977). Seed germination in response to diurnal fluctuations of temperature. *Nature*, **267**, 147–9.

Thompson, P. A. (1980). Germination strategy of a woodland grass: *Milium effusum* L. *Annals of Botany*, **46**, 593–602.

Thornton, M. L. (1966). Seed dormancy in buffalograss (*Buchlöe dactyloides*). *Proceedings of the Association of Official Seed Analysts*, **56**, 120–3.

Thurston, J. M. (1966). Survival of seeds of wild oats (*Avena fatua* L. and *Avena ludoviciana* Dur.) and charlock (*Sinapis arvensis* L.) in soil under leys. *Weed Research*, **6**, 67–80.

Thurston, J. M. (1951). A comparison of the growth of wild and of cultivated oats in manganese-deficient soils. *Annals of Applied Biology*, **38**, 289–302.

Tolstead, W. L. (1941). Germination habits of certain sand-hill plants in Nebraska. *Ecology*, **22**, 393–7.

Toole, E. H. & Brown, E. (1946). Final results of the Duvel buried seed experiment. *Journal of Agricultural Research*, **72**, 201–10.

Toole, E. H. & Toole, V. K. (1940). Germination of seed of goosegrass, *Eleusine indica*. *Agronomy Journal*, **32**, 320–1.

Toole, E. H. & Toole, V. K. (1941). Progress of germination of seed of *Digitaria* as influenced by germination temperature and other factors. *Journal of Agricultural Research*, **63**, 65–90.

Toole, V. K. (1939). Germination of the seeds of poverty grass, *Danthonia spicata*. *Agronomy Journal*, **31**, 954–65.

Toole, V. K. (1940). Germination of seeds of vine-mesquite, *Panicum obtusum*, and plains bristle-grass, *Setaria macrostachya*. *Agronomy Journal*, **32**, 503–12.

Toole, V. K. (1941). Factors affecting the germination of various dropseed grasses (*Sporobolus* spp.). *Journal of Agricultural Research*, **62**, 691–715.

Torssell, B. W. R. & McKeon, G. M. (1976). Germination effects on pasture composition in a dry monsoonal climate. *Journal of Applied Ecology*, **13**, 593–603.

Tothill, J. C. (1969). Soil temperatures and seed burial in relation to the performance of *Heteropogon contortus* and *Themeda australis* in burnt native woodland pastures in eastern Queensland. *Australian Journal of Botany*, **17**, 269–75.

Tothill, J. C. (1977). Seed germination studies with *Heteropogon contortus*. *Australian Journal of Ecology*, **2**, 477–84.

Twentyman, J. D. (1974). Environmental control of dormancy and germination in the seeds of *Cenchrus longispinus* (Hack.) Fern. *Weed Research*, **14**, 1–11.

Tyler, B., Borrill, M. & Chorlton, K. (1978). Studies in *Festuca*. X. Observations on germination and seedling cold tolerance in diploid *Festuca pratensis* and tetraploid *F. pratensis* var. *apennina* in relation to their altitudinal distribution. *Journal of Applied Ecology*, **15**, 219–26.

Ungar, I. A. (1974). The effect of salinity and temperature on seed germination and growth of *Hordeum jubatum*. *Canadian Journal of Botany*, **52**, 1357–62.

Ungar, I. A. & Woodell, S. R. J. (1993). The relationship between the seed bank and species composition of plant communities in two British salt marshes. *Journal of Vegetation Science*, **4**, 531–6.

Uphof, J. C. Th. (1938). Cleistogamic flowers. *The Botanical Review*, **4**, 21–49.

Vanden Born, W. H. (1971). Green foxtail: Seed dormancy, germination and growth. *Canadian Journal of Plant Science*, **51**, 53–9.

van Esso, M. L., Ghersa, C. M. & Soriano, A. (1986). Cultivation effects on the dynamics of a Johnson grass seed population in the soil profile. *Soil and Tillage Research*, **6**, 325–35.

van Rooden, J., Akkermans, L. M. A & van der Veen, R. (1970). A study on photoblastism in seeds of some tropical weeds. *Acta Botanica Neerlandica*, **19**, 257–64.

Veenendaal, E. M. & Ernst, W. H. O. (1991). Dormancy patterns in accessions of caryopses from savanna grass species in south eastern Botswana. *Acta Botanica Neerlandica*, **40**, 297–309.

Vegis, A. (1964). Dormancy in higher plants. *Annual Review of Plant Physiology*, **15**, 185–215.

Villiers, T. A. (1971). Cytological studies in dormancy. I. Embryo maturation during dormancy in *Fraxinus excelsior*. *New Phytologist*, **70**, 751–60.

Vlahos, S. & Bell, D. T. (1986). Soil seed-bank components of the northern jarrah forest of western Australia. *Australian Journal of Ecology*, **11**, 171–9.

Voight, R. L., Gardner, C. O. & Webster, O. J. (1966). Inheritance of seed size in sorghum, *Sorghum vulgare* Pers. *Crop Science*, **6**, 582–6.

Wagner, R. H. (1964). The ecology of *Uniola paniculata* L. in the dune-strand habitat of North Carolina. *Ecological Monographs*, **34**, 79–96.

Wallgren, B. E. & Aamisepp, A. (1977). Biology and control of *Alopecurus myosuroides* Huds. and *Apera spica-venti* L. *Proceedings of the European Weed Research Society Symposium*, 1977, 229–41.

Washitani, I. & Masuda, M. (1990). A comparative study of the germination characteristics of seeds from a moist tall grassland community. *Functional Ecology*, **4**, 543–57.

Watanabe, Y. & Hirokawa, F. (1975a). Ecological studies on the germination and emergence of annual weeds. 4. Seasonal changes in dormancy status of viable seeds in cultivated and uncultivated soil. *Weed Research, Japan*, **19**, 20–4. (in Japanese with English summary).

Watanabe, Y. & Hirokawa, F. (1975b). Requirement of temperature conditions in germination of annual weed seeds and its relation to seasonal distribution of emergence in the field. *Proceedings of the Fifth Asian-Pacific Weed Science Society Conference*, 5–11 October 1975, Tokyo, Japan, pp. 38–41.

Watt, L. A. (1974). The effect of water potential on the germination behaviour of several warm season grass species, with special reference to cracking black clay soils. *Journal of Soil Conservation of New South Wales*, **30**, 28–41.

Watt, L. A. (1978). Some characteristics of the germination of Queensland blue grass on cracking black earths. *Australian Journal of Agricultural Research*, **29**, 1147–55.

Watt, L. A. (1982). Germination characteristics of several grass species as affected by limiting water potentials imposed through a cracking black clay soil. *Australian Journal of Agricultural Research*, **33**, 223–31.

Wellington, P. S. & Hitchings, S. (1965). Germination and seedling establishment of blackgrass (*Alopecurus myosuroides* Huds.). *Journal of the National Institute of Agricultural Botany*, **10**, 262–73.

Wenger, L. E. (1941). Soaking buffalo grass (*Buchlöe dactyloides*) seed to improve its germination. *Agronomy Journal*, **33**, 135–41.

Went, F. W. (1948). Ecology of desert plants. I. Observations on germination in the Joshua Tree National Monument, California. *Ecology*, **29**, 242–53.

West, S. H. & Marousky, R. (1989). Mechanism of dormancy in Pensacola bahiagrass. *Crop Science*, **29**, 787–91.

Westra, R. N. & Loomis, W. E. (1966). Seed dormancy in *Uniola paniculata*. *American Journal of Botany*, **53**, 407–11.

Whittington, W. J., Hillman, J., Gatenby, S. M., Hooper, B. E & White, J. C. (1970). Light and temperature effects on the germination of wild oats. *Heredity*, **25**, 641–50.

Wiesner, L. E. & Kinch, R. C. (1964). Seed dormancy in green needlegrass. *Agronomy Journal*, **56**, 371–3.

Williams, E. D. (1971). Germination of seeds and emergence of seedlings of *Agropyron repens* (L.) Beauv. *Weed Research*, **11**, 171–81.

Williams, E. D. (1983a). Effects of temperature fluctuation, red and far-red light and nitrate on seed germination of five grasses. *Journal of Applied Ecology*, **20**, 923–35.

Williams, E. D. (1983b). Effects of temperature, light, nitrate and pre-chilling on seed germination of grassland plants. *Annals of Applied Biology*, **103**, 161–72.

Wilson, M. V., Ingersoll, C. A. & Roush, M. L. (1989). Measuring seed movement in soil. *Bulletin of the Ecological Society of America*, **70**(2), 300–1. (abstract).

Wisheu, I. C. & Keddy, P. A. (1991). Seed banks of a rare wetland plant community: Distribution patterns and effects of human-induced disturbance. *Journal of Vegetation Science*, **2**, 181–8.

Wolters, G. L. (1970). Breaking dormancy of longleaf *Uniola* seeds. *Journal of Range Mangement*, **23**, 178–80.

Wright, L. N. (1973). Seed dormancy, germination environment, and seed structure of Lehmann lovegrass, *Eragrostis lehmanniana* Nees. *Crop Science*, **13**, 432–5.

Wright, L. N. (1976). Recurrent selection for shifting gene frequency of seed weight in *Panicum antidotale* Retz. *Crop Science*, **16**, 647–9.

Wurzburger, J. & Koller, D. (1976). Differential effects of the parental photothermal environment on development of dormancy in caryopses of *Aegilops kotschyi*. *Journal of Experimental Botany*, **27**, 43–8.

Young, J. A., Eckert, R. E., Jr. & Evans, R. A. (1981*a*). Temperature profiles for germination of blue-bunch and beardless wheatgrasses. *Journal of Range Management*, **34**, 84–9

Young, J. A., Emmerich, F. L. & Patten, B. (1990). Germination of seeds of Columbia needlegrass. *Journal of Seed Technology*, **14**, 94–100.

Young, J. A. & Evans, R. A. (1980). Germination of desert needlegrass. *Journal of Seed Technology*, **5**, 40–6.

Young, J. A., Evans, R. A & Eckert, R. E., Jr. (1968). Germination of medusahead in response to temperature and afterripening. *Weed Science*, **16**, 92–5.

Young, J. A., Evans, R. A, Eckert, R. E., Jr. & Ensign, R. D. (1981b). Germination-temperature profiles for Idaho and sheep fescue and Canby bluegrass. *Agronomy Journal*, **73**, 716–20.

Young, J. A., Evans, R. A & Kay, B. L. (1975). Germination of italian ryegrass seeds. *Agronomy Journal*, **67**, 386–9.

Young, J. A., Evans, R. A, Raguse, C. A & Larson, J. R. (1981c). Germinable seeds and periodicity of germination in annual grasslands. *Hilgardia*, **49**, 1–37.

Zhang, J. & Maun, M. A. (1989). Seed dormancy of *Panicum virgatum* L. on the shoreline sand dunes of Lake Erie. *The American Midland Naturalist*, **122**, 77–87.

3

Seed dispersal and seedling establishment in grass populations

GREGORY P. CHEPLICK

Whether as native components of a regional flora or as weedy invaders in agricultural fields, their widespread distribution on Earth implies that grasses are successful in the colonization of a diversity of habitats. Critical to this success are various forms of asexual reproduction such as clonality or apomixis (Clark & Fisher, 1987; Chapman, 1992). In addition, many grasses possess morphological modifications of sexual reproductive structures that enhance the dispersal of caryopses (Davidse, 1987; Clayton, 1990) or promote successful seedling establishment (Peart, 1979, 1981). In his classic book *The Dispersal of Plants Throughout the World*, Ridley (1930) stated that the common reed grass *Phragmites communis* was 'the most widely distributed species of all flowering plants in the world' and noted that it was one of the first grasses to reach Krakatau after the island's vegetation had been destroyed by a volcano. He attributed the species' success to its adaptation for dispersal by wind, claiming that 'the spikelets can evidently fly 25 miles [40 km] without falling.'

Additional anecdotal evidence for the long-distance transport of grasses based on geographical patterns of distribution including islands can be found in many other examples provided by Ridley (1930) and van der Pijl (1982). However, such long-distance dispersal is likely to be rare and dependent on catastrophic events like severe tropical storms. This review will focus on processes occurring at the scale of the population and it will be shown that the usual dispersal potential for members of the grass family actually may be quite modest.

Davidse (1987) provided a systematic overview of the characteristic features of spikelets and infloresences adapted to a diversity of dispersal agents, using examples mostly from three tribes of the Poaceae. Hence, in this chapter only brief consideration will be given to the morphological adaptations that promote seed dispersal; rather, I will focus on the

84

Table 3.1. *The dispersal unit in grasses*

Data show the number and percentage of grass genera that
have the indicated dispersal unit.

Dispersal unit	No. of genera	Percentage
Seed/caryopsis	5	0.8
Floret	296	45.5
Floret plus accessory sterile lemmas	62	9.5
Spikelet	136	20.9
Rachis internode	89	13.7
Inflorescence branches	50	7.7
Inflorescence	13	2.0

Note:
Adapted from Clayton (1990).

processes of seed dispersal and seedling establishment as they occur within
grass populations. Topics to be covered include the units and agents of dispersal, dispersal distances, antitelechory, and the primary determinants of
seedling establishment.

The units of dispersal

Because it is functionally comparable to the seed of other plant families,
the one-seeded grass fruit (caryopsis) will here be referred to simply as a
'seed'. Naked seeds rarely function as the dispersal unit in grasses (Davidse,
1987; Clayton, 1990); thus, the term 'diaspore' is best used to describe the
unit of the plant that is actually dispersed (Howe & Smallwood, 1982;
Chapman & Peat, 1992). In the grasses, the diaspore is most commonly a
seed enclosed by various spikelet bracts (lemma, palea, glumes).

Clayton (1990) categorized grass genera in relation to dispersal unit
(Table 3.1). The most frequent diaspore was the floret – a seed surrounded
by a lemma and palea. In some grasses, the floret was enclosed by additional sterile lemmas, while in others the entire spikelet was dispersed.
Together, dispersal of florets and spikelets comprised 76% of the genera
surveyed (Table 3.1). Although quantitative data are not available, it is
likely that intraspecific variation in the diaspore unit can occur. For
example, in some awned species with adhesive dispersal, both individual
florets or entire spikelets might readily attach to animal fur. In other
grasses, wind-blown inflorescences serve as the dispersal unit, although

Table 3.2. *The dispersal agents of the grasses and the associated adaptive features of the grass plant, spikelet, or inflorescence*

Dispersal agent	Associated adaptive features
I. Passive dispersal	
A. Adhesion	Awns, hairs, spines, or sticky secretions on spikelet parts; sharp or hairy callus; sterile spikelets or inflorescence branches with bristles
B. Wind	Long, silky hairs on spikelet parts, including awns; lightweight seeds; thin, membraneous spikelet bracts; plume-like callus; open inflorescences elevated above foliage; disarticulating inflorescences ('tumbleweeds')
C. Water	Aerenchymatous tissue in spikelet or inflorescence parts; buoyant seeds and spikelet parts
II. Active dispersal	
D. Ingestion (birds, mammals)	Smooth, indurate spikelet parts; oil production (birds)[a]; inflorescences within foliage mammals)[b]
E. Transport (ants)	Elaiosomes[c]

Notes:
[a] Davidse & Morton (1973).
[b] Janzen (1994).
[c] Davidse (1987).

some florets (with seeds) may disperse prior to disarticulation of the inflorescence (Ridley, 1930; Rabinowitz & Rapp, 1979).

The agents of dispersal

In the Poaceae, the agents of dispersal may be classified as passive or active, depending on whether or not animals are involved in ingestion or active transport of the diaspores. It is recognized that the distinction between active and passive dispersal may not always be clear: for example, adhesion is really passive dispersal by an active agent (Begon, Harper & Townsend, 1990). When categorized in this way, adhesion, wind, and water constitute passive modes while ingestion and transport by animals constitute active modes of dispersal.

It should be noted when considering the dispersal agents of grasses that there may be overlap among dispersal modes in the morphological features of the associated reproductive structures. For example, the presence of long, silky hairs on spikelet parts could represent adaptations to dispersal by either wind or adhesion (Table 3.2). A cursory examination of diaspore morphology certainly does not always lead to a definitive classification as to dispersal agent for any group of plants (Jurado, Westoby & Nelson, 1991). Some grasses show no obvious morphological features that would implicate a particular dispersal agent (Davidse, 1987; Jurado *et al.*, 1991; Cousens & Mortimer, 1995).

It is probably no accident that there are structural features common to both wind and adhesively dispersed grasses as this has been noted in other plant families (Sorensen, 1986). Indeed, wind dispersal may be ancestral to adhesive dispersal (Stebbins, 1974). When comparing habitats occupied by adhesive versus non-adhesive plants, Sorensen (1986) noted that adhesive species occurred more often in disturbed, dry habitats. Also, lightweight fruits or seeds may be selectively favoured in plants with adhesive dispersal. Certainly the grasses, which commonly inhabit dry or disturbed environments and which possess relatively lightweight diaspores (Leishman, Westoby & Jurado, 1995), may be expected to be well represented among adhesively dispersed plants.

However, quantitative data on dispersal modes in regional grass flora are relatively rare. In part, this may be responsible for the lack of agreement on which dispersal agents are most important to the grass family. For example, both wind (Ridley, 1930) and ingestion (Clayton, 1990) have been described as being the most common dispersal agents in the Poaceae.

It is likely that the relative importance of various dispersal agents depends on the ecosystem or community type examined (Hodgson & Grime, 1990). Two community level studies are considered here. Jurado *et al.* (1991) examined dispersal in relation to diaspore mass and growth form in 229 species, 57 of which were grasses, in the dry, open habitats of central Australia. Although some grasses were categorized as 'unassisted' with no apparent dispersal structure, the predominant dispersal agent for the remaining grasses (60%) was wind (Fig. 3.1). They noted that, relative to other species, the graminoid species 'tended to be over-represented in the wind dispersal classes.'

In contrast to the Australian research, the second study showed that in cerrado vegetation in central Brazil the majority (46%) of the 28 grasses examined were dispersed by animal ingestion (Silberbauer-Gottsberger, 1984). Wind and adhesively dispersed grasses also were found, but these were not as common (Fig. 3.1). It is possible that the more forested and savanna-like nature of the cerrado vegetation favoured closer coevolution of plants with mammals and birds compared with the arid communities in central Australia. Indeed, Willson, Rice & Westoby (1990) have reported that vertebrate-dispersed plants tend to predominate in temperate forests

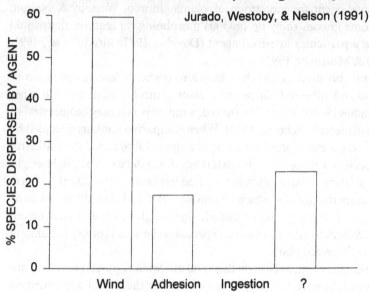

Fig. 3.1. The percentage of grass species dispersed by various dispersal agents in two plant communities: central Australia (Jurado *et al.*, 1991) and cerrado vegetation in central Brazil (Silberbauer-Gottsberger, 1984).

relative to other communities. Wind dispersal may be more common in disturbed, temporary habitats (Hodgson & Grime, 1990).

The primary animal groups involved in the active dispersal of grasses are mammals, birds, and ants. Mammals and birds that ingest viable diaspores presumably pass some fraction of the seeds unharmed through their digestive tracts (van der Pijl, 1982; Janzen, 1984). Quinn *et al.* (1994) have used cattle as a surrogate for buffalo and reported enhanced germination of buffalograss (*Buchloe dactyloides*) diaspores after passage through the animals. Smooth, indurate spikelet parts resistant to digestion may be associated with mammalian dispersal (Table 3.2). In a plant community in central Spain, successful germination of the seeds of 15 grass species was reported from dung samples of rabbit, deer, and cattle (Malo & Suarez, 1995). By far the species with the greatest number of dung-germinated seeds in that study was *Poa annua* ($\bar{x}=193$ seeds per 3-g dung sample for four herbivores). Davidse & Morton (1973) described a neotropical grass genus (*Lasiacis*) adapted to bird dispersal. Their spikelets are black, produce oils, and possess caryopses protected by tightly enclosed lemmas and paleas. Species with seeds transported by ants usually bear an oil-producing elaiosome; excellent examples from the grass family can be found in Davidse (1987).

Dispersal distances

Although it is likely that long-distance transport has played a role in the successful colonization of geographically widespread regions by grasses (Ridley, 1930; Campbell, 1983), most seeds within grass populations are dispersed no more than a few metres from parental sources (Watkinson, 1978; Rabinowitz & Rapp, 1981; Carey & Watkinson, 1993; Cousens & Mortimer, 1995). However, most of the available data are for wind-dispersed species and dispersal distances often have been measured under artificial conditions.

Ernst, Veenendaal & Kebakile (1992) experimentally estimated the dispersal potential of 11 savanna grasses of southern Africa by dropping diaspores in still air from a range of heights that corresponded to infructescence heights recorded in the field. Then they used a simple ballistic formulation to calculate distances to which diaspores could be carried given wind speeds of 2.5–10 m sec^{-1} (Table 3.3). For low infructescence heights and low wind speeds, dispersal distances were generally less than 1 m. The maximum distance measured was 13 m for the wind-dispersed *Enneapogon cenchroides* at the highest wind speed (Table 3.3). An additional analysis of the disper-

Table 3.3. *Dispersal distances for 11 grasses of the African savanna*

Species	Dispersal agent	Infructescence height (m)[a]	Dispersal distance (m)[a]
Aristida congesta	Adhesion	0.2–0.5	0.23–2.31
Cenchrus ciliaris	Adhesion	0.3–0.9	0.60–7.26
Chloris virgata	Wind	0.3–0.5	0.83–5.56
Enneapogon cenchroides	Wind	0.3–0.8	1.23–13.11
Eragrostis rigidior	Gravity	0.3–1.1	0.69–10.19
Panicum maximum	Gravity	0.6–1.4	0.85–7.95
Schmidtia pappaphoroides	Adhesion	0.3–1.1	0.52–7.59
Setaria verticillata	Gravity/adhesion	0.5–0.8	0.51–3.28
Tragus berteronianus	Adhesion	0.1–0.3	0.19–2.24
Urochloa panicoides	Gravity (ingestion?)	0.4–0.7	0.43–3.16
Urochloa mosambicensis	Gravity (ingestion?)	0.4–0.7	0.64–4.46

Note:
[a] Ranges are provided. Transport distances are based on wind velocities from 2.5 to 10 m s^{-1}.
Source: Adapted from Ernst, Veenendaal & Kebakile (1992).

sal of diaspores around a parent plant of *Setaria verticillata* revealed that 77% of the diaspores were dispersed within 50 cm of the parent (Ernst *et al.*, 1992).

Dispersibility of seven prairie grasses from central North America was estimated by dropping diaspores in still air from a height of 1 m (Rabinowitz & Rapp, 1981). The maximum lateral movement observed was between 0.11 m and 0.46 m, depending on species. The dispersal of *Sphenopholis obtusata* was investigated for diaspores with or without glumes; interestingly, more diaspores with glumes travelled to greater distances. Ridley (1930) referred to spikelet bracts that enhanced wind dispersal of the enclosed seeds as 'glume wings', but he provided no quantitative data.

More realistic studies of grass seed dispersal under field conditions have also revealed transport distances to be modest, even when secondary dispersal (i.e. lateral movement of diaspores across the soil surface) is taken into account (Mortimer, 1974; Watkinson, 1978; O'Toole & Cavers, 1983; Cousens & Mortimer, 1995). In the wind-dispersed annual *Vulpia fasciculata*, the maximum distance diaspores fell from the parent was 36 cm, with 79% falling within 10 cm of the parent plant (Watkinson, 1978). Even with secondary dispersal, the maximum dispersal distance was less than 1 m. A more recent study of the congeneric species *V. ciliata* showed that the mean dispersal distance under field conditions was only 6 cm (Carey & Watkinson, 1993). As might be expected, for both species infructescence height was correlated with dispersal distance.

Field experiments with the annual grass weed *Panicum miliaceum* revealed usual dispersal distances of less than 1 m (McCanny & Cavers, 1987). For *Agrostis hiemalis*, a perennial grass with tumbleweed dispersal, an average seed travelling by this mode moved three times farther than a seed falling from a stationary parent (Rabinowitz & Rapp, 1979). However, even with tumbleweed dispersal, the geometric mean distance seeds moved was 1.4 m and 95% fell within 9 m of the parental source.

Antitelechory

In common with other herbaceous species where seed shadows have been quantified (Levin & Kerster, 1974; Levin, 1981; Willson, 1993), a number of grasses appear to have limited dispersibility within natural populations. In fact, a proportion of the seeds may be 'gravity dispersed' or retained by the parent plant by a number of different mechanisms. Antitelechory has been defined as the hindrance of dispersal by the placement of seeds at, near, or below the soil surface, or by morphological characteristics of the seeds, diaspores, or mother plant (Ellner & Shmida, 1981). Using this definition, the life history of a variety of grasses show antitelechoric features.

The most extreme example of the antitelechoric strategy can be found in amphicarpic grasses that mature some proportion of their seeds on subterranean culms or rhizomes. The best known example is peanutgrass (*Amphicarpum purshii*), an annual with subterranean seeds which are produced on culms no more than a few centimetres long (Cheplick, 1987, 1994). Thus, seedlings arising from these seeds are retained within the immediate vicinity of the parent that produced them. Subterranean seeds have also been reported in seven other species of the Poaceae (Campbell *et al.*, 1983).

Besides amphicarpy, there are other less obvious, but more common, mechanisms by which dispersal may be precluded in grasses. At least 22 species produce 'cleistogenes' (Chase, 1908) which are cleistogamous spikelets enclosed by leaf sheaths on the lowermost culm nodes (Campbell *et al.*, 1983). The resulting axillary seeds are not usually dispersed and often remain enclosed within the lowermost leaf sheaths for some time after parent culms have senesced. Because both cleistogenes and subterranean seeds remain in a microhabitat where the parent grew and reproduced, and in addition may receive some protection from fire, grazers, or environmental extremes, these seeds may be more likely to produce established seedlings (Dyksterhuis, 1945; Campbell *et al.*, 1983; Clay, 1983; Cheplick, 1987, 1994; Cheplick & Quinn, 1988).

Other cleistogamous grasses produce additional axillary inflorescences enclosed by leaf sheaths on the middle or uppermost culm nodes. In their tabulation, Campbell *et al.* (1983) stated that 118 species in 41 genera showed this form of cleistogamy. In *Triplasis purpurea* and *Sporobolus vaginiflorus*, two annual grasses with sheath-enclosed cleistogamous panicles at all culm nodes, seeds in the upper nodes can be dispersed by wind as culms break apart. However, the lowermost culm nodes with their seeds remain intact *in situ* over winter and thus do not disperse (Cheplick, 1992, 1996*a,b*).

These antitelechoric life history features have some intriguing ecological consequences that can offer clues as to why and how such a system could evolve. There are negative density-dependent consequences (sibling competition) predicted for seedlings that arise from seeds that are not dispersed in space or time (Cheplick, 1992; Willson, 1992; Venable & Brown, 1993). In *Sporobolus vaginiflorus*, the seeds in the lower culm nodes germinate directly from the senesced parent tillers in spring, resulting in a high-density zone of competing siblings centred around the original parent base. This zone of sibling competition severely reduces the growth and reproduction of individuals and thus has significant fitness consequences (Cheplick, 1992, 1993).

Casual observations have revealed that sibling competition also may occur in *Triplasis purpurea* which has a life history that is strikingly similar to *S. vaginiflorus* (Cheplick, 1996b). An analysis of tiller lengths in a population of *T. purpurea* on Staten Island, New York shows why sibling seedling densities in spring are greatest at the locations where the maternal parent originally stood (Fig. 3.2). In late autumn when the growing season is over, the majority of tillers are short, being composed of only a few nodes containing sheath-enclosed seeds. Thus, a few nodes (with seeds) of every tiller are within 5 cm of a parent base; fewer and fewer tillers extend out to increasingly greater distances (Fig. 3.2). Because the lower 2–4 nodes remain intact over winter, if most sheath-enclosed seeds at these nodes germinate the following spring, a circular zone of competing siblings will arise.

Recent laboratory experiments have suggested, however, that the non-dispersed seeds in the lowermost nodes of both *T. purpurea* and *S. vaginiflorus* have lower and slower germination than the potentially dispersible seeds in the uppermost nodes (Cheplick, 1996*a*; Fig. 3.3). This position-dependent germination pattern shared by the two grasses concords with the

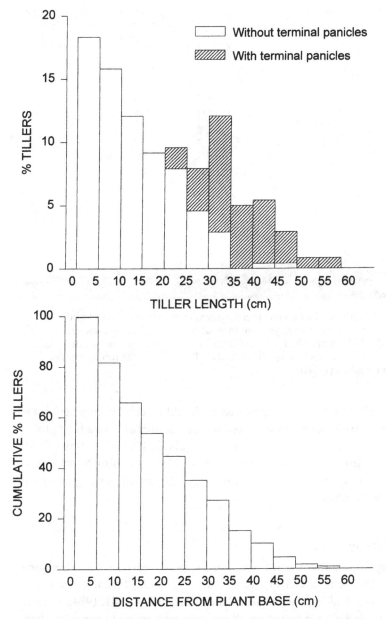

Fig. 3.2. Frequency distribution of tiller lengths (top) and the cumulative percentage of tillers extending to various distances from the parental plant base (bottom) in a population of *Triplasis purpurea* on Staten Island, New York. All tillers contained axillary cleistogamous panicles; tillers with additional exserted terminal panicles are indicated by hatching (top).

Fig. 3.3. Germination of intact stem fragments representing distinct nodes of *Triplasis purpurea* after 16 days in an incubator set on a diurnal cycle of 25/15 °C, 12-hr light/12-hr dark. The leftmost fragment represents the most basal node; the rightmost seedlings germinated from seeds collected from the terminal panicle. See also Cheplick (1996a).

theoretical prediction that germination should be delayed or staggered in time for poorly dispersed seeds because their seedlings have a high probability of experiencing sibling competition (Ellner, 1986; Silvertown, 1988; Venable & Brown, 1988; Nilsson *et al.*, 1994). The mechanism by which germination in these grasses is differentially regulated among nodes is presently unknown.

Role of the awn

Probably the best examples of how morphological features of spikelet parts can be adaptive in seed dispersal and seedling establishment are provided in a series of studies done in Australia by Peart (1979, 1981, 1984). For five grasses subjected to experiments where awns were excised from some diaspores, he showed the importance of the awn in locating and lodging in appropriate microsites on a cracked-clay surface (Peart, 1979). For *Aristida vagans* and *Microlaena stipoides*, the rigid awns orient the diaspore during its fall from the parent, thereby increasing the chances that it will land in an upright position (Peart, 1981). Such orientation increases the probability of germination and successful seedling establishment. Other experiments dramatically illustrate the role of the hygroscopically active awn in seedling establishment: for all four grasses examined, excision of the awn significantly reduced the establishment of seedlings on almost all of the four soil surface types studied (Peart, 1979).

The presence or absence of the awn may even explain patterns of species distribution. After an examination of the grasses distributed among 36 sites in south-eastern Queensland, it was found that species with hygroscopically active awns tended to predominate on soils that characteristically develop cracks or loose, crumbly surfaces when dry (Peart & Clifford, 1987). Thus, the surface properties of such soils provided microsites in which awned diaspores could easily lodge and presumably establish seedlings. Additional examples of the importance of spikelet hairs and awns in grass seed entrapment and seedling establishment are provided by Chambers (1995) for an alpine ecosystem.

Seedling establishment

Between seed dispersal and seedling recruitment into a population, a variety of processes occur including seed germination and seed/seedling mortality (Harper, 1977; Cook, 1979; Cavers, 1983; Fenner, 1987; Houle, 1995). Because the ecology of seed germination and dormancy in grasses is thoroughly explored by Baskin & Baskin (this volume), here the focus will be primarily on the environmental factors that impact grass seedlings after germination.

Relevant environmental factors

As might be expected, many of the environmental factors affecting seedling establishment in the grasses are the same as those that affect establishment in any plant population (Dowling, Clements & McWilliam, 1971; Sheldon, 1974; Harper, 1977; Fenner, 1987). However, some of the factors have been better characterized for grasses compared with other plants. Also, specific characterisitics of the unique grass diaspore such as the awn have been found to play an important role in seedling establishment (see above). This section will highlight the major environmental determinants of seedling establishment in grass populations. There are many studies that have examined seedling establishment in grasses, especially for economically important species like crops and forages; here, only a few studies will be examined

Table 3.4. *The fine-scale, primary factors and the associated larger-scale, primary environmental determinants that affect the seedling establishment of grasses*

Primary factors	Primary determinants
I. Abiotic factors	
Local edaphic conditions (moisture, nutrients, aeration . . .)	Rainfall patterns and distribution; topography; soil type
Light	Daylength; vegetation structure/type; disturbance levels
Temperature	Regional climate; vegetation structure/type
II. Biotic factors	
Competition	Population density; vegetation structure/ type; disturbance levels
Neighbours (e.g. allelopathy, nurse plant effects)	Population density; vegetation structure/ type
Herbivory	Insect/mammal abundance; structural and chemical defences
Litter	Vegetation structure/type; decomposition rates; fire frequency
Pathogens/mutualists/other symbionts	Structural and chemical defences; environmental conditions (moisture, temperature . . .); availability of inocula
Maternal effects (e.g. seed size heteromorphism)	Environmental conditions present during seed maturation; position of seed maturation on maternal plant

to provide a broad overview of how the grass diaspore and the resulting seedling interact with the local microenvironment during the establishment process.

The primary factors affecting seedling establishment can be classified as abiotic or biotic (Table 3.4). At the scale of the seedling, typical factors include soil conditions (moisture, nutrients, aeration), light, and temperature. The primary large-scale determinants of the abiotic microenvironment include rainfall patterns, soil type, topography, regional climate, and vegetation structure. As a general ecological force, 'disturbance' at any scale can be listed as a primary determinant for almost any abiotic factor because of its role in changing local microenvironmental conditions (Table 3.4).

Biotic factors that can impact the establishment of grass seedlings include competition, neighbours, allelopathy, herbivores, litter, symbiotic associates, and maternal effects. Primary determinants of the biotic microenvironment of a seedling depend greatly on the factor considered, but typical ones might include the density of conspecific individuals and the surrounding vegetation, disturbance levels, fire frequency, and herbivore abundance (Table 3.4). As noted in the previous section on antitelechory, seed germination that depends on the position of maturation on the parent provides an example of a maternal effect that could subsequently influence seedling establishment patterns in the field. The importance of plant–plant interactions to seedling establishment are considered by Lauenroth & Aguilera (this volume).

Safe sites

The microsites in which the probability of seed germination followed by successful seedling establishment is greatest can be termed 'safe sites' (Harper, 1977). It should be noted that a site that is 'safe' for a seed is not necessarily conducive to seedling establishment (Schupp, 1995). However, for many field studies seed germination and seedling establishment have not been separately examined and the two processes are difficult to unravel. There is broad overlap in the environmental factors affecting both seeds and seedlings (e.g. soil moisture). The successful establishment of a particular seedling must be preceded by the germination of a particular seed!

Fowler (1988) has examined some of the factors that help define the safe sites required by *Aristida longiseta* and *Bouteloua rigidiseta*, two perennial grasses native to North America. There was tremendous variation in the relative degree to which a patch was favourable for seedling establishment between years, among census intervals, and among germination cohorts. Interestingly, seedlings had higher chances of surviving if they had one or more seedlings or juveniles of either species within 2 cm of them. This apparent positive effect of neighbours on seedling performance was most likely due to a correlation between microsite favourableness and the presence of an existing parent, rather than any type of direct facilitation (Fowler, 1988). The timing of germination and litter also affected seedling establishment: seedlings germinating earlier in autumn were more successful, but litter reduced the survival rates of seedlings of both species.

The negative impact of litter on seedlings reported by Fowler (1988) is unexpected, given the ability of litter to increase water availability by reducing evaporation (Facelli & Pickett, 1991). In an earlier study of *A. longiseta* and *B. rigidiseta*, germination was increased by a thin cover of litter

(Fowler, 1986). Emergence of seedlings from shallowly buried seeds of *Amphicarpum purshii* has been shown to be increased by litter (Cheplick & Quinn, 1987). Because water has proven to be very important to both seed germination and seedling establishment in grasses (Dowling *et al.*, 1971; Mack & Pyke, 1984; Fowler, 1986; Wester, Dahl & Cotter, 1986; Simpson, 1990; Potvin, 1993; Lauenroth *et al.*, 1994), any environmental factor like litter that improves the availability of water is likely to be a component of a seed's safe site. However, it should be noted that thick layers of litter can act as a mechanical barrier (Facelli & Pickett, 1991) and may curtail the establishment of seedlings in some habitats (Fowler, 1986; Bergelson, 1990).

Habitat disturbance has been found to be another important factor in the generation of safe sites for grass seed germination and seedling establishment. Whether natural or human-caused, any disturbance that removes intact vegetation tends to promote the seedling establishment of grasses (e.g. Bullock *et al.*, 1995). Experiments with the amphicarpic annual *Amphicarpum purshii* have shown that fire may create open conditions more conducive to germination and seedling establishment by removing perennial vegetation (Cheplick & Quinn, 1988). For the perennial bunchgrass *Danthonia sericea*, seedling survival was greatest in the field in quadrats where all aboveground vegetation had been removed by the investigator (Moloney, 1990).

Probably the most realistic examples of how disturbance affects seedling establishment come from those studies where investigators considered the interaction of other environmental factors with disturbance by experimentally manipulating them and/or by examining sites that differed in such factors (Moloney, 1990; Potvin, 1993; Aguilera & Lauenroth, 1995; Burke & Grime, 1996). In a study of three grasses on three sites in the Nebraska Sandhills, USA (*Andropogon hallii*, *Sorghastrum nutans*, and *Panicum virgatum*), Potvin (1993) established four treatments: (1) soil disturbed, (2) soil undisturbed, irrigated, (3) soil disturbed, irrigated, and (4) undisturbed. A rototill was used for the disturbance treatments. Although there was considerable variation among sites, most mortality occurred during seedling establishment and was primarily due to desiccation. There were no surviving seedlings of any species at any site in undisturbed plots by the end of the first growing season and the greatest seedling survivorship occurred in the plots where the soil was both disturbed and irrigated (Potvin, 1993).

A series of experiments by Aguilera & Lauenroth (1993, 1995) with the perennial bunchgrass *Bouteloua gracilis* provide additional evidence of the role of gap disturbances and microsite environment in seedling establishment under field conditions. By excluding the roots of neighbouring conspecific adults in some plots, they showed that emergence and survival of seedlings was greatest when adult plants were prevented from interacting with the seedlings (Aguilera & Lauenroth, 1993). The removal of established plants clearly favoured seedling establishment and any form of disturbance (clipping, herbicide, shading) tended to enhance seedling establishment, survivorship, and size (Aguilera & Lauenroth, 1995).

One final, more subtle, biotic factor needs to be recognized in light of its ability to affect grass seeds and seedlings. Symbiotic microorganisms, such as the endophytic fungi that are found in so many species (Clay, 1990 and this volume), have the potential to influence seedling establishment in grass populations. Endophytic fungi have been shown to increase germination and seedling growth of both tall fescue (*Festuca arundinacea*) and perennial ryegrass (*Lolium perenne*) under controlled environmental conditions (Clay, 1987). Compared with uninfected seedlings, endophyte infection also tended to enhance the competitive ability of infected seedlings in glasshouse experiments (Marks, Clay & Cheplick, 1991). Although additional research is clearly required before these results can be extrapolated to natural populations, the possibility that symbionts may have a subtle effect on the establishment of grasses should not be overlooked.

Summary and future research directions

Seed dispersal and seedling establishment are ecological processes critical to the colonization and persistence of grass populations. While much data are available regarding many of the factors affecting seedling establishment, less information can be found on dispersal systems, coevolution of grasses with dispersal agents, and dispersal distances. Future research should consider the impact of dispersal on population genetic structure and adaptation to local habitats. For the mostly wind-dispersed species that have been examined, most seeds are dispersed no more than a few metres from parental sources. The primary modes of dispersal are wind and ingestion by mammals or birds. Morphology of the spikelet or inflorescence can provide clues as to the dispersal agent; however, grasses may have more than one agent (e.g. wind and adhesion) and there can be overlap among dispersal modes in the morphological features of the associated reproductive structures. More information will be needed on the coevolutionary relationship between grasses and biotic dispersal agents before generalizations can be made about the influence of animal dispersers on the morphological features of grass diaspores.

Some grasses inhibit seed dispersal by the placement of seeds at, near, or below the soil surface, or by particular characteristics of the seed, diaspore, or mother plant (antitelechory). Although a number of cleistogamous grasses show antitelechoric life history features, the adaptive advantages of limited dispersal are unclear. Clearly, the genetic structure, adaptation, and microevolution of populations is influenced by dispersal ability (Hamrick & Loveless, 1986; Richards, 1990; Williams, 1994), but to date, the ecological and evolutionary consequences of restrictive dispersal systems have not been well documented.

The spikelet awn of some grass diaspores has been shown to be of critical importance in lodging in appropriate microsites, and for orienting the diaspore in a manner conducive to seedling establishment. For many grass seeds, a safe site for germination and establishment has adequate moisture, a thin cover of litter, and is relatively free of other vegetation. Disturbance clearly plays an important role in providing safe sites by creating openings in closed perennial communities (Bullock et al., 1995; Burke & Grime, 1996). Future workers should consider the magnitude, timing, and frequency of disturbance when exploring the dynamics of grass populations in successional communities. The spatial and temporal distribution of safe sites within a community for grass colonization and establishment is generally not known and represents a critical area for future research.

Despite their probable importance to life histories, there is much that is not known about the dispersal and establishment phases of most grasses. This is unfortunate because more knowledge on the processes of seed dispersal and seedling establishment could shed light on how troublesome grass weeds colonize areas and establish populations (Cousens & Mortimer, 1995), and contribute much to the understanding of how genetic variation is distributed within and between populations. In addition, such information could provide management tools for ecosystems where grasses predominate (Chambers & MacMahon, 1994) and increase the understanding of successional processes in many natural and disturbed communities.

Acknowledgements

Sincere thanks are extended to Polly Cheplick for help in collecting the data reported in Fig. 3.2, and to M. L. Cain and C. C. Baskin for providing helpful suggestions on improving the manuscript. The *Triplasis* research was supported by a grant from the City University of New York PSC-CUNY Research Award Program.

References

Aguilera, M. O. & Lauenroth, W. K. (1993). Seedling establishment in adult neighbourhoods – intraspecific constraints in the regeneration of the bunchgrass *Bouteloua gracilis, Journal of Ecology,* **81**, 253–61.

Aguilera, M. O. & Lauenroth, W. K. (1995). Influence of gap disturbances and type of microsites on seedling establishment in *Bouteloua gracilis. Journal of Ecology,* **83**, 87–97.

Begon, M., Harper, J. L. & Townsend, C. R. (1990). *Ecology: Individuals, Populations and Communities.* London: Blackwell Scientific.

Bergelson, J. (1990). Life after death: site pre-emption by the remains of *Poa annua. Ecology,* **71**, 2157–65.

Bullock, J. M., Clear Hill, B., Silvertown, J. & Sutton, M. (1995). Gap colonization as a source of grassland community change: effects of gap size and grazing on the rate and mode of colonization by different species. *Oikos,* **72**, 273–82.

Burke, M. J. W. & Grime, J. P. (1996). An experimental study of plant community invasibility. *Ecology,* **77**, 776–90.

Campbell, C. S. (1983). Wind dispersal of some North American species of *Andropogon* (Gramineae). *Rhodora,* **85**, 65–72.

Campbell, C. S., Quinn, J. A., Cheplick, G. P. & Bell, T. J. (1983). Cleistogamy in grasses. *Annual Review of Ecology and Systematics,* **14**, 411–41.

Carey, P. D. & Watkinson, A. R. (1993). The dispersal and fates of seeds of the winter annual grass *Vulpia ciliata. Journal of Ecology,* **81**, 759–67.

Cavers, P. B. (1983). Seed demography. *Canadian Journal of Botany,* **61**, 3578–90.

Chambers, J. C. (1995). Relationships between seed fates and seedling establishment in an alpine ecosystem. *Ecology,* **76**, 2124–33.

Chambers, J. C. & MacMahon, J. A. (1994). A day in the life of a seed: movements and fates of seeds and their implications for natural and managed systems. *Annual Review of Ecology and Systematics,* **25**, 263–92.

Chapman, G. P. (1992). Apomixis and evolution. In *Grass Evolution and Domestication,* ed. G. P. Chapman, pp. 138–55. Cambridge: Cambridge University Press.

Chapman, G. P. & Peat, W. E. (1992). *An Introduction to the Grasses (Including Bamboos and Cereals).* Wallingford: CAB International.

Chase, A. (1908). Notes on cleistogamy in grasses. *Botanical Gazette,* **45**, 135–6.

Cheplick, G. P. (1987). The ecology of amphicarpic plants. *Trends in Ecology and Evolution,* **2**, 97–101.

Cheplick, G. P. (1988). Influence of environment and population origin on survivorship and reproduction in reciprocal transplants of amphicarpic peanutgrass (*Amphicarpum purshii*). *American Journal of Botany,* **75**, 1048–56.

Cheplick, G. P. (1992). Sibling competition in plants. *Journal of Ecology,* **80**, 567–75.

Cheplick, G. P. (1993). Reproductive systems and sibling competition in plants. *Plant Species Biology,* **8**, 131–9.

Cheplick, G. P. (1994). Life history evolution in amphicarpic plants. *Plant Species Biology,* **9**, 119–31.

Cheplick, G. P. (1996*a*). Do seed germination patterns in cleistogamous annual grasses reduce the risk of sibling competition? *Journal of Ecology,* **84**, 247–55.

Cheplick, G. P. (1996*b*). Cleistogamy and seed heteromorphism in *Triplasis purpurea* (Poaceae). *Bulletin of the Torrey Botanical Club,* **123**, 25–33.

102   G. P. Cheplick

Cheplick, G. P. & Quinn, J. A. (1987). The role of seed depth, litter, and fire in the seedling establishment of amphicarpic peanutgrass (*Amphicarpum purshii*). *Oecologia*, **73**, 459–64.

Cheplick, G. P. & Quinn, J. A. (1988). Subterranean seed production and population responses to fire in *Amphicarpum purshii* (Gramineae). *Journal of Ecology*, **76**, 263–73.

Clark, L. G & Fisher, J. B. (1987). Vegetative morphology of grasses: shoot and roots. In *Grass Systematics and Evolution*, ed. T. R. Soderstrom, K. W. Hilu, C. S. Campbell, & M. E. Barkworth, pp. 37–45. Washington, D.C.: Smithsonian Institution Press.

Clay, K. (1983). Variation in the degree of cleistogamy within and among species of the grass *Danthonia*. *American Journal of Botany*, **70**, 835–43.

Clay, K. (1987). Effects of fungal endophytes on the seed and seedling biology of *Lolium perenne* and *Festuca arundinacea*. *Oecologia*, **73**, 358–62.

Clay, K. (1990). Fungal endophytes of grasses. *Annual Review of Ecology and Systematics*, **21**, 275–97.

Clayton, W. D. (1990). The spikelet. In *Reproductive Versatility in the Grasses*, ed. G. P. Chapman, pp. 32–51. Cambridge: Cambridge University Press.

Cook, R. E. (1979). Patterns of juvenile mortality and recruitment in plants. In *Topics in Plant Population Biology*, ed. O. T. Solbrig, S. Jain, G. B. Johnson, & P. H. Raven, pp. 207–31. Columbia University Press, New York.

Cousens, R. & Mortimer, M. (1995). *Dynamics of Weed Populations*. Cambridge: Cambridge University Press.

Davidse, G. (1987). Fruit dispersal in the Poaceae. In *Grass Systematics and Evolution*, ed. T. R. Soderstrom, K. W. Hilu, C. S. Campbell & M. E. Barkworth, pp. 143–55. Washington, D.C.: Smithsonian Insitution Press.

Davidse, G. & Morton, E. (1973). Bird-mediated fruit dispersal in the tropical grass genus *Lasiacis* (Gramineae: Paniceae). *Biotropica*, **5**, 162–7.

Dowling, P. M., Clements, R. J. & McWilliam, J. R. (1971). Establishment and survival of pasture species from seeds sown on the soil surface. *Australian Journal of Agriculture Research*, **22**, 61–74.

Dyksterhuis, E. J. (1945). Axillary cleistogenes in *Stipa leucotricha* and their role in nature. *Ecology*, **26**, 195–9.

Ellner, S. (1986). Germination dimorphisms and parent–offspring conflict in seed germination. *Journal of Theoretical Biology*, **123**, 173–85.

Ellner, S. & Shmida, A. (1981). Why are adaptations for long-range seed dispersal rare in desert plants? *Oecologia*, **51**, 133–44.

Ernst, W. H. O., Veenendaal, E. M. & Kebakile, M. M. (1992). Possibilities for dispersal in annual and perennial grasses in a savanna in Botswana. *Vegetatio*, **102**, 1–11.

Facelli, J. M. & Pickett, S. T. A. (1991). Plant litter: its dynamics and effects on plant community structure. *The Botanical Review*, **57**, 1–32.

Fenner, M. (1987). Seedlings. *New Phytologist*, **106** (Suppl.), 35–47.

Fowler, N. L. (1986). Microsite requirements for germination and establishment of three grass species. *American Midland Naturalist*, **115**, 131–45.

Fowler, N. L. (1988). What is a safe site?: neighbor, litter, germination date, and patch effects. *Ecology*, **69**, 947–61.

Hamrick, J. L. & Loveless, M. D. (1986). The influence of seed dispersal mechanisms on the genetic structure of plant populations. In *Frugivores and Seed Dispersal*, ed. A. Estrada & T. H. Fleming, pp. 211–23. Dordrecht: Dr. W. Junk Publishers.

Harper, J. L. (1977). *Population Biology of Plants*. London: Academic Press.

Hodgson, J. G. & Grime, J. P. (1990). The role of dispersal mechanisms, regenerative strategies and seed banks in the vegetation dynamics of the British landscape. In *Species Dispersal in Agricultural Habitats*, ed. R. G. H. Bunce & D. C. Howard, pp. 65–81. London: Belhaven Press.

Houle, G. (1995). Seed dispersal and seedling recruitment: the missing link(s). *Ecoscience*, **2**, 238–44.

Howe, H. F. & Smallwood, J. (1982). Ecology of seed dispersal. *Annual Review of Ecology and Systematics*, **13**, 201–8.

Janzen, D. H. (1984). Dispersal of small seeds by big herbivores: foliage is the fruit. *American Naturalist*, **123**, 338–53.

Jurado, E., Westoby, M. & Nelson, D. (1991). Diaspore weight, dispersal, growth form and perenniality of central Australian plants. *Journal of Ecology*, **79**, 811–30.

Lauenroth, W. K., Sala, O. E., Coffin, D. P. & Kirchner, T. B. (1994). The importance of soil water in the recruitment of *Bouteloua gracilis* in the shortgrass steppe. *Ecological Applications*, **4**, 741–9.

Leishman, M. R., Westoby, M. & Jurado, E. (1995). Correlates of seed size variation: a comparison among five temperate floras. *Journal of Ecology*, **83**, 517–30.

Levin, D. A. (1981). Dispersal versus gene flow in plants. *Annals of the Missouri Botanical Garden*, **68**, 233–53.

Levin, D. A. & Kerster, H. W. (1974). Gene flow in seed plants. *Evolutionary Biology*, **7**, 139–220.

Mack, R. N. & Pyke, D. A. (1984). The demography of *Bromus tectorum*: the role of microclimate, grazing and disease. *Journal of Ecology*, **72**, 731–48.

Malo, J. E. & Suarez, F. (1995). Herbivorous mammals as seed dispersers in a Mediterranean *dehesa*. *Oecologia*, **104**, 246–55.

Marks, S., Clay, K. & Cheplick, G. P. (1991). Effects of fungal endophytes on interspecific and intraspecific competition in the grasses *Festuca arundinacea* and *Lolium perenne*. *Journal of Applied Ecology*, **28**, 194–204.

McCanny, S. J. & Cavers, P. B. (1987). The escape hypothesis: a test involving a temperate, annual grass. *Oikos*, **49**, 67–76.

Moloney, K. A. (1990). Shifting demographic control of a perennial bunchgrass along a natural habitat gradient. *Ecology*, **71**, 1133–43.

Mortimer, A. M. (1974). Studies of germination and establishment of selected species with special reference to the fates of seeds. Ph.D. Thesis, University of Wales.

Nilsson, P., Fagerstrom, T., Tuomi, J. & Astrom, M. (1994). Does seed dormancy benefit the mother plant by reducing sib competition? *Evolutionary Ecology*, **8**, 422–30.

O'Toole, J. T. & Cavers, P. B. (1983). Input to seed banks of proso millet (*Panicum miliaceum*) in southern Ontario. *Canadian Journal of Plant Science*, **63**, 1023–30.

Peart, M. H. (1979). Experiments on the biological significance of the morphology of seed-dispersal units in grasses. *Journal of Ecology*, **67**, 843–63.

Peart, M. H. (1981). Further experiments on the biological significance of the morphology of seed-dispersal units in grasses. *Journal of Ecology*, **69**, 425–36.

Peart, M. H. (1984). The effects of morphology, orientation and position of grass diaspores on seedling survival. *Journal of Ecology*, **72**, 437–53.

Peart, M. H. & Clifford, H. T. (1987). The influence of diaspore morphology and soil-surface properties on the distribution of grasses. *Journal of Ecology*, **75**, 569–76.

Potvin, M. A. (1993). Establishment of native grass seedlings along a topographic/moisture gradient in the Nebraska Sandhills. *American Midland Naturalist*, **130**, 248–61.

Quinn, J. A., Mowrey, D. P., Emanuele, S. M. & Whalley, R. D. B. (1994). The 'Foliage is the Fruit' hypothesis: *Buchloe dactyloides* (Poaceae) and the shortgrass prairie of North America. *American Journal of Botany*, **81**, 1545–54.

Rabinowitz, D. & Rapp, J. K. (1979). Dual dispersal modes in hairgrass, *Agrostis hiemalis* (Walt.) B.S.P. (Gramineae). *Bulletin of the Torrey Botanical Club*, **106**, 32–6.

Rabinowitz, D. & Rapp, J. K. (1981). Dispersal abilities of seven sparse and common grasses from a Missouri prairie. *American Journal of Botany*, **68**, 616–24.

Richards, A. J. (1990). The implications of reproductive versatility for the structure of grass populations. In *Reproductive Versatility in the Grasses*, ed. G. P. Chapman, pp. 131–53. Cambridge: Cambridge University Press.

Ridley, H. N. (1930). *The Dispersal of Plants Throughout the World*. London: L. Reeve & Co., Ltd.

Schupp, E. W. (1995). Seed–seedling conflicts, habitat choice, and patterns of plant recruitment. *American Journal of Botany*, **82**, 399–409.

Sheldon, J. C. (1974). The behaviour of seeds in soil. III. The influence of seed morphology and the behaviour of seedlings on the establishment of plants from surface-lying seeds. *Journal of Ecology*, **62**, 47–66.

Silberbauer-Gottsberger, I. (1984). Fruit dispersal and trypanocarpy in Brazilian cerrado grasses. *Plant Systematics and Evolution*, **147**, 1–27.

Silvertown, J. (1988). The demographic and evolutionary consequences of seed dormancy. In *Plant Population Ecology*, ed. A. J. Davy, M. J. Hutchings & A. R. Watkinson, pp. 205–19. Oxford: Blackwell Scientific Publications.

Simpson, G. M. (1990). *Seed Dormancy in Grasses*. Cambridge: Cambridge University Press.

Sorensen, A. E. (1986). Seed dispersal by adhesion. *Annual Review of Ecology and Systematics*, **17**, 443–63.

Stebbins, G. L. (1974). *Flowering Plants: Evolution Above the Species Level*. Cambridge, MA: Harvard University Press.

van der Pijl, L. (1982). *Principles of Dispersal in Higher Plants*, 3rd edn. New York: Springer-Verlag.

Venable, D. L. & Brown, J. S. (1988). The selective interactions of dispersal, dormancy, and seed size as adaptations for reducing risk in variable environments. *American Naturalist*, **131**, 360–84.

Venable, D. L. & Brown, J. S. (1993). The population-dynamic functions of seed dispersal. *Vegetatio*, **107/108**, 31–55.

Watkinson, A. R. (1978). The demography of a sand dune annual: *Vulpia fasciculata*. III. The dispersal of seeds. *Journal of Ecology*, **66**, 483–98.

Wester, D. B., Dahl, B. E. & Cotter, P. F. (1986). Effects of pattern and amount of simulated rainfall on seedling dynamics of weeping lovegrass and Kleingrass. *Agronomy Journal*, **78**, 851–5.

Williams, C. F. (1994). Genetic consequences of seed dispersal in three sympatric forest herbs. II. Microspatial genetic structure within populations. *Evolution*, **48**, 1959–72.

Willson, M. F. (1992). The ecology of seed dispersal. In *Seeds: The Ecology of*

Regeneration in Plant Communities, ed. M. Fenner, pp. 61–85. Tuscon, AZ: Tuscon University Press.

Willson, M. F. (1993). Dispersal mode, seed shadows, and colonization patterns. *Vegetatio*, **107/108**, 261–80.

Willson, M. F., Rice, B. & Westoby, M. (1990). Seed dispersal spectra: comparison of temperate plant communities. *Journal of Vegetation Science*, **1**, 547–62.

4

Clonal biology of caespitose grasses

DAVID D. BRISKE AND JUSTIN D. DERNER

Introduction

Graminoids comprise one of the largest subgroups of clonal plants among terrestrial angiosperms (Tiffney & Niklas, 1985). Caespitose graminoids represent a unique growth form that is characterized by the compact spatial arrangement of ramets within individual clones and the absence of rhizomes or stolons. Caespitose graminoids occur on all continents from the high Arctic to the Sub-Antarctic and are distributed over a wide range of precipitation zones (Leith, 1978; Walter, 1979). This growth form is particularly dominant in the grassland biome which occupies 24 million km^2 including tropical and temperate grasslands, savannas and shrub steppe (Leith, 1978).

The wide distribution and dominance expressed by caespitose graminoids is somewhat surprising given that the majority of modern monocot families are rhizomatous (Tiffney & Niklas, 1985). Rhizomes are modified horizontal ramets located belowground that represent a mode of clonal growth well suited for ramet dispersal, effective resource acquisition, and resource storage (Grace, 1993). In contrast, caespitose graminoids possess several attributes that could potentially limit their success, including intense intraclonal competition for nutrients (Hartnett, 1993) and for photosynthetically active radiation (Caldwell et al., 1983; Ryel, Beyschlag & Caldwell, 1993, 1994), and a limited ability to access heterogeneously distributed resources (Van Auken, Manwaring & Caldwell, 1992). This poses the question, 'what structural and/or functional attributes contribute to the ecological success of caespitose graminoids without the benefits conferred by rhizomes?' The adaptive value of caespitose clones must be sufficient to offset the benefits associated with rhizomes or this growth form would not be so successful (e.g. Pedersen & Tuomi, 1995). Because they possess minimal plasticity for ramet placement compared with rhizomatous and stoloniferous (produce modified horizontal ramets aboveground) species (e.g. de Kroon & van Groenendael, 1990), caespitose graminoids may be an ideal growth form in which to evaluate ecological success conferred by clonality.

A greater understanding of the processes and mechanisms influencing and/or regulating ramet demography within individual clones would increase insight into the ecological success of caespitose graminoids. For example, are ramets within individual clones interdependent or independent? What mechanism(s) regulates ramet densities, including recruitment and/or mortality, within clones? Do caespitose clones represent an alternative strategy to active foraging characteristic of rhizomatous and stoloniferous species? How does the spatial arrangement of ramets within individual clones influence growth efficiency? Answers to these questions are essential for a thorough understanding of the structure, function, and competitive ability expressed by this important and widely distributed group of clonal plants. Unfortunately, the biology of caespitose clones has received minimal attention compared with resource foraging by rhizomatous and stoloniferous species. However, the concept of foraging strategies within clonal plants has recently been broadened to incorporate the caespitose growth form (Hutchings & de Kroon, 1994; de Kroon & Hutchings, 1995).

The goals of this chapter are to evaluate mechanisms potentially capable of regulating ramet demography within caespitose clones and to assess mechanisms conferring ecological success to caespitose grasses. These two goals are related in that intraclonal regulation of ramet recruitment and density may optimize the growth efficiency of individual clones. Specific objectives are to:

1. establish the architectural constraints that define the caespitose growth form;
2. evaluate the status of mechanisms capable of regulating intraclonal ramet demography, including apical dominance, resource competition, physiological integration, the red:far-red ratio of solar radiation, and equitable resource acquisition among ramets within clones;
3. assess the relative magnitude and influence of intraclonal ramet competition;
4. survey the ability of caespitose clones to consolidate and monopolize resources within their immediate environment as a potential mechanism for their ecological success.

Architecture and demography of caespitose grasses

Clonal architecture

A thorough comparison of caespitose and rhizomatous or stoloniferous grasses requires some knowledge of the developmental morphology contributing to architectural variations among them. In grasses, the fundamental growth unit is the phytomer which consists of a blade, sheath, node, internode, and axillary bud (Etter, 1951; Langer, 1972; Dahl & Hyder, 1977). Ramets comprise a series of phytomers successively differentiated from individual apical meristems (White, 1979; Briske, 1991). This pattern of developmental morphology determines that grass clones are composed of an assemblage of phytomers organized within a variable number of ramets. Morphological variation of individual ramets is a consequence of the number and size of phytomers which comprise them. For example, variation in ramet architecture among grasses of various heights results from a modification in the size and/or number of phytomers determining cumulative ramet height and leaf area. Following a period of juvenile development, ramets are potentially capable of initiating subsequent ramets from axillary buds.

The capacity for physiological integration between juvenile and parental ramets is established at the time of ramet development. Inter-ramet vascular connections are formed soon after bud differentiation beginning with vascular traces of the prophyll (Hitch & Sharman, 1968a,b; Bell, 1976). Basal differentiation of these vascular traces eventually allows them to join the vascular system of the parental ramet, thus conferring the ability for inter-ramet resource allocation. Physiological processes governing resource availability and sink strengths determine the subsequent patterns of intra- and inter-ramet resource allocation once the vascular system has formed (Geiger, 1979). Inter-ramet carbon allocation patterns in grasses have been summarized by Pitelka & Ashmun (1985; also see Welker & Briske, 1992).

Architectural variation among plants is often determined by a small number of attributes including branch angles, internode length, and the probability of bud growth (Harper, 1981). Spatial arrangement of ramets within clones is a major determinant of architectural variation among grass growth forms (i.e. caespitose, rhizomatous and stoloniferous species), and is dependent upon the pattern of ramet development (Fig. 4.1). Intravaginal ramet development within the subtending leaf sheaths results in minimal inter-ramet distances and defines the caespitose (tussock, bunchgrass, or phalanx) growth form (White, 1979; Briske, 1991). Extravaginal ramet development proceeds laterally through the subtending

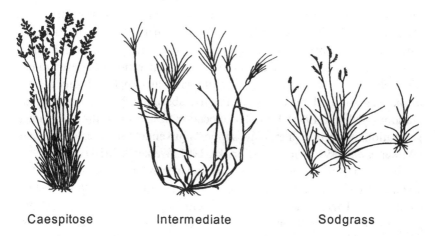

Caespitose   Intermediate   Sodgrass

Fig. 4.1. Architectural variation within the grass growth form originates from the pattern of juvenile ramet emergence. The caespitose growth form originates from intravaginal ramet development within surrounding leaf sheaths while extravaginal ramet development through the subtending leaf sheaths contributes to a more diffuse ramet arrangement represented by the intermediate growth form. Extravaginal ramet development is a prerequisite for rhizome and stolon development characteristic of the sodgrass growth form. (From Briske, 1991).

leaf sheath contributing to a greater inter-ramet distance within clones. This pattern of ramet development defines the intermediate growth form. Extravaginal ramet development is a prerequisite for development of the sodgrass (creeping, spreading, or guerrilla) growth form and inter-ramet distances may be further accentuated by the development of rhizomes and stolons. Architectural variation also determines how clones exploit their environment and interact with neighbours (Harper, 1981).

Clonal demography

Clonal growth and continued site occupation in perennial grasses results from a condition termed meristem dependence (Tomlinson, 1974). Active meristems are continually required to produce juvenile ramets to offset mortality losses associated with the relatively short longevity of these structures (≤ 2 years; Langer, 1972; Briske & Richards, 1995). Successive ramet recruitment produces a series of connected generations referred to as ramet hierarchies or families (Langer, 1972). The number of ramet generations comprising a hierarchy is determined by the rate of ramet recruitment and ramet longevity as influenced by genetic and environmental constraints. In the case of the C_4 caespitose grass, *Schizachyrium scoparium*, ramet hierarchies comprise three connected ramet generations (Welker, Briske & Weaver, 1991; Williams & Briske, 1991). Ramet hierarchies are restricted to three generations because the oldest ramet generation dies and decomposes prior to development of the quaternary ramet generation. However, ramet hierarchies have been shown to separate without ramet mortality in *Poa alpina*, suggesting the occurrence of active abscission between ramet generations (Wilhalm, 1995). Resource allocation from both of the older ramet generations within a hierarchy supports juvenile ramet establishment (Welker *et al.*, 1991; Williams & Briske, 1991; Welker & Briske, 1992).

The numbers of ramets per hierarchy and hierarchies per clone define the size and architectural configuration of caespitose clones. With increasing clone size and age, ramet hierarchies become separated as the initial ramet generations die and decompose (Gatsuk *et al.*, 1980; Olson & Richards, 1988). The hollow crown phenomena characteristic of many long-lived perennial caespitose grasses is very likely a natural consequence of the architectural development of clones and not a symptom of stress or disturbance (e.g. Briske, 1991; Danin & Orshan, 1995). Disproportionate ramet recruitment at the clone periphery eventually reduces axillary bud availability within the clone interior and limits ramet recruitment (Butler & Briske, 1988; Olson & Richards, 1988). The interior regions of clones may not be recolonized because of insufficient plasticity for ramet placement in this location.

Annual ramet replacement theoretically confers clones with potential immortality (Watkinson & White, 1986). Individual genets of *Carex curvula*, *Festuca rubra*, *F. ovina*, and *Holcus mollis* have been estimated to attain great longevities, perhaps exceeding 1000 years (Steinger, Körner & Schmid, 1996), and occupy large areas (Harberd, 1961, 1962, 1967). However, the few age estimates available for North American caespitose grasses indicate that maximum clone longevity does not exceed 50 years (Briske & Richards, 1995). Comparable evaluations of caespitose grasses in Kazakhstan, including *Festuca*, *Koeleria*, and *Stipa* spp., suggest maximum clone longevities of 30–80 years (Vorontzova & Zaugolnova, 1985; Zhukova & Ermakova, 1985). Estimates of relatively short life spans may partially result from the architectural development of caespitose clones and the charting procedures frequently used to monitor clone survival. Clonal expansion and subsequent fragmentation may yield estimates of premature clone mortality even though the genet continues to survive in the form of one or more remnants of the original clone (e.g. West, Rea & Harniss, 1979; Cain, 1990; Lord, 1993). Chronological estimates of the architectural development of *Deschampsia caespitosa* in northern Europe suggest that

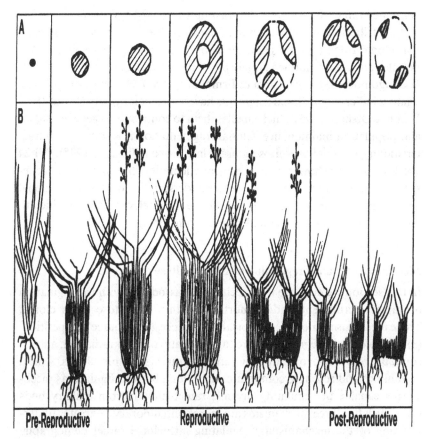

Fig. 4.2. Age estimates of the architectural development of *Deschampsia caespitosa* in northern Europe suggest that 35–60 years is required for clones to proceed from seedlings to senescence. Development of hollow crowns occurs in the reproductive stage and clonal fragmentation occurs in the post-reproductive stage. (From Gatsuk *et al.*, 1980).

35–60 years is required for clones to progress from seedlings to senescent clones (Gatsuk *et al.*, 1980; Zhukova & Ermakova, 1985) (Fig. 4.2). The pre-reproductive, reproductive, and post-reproductive stages require approximately 5–10, 15–30, and 15–25 years to complete, respectively. Clones develop hollow crowns during the reproductive stage and may fragment into as many as 20 units during the post-reproductive stage. The progression of clonal development is assumed to be both species and habitat specific (Vorontzova & Zaugolnova, 1985; Zhukova & Ermakova, 1985; Wilhalm, 1995).

Although each clone fragment is potentially free living and capable of continued ramet recruitment, contrasting views exist concerning their fate and contribution to population maintenance. One interpretation indicates that clonal fragments are relatively short-lived and that caespitose grass populations are maintained largely by reproduction from seed (Vorontzova & Zaugolnova, 1985; Zhukova & Ermakova, 1985). However, other investigators indicate that clonal fragments have a greater longevity (Lord, 1993; Danin & Orshan, 1995) which enables them to contribute to genet existence and population maintenance. Clonal fragmentation is a common characteristic of perennial caespitose grasses in Europe (Wilhalm, 1995). All 24 species evaluated showed evidence of fragmentation into smaller units consisting of several ramets each. A greater understanding of the fate of clonal fragments is required because it may be the most relevant level at which to investigate the ecology of clonal plants (Cain, 1990).

Mechanisms of ramet regulation within caespitose clones

In this section we evaluate five mechanisms associated with the regulation of ramet recruitment and/or mortality within caespitose grass clones. These mechanisms include: (1) apical dominance, (2) resource competition, (3) physiological integration, (4) red:far-red ratio of solar radiation, and (5) equitable resource acquisition by ramet hierarchies within clones. Apical dominance and resource competition have long been recognized as mechanisms capable of regulating juvenile ramet growth from axillary buds while the latter three mechanisms are more contemporary. A greater understanding of the mechanism(s) regulating intraclonal ramet demography would increase insight into the structure and function of individual clones, and more clearly establish whether ramet interdependence or independence is the prevalent mode of intraclonal function.

Apical dominance

Apical dominance describes the physiological regulation of axillary bud growth by the apical meristem region (Phillips, 1975; Cline, 1991). The physiological mechanism termed the direct hypothesis of auxin action was initially proposed by Thimann & Skoog (1933) shortly following the discovery of the plant hormone auxin. This mechanism indicates that auxin (indoleacetic acid), produced in the apical meristem and young leaves, directly inhibits axillary bud growth. Apical meristem destruction or removal is assumed to release axillary buds from hormonal inhibition and stimulate ramet initiation by eliminating the source of auxin (Fig. 4.3). The

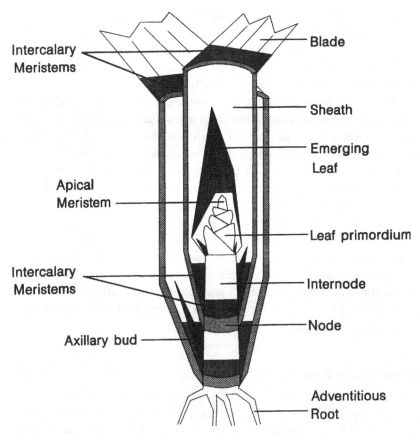

Fig. 4.3. Longitudinal view of a ramet base (crown) illustrating the apical meristem and an emerging leaf and ramet. A single axillary bud is differentiated with each phytomer from the apical meristem. (From Murphy & Briske, 1992).

direct hypothesis of auxin action continues to be a predominant interpretation of the physiological mechanism of apical dominance by grassland ecologists and resource managers (Murphy & Briske, 1992). However, the direct hypothesis was abandoned by plant physiologists during the 1950s because of experimental and interpretive inconsistencies and the demonstrated involvement of a second hormone, cytokinin. A hypothesis based on the auxin:cytokinin ratio has replaced the direct hypothesis as the current hormonally based interpretation of apical dominance (Cline, 1991; Murphy & Briske, 1992). This hypothesis indicates that auxin produced in the apical meristem region blocks the synthesis or utilization of cytokinin within axillary buds, inhibiting their growth (see Hutchings & Mogie,

1990). Despite wide acceptance, numerous issues remain unresolved concerning this hypothesis, suggesting that it may also be an incomplete interpretation of the physiological mechanism of apical dominance.

The ability of apical dominance to explain ramet recruitment in grasses is less consistent than generally recognized, regardless of the underlying physiological mechanisms. Apical meristem removal does not consistently promote ramet initiation in grasses and ramet initiation may occur in plants with intact apical meristems (Murphy & Briske, 1992). However, forage grasses adapted to mesic, fertile environments (e.g. *Lolium perenne*) frequently do increase ramet initiation in response to plant defoliation and grazing (e.g. Grant, Barthram & Torvell, 1981). The mechanism(s) associated with these contrasting responses is unknown. The large number of potentially intervening variables, including environmental conditions, species-specific responses, stage of phenological development, and frequency and intensity of defoliation, also minimizes the likelihood of consistent ramet initiation in response to defoliation. These inconsistencies indicate that the traditional concept of apical dominance is an overly restrictive interpretation of ramet regulation in perennial grasses.

Resource competition

Resource competition exerts a substantial influence on ramet recruitment in grass clones (Briske & Butler, 1989; Hartnett, 1993). Presumably competition influences resource availability and, therefore, ramet recruitment and/or mortality. Ramet recruitment is strongly influenced by both intraclonal and interclonal competition in populations of caespitose grasses (Briske & Butler, 1989; Cheplick & Salvadori, 1991). Clone size and distribution also mediate competitive interactions and influence ramet initiation and clonal expansion (Briske & Anderson, 1990). A high density of small *Schizachyrium scoparium* clones exhibited greater relative increases in ramet density and basal area expansion than did a comparable number of ramets arranged in a low density of large clones. These responses apparently were mediated through a specific regulation mechanism, as opposed to increased efficiency of resource acquisition, because annual shoot biomass production was comparable for all three combinations of clone size and ramet distribution evaluated. Greater ramet recruitment from the high density of small clones was very likely a function of the greater clonal periphery associated with a large number of small clones (Briske & Anderson, 1990). The majority of juvenile ramet recruitment occurs on the periphery, rather than the interior, of individual clones (Olson & Richards, 1988; Briske & Butler, 1989).

Resource availability has been associated with regulation of ramet recruitment based on a correlative relationship between the rate and magnitude of ramet recruitment and resource availability. However, it is impossible to conclude from such evidence that ramet recruitment is regulated by resource availability as opposed to being one component of an overall growth increase (Murphy & Briske, 1992). Higher nutrient concentrations in axillary buds released from inhibition in comparison with their inhibited counterparts do not necessarily imply a causal relationship. Greater nutrient concentrations in growing buds may more accurately reflect the result, rather than the cause, of axillary bud growth (Rubinstein & Nagao, 1976). Consequently, it is difficult to deny the involvement of resource availability in the regulation of ramet recruitment, but it has yet to be established that ramet recruitment is specifically regulated by resource availability (Murphy & Briske, 1992).

Physiological integration

Physiological integration among ramets has been proposed to function as a mechanism capable of regulating ramet recruitment within clones (Hutchings, 1979; Pitelka, 1984; Hutchings & Bradbury, 1986). Resource integration among ramets within clones presumably provides a mechanism to: (1) equitably distribute resources among ramets, (2) minimize inter-ramet competition, and (3) optimize the efficiency of resource acquisition from the local environment (Caraco & Kelly, 1991). Although the occurrence of inter-ramet resource allocation is well documented in grasses (Pitelka & Ashmun, 1985), substantial evidence has accumulated to indicate that resource integration does not occur among all ramets within caespitose clones (Briske & Butler, 1989; de Kroon & Kwant, 1991).

Experiments conducted with three dominant grass species (*Panicum virgatum, Schizachyrium scoparium* and *Bouteloua gracilis*) along an east–west environmental gradient in the North American Great Plains demonstrate that physiological integration is confined to individual ramet hierarchies, rather than throughout all ramets within clones (Derner & Briske, 1998). A vast majority of the stable isotope of nitrogen (^{15}N) introduced into individual parental ramets remained within the labelled ramet hierarchies, rather than being allocated to associated ramet hierarchies within the clone. These findings confirm that intraclonal integration is not complete and support previous hypotheses drawn from ^{15}N experiments with

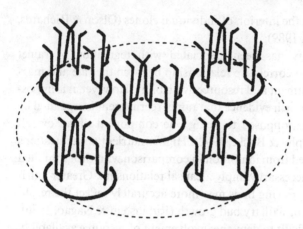

Fig. 4.4. Clones of temperate, caespitose grasses are organized as assemblages of autonomous ramet hierarchies, rather than as a sequence of completely integrated ramets. The benefits of physiological integration are restricted to individual ramet hierarchies (solid circles), which consist of approximately three connected ramet generations, while interhierarchical competition occurs for soil resources accumulated directly beneath the basal area of individual clones (dashed circle).

containerized *S. scoparium* clones and experiments severing vascular connections among ramets within established clones in the field (Welker *et al.*, 1987, 1991; Williams & Briske, 1991).

Partial resource integration can be explained on the basis of the developmental architecture of caespitose clones. Most investigations of interramet resource allocation in grasses have been conducted with young plants established from seed (e.g. Pitelka & Ashmun, 1985). The potential for complete resource integration exists in young plants because all ramets possess vascular connections with the seminal ramet produced from the embryo. This provides complete vascular continuity throughout the entire clone. However, the seminal ramet dies during the second or third growing season, disrupting complete vascular continuity within older clones (Briske & Butler, 1989). Annual grasses are anticipated to maintain complete vascular continuity because longevity of the seminal ramet equals that of the clone.

Resource allocation among only those ramets within individual ramet hierarchies indicates that ramet hierarchies function autonomously within clones, rather than as a sequence of completely integrated ramets (Fig. 4.4). Consequently, the benefits of physiological integration are restricted to these connected generations. Therefore, ramet hierarchies comprise the physiologically integrated individual in this growth form (*sensu* Watson &

Casper, 1984). Low levels of resource allocation observed between ramet hierarchies may result from the occurrence of mycorrhizal connections among root systems of ramet hierarchies (Newman, 1988; Fischer Walter *et al.*, 1996). Documentation of partial clonal integration within caespitose grasses requires that an alternative mechanism of intraclonal ramet regulation be identified and investigated.

Equitable resource acquisition among ramet hierarchies

Competitive interactions among ramet hierarchies within clones, rather than physiological integration among them, may potentially regulate ramet recruitment and density. A broadly distributed root system associated with each ramet hierarchy may potentially enable all ramet hierarchies to access resources equitably from small patches within the immediate vicinity of individual clones. Therefore, a small-scale resource limitation or pulse would equitably constrain or promote growth of all ramet hierarchies within a clone, rather than only the hierarchies nearest the resource patch. Consequently, equitable resource acquisition among ramet hierarchies may minimize intraclonal competition, promote ramet interdependence, and enhance growth efficiency of individual clones.

Experiments conducted with three dominant grass species along an east–west environmental gradient in the North American Great Plains documented that ramet hierarchies within individual clones did not exhibit equitable resource acquisition. Placement of ^{15}N in the soil at three locations peripheral to clones demonstrated that ramets in closer proximity to the nitrogen pulse exhibited significantly greater nitrogen acquisition than did those further away from the nitrogen pulse (Derner & Briske, 1998). Vascular connections within clones were severed perpendicular to the location of the ^{15}N pulse to ensure that ramets acquired ^{15}N by root absorption, rather than by physiological integration from associated hierarchies. Inequitable resource acquisition among ramet hierarchies demonstrates that hierarchies are independent and that intense competition potentially occurs among hierarchies within clones. Therefore, equitable resource acquisition is not a viable mechanism of intraclonal ramet regulation.

Red:far-red radiation ratio

Perception of an environmental signal, rather than resource availability or acquisition, may regulate intraclonal ramet recruitment and density.

Depressions in the red:far-red ratio (R:FR; 660±5 nm:730±5 nm) of solar radiation from values typical of sunlight have been proposed as a potential mechanism capable of regulating ramet initiation within perennial grasses (Deregibus *et al.*, 1985; Casal, Deregibus & Sanchez, 1985). A suppression of ramet initiation in response to the R:FR ratio was first demonstrated with seedlings of *Lolium perenne* and *L. multiflorum* in a controlled environment (Deregibus, Sanchez & Casal, 1983) and *Paspalum dilatatum* plants in the field (Deregibus *et al.*, 1985). However, in spite of the rapidly growing acceptance of the R:FR hypothesis (Casal *et al.*, 1985, 1987*a,b*; Casal, 1988; Skinner & Simmons, 1993), additional understanding of three important processes is required to determine accurately the ecological significance of the R:FR as an environmental signal capable of regulating ramet initiation and growth in caespitose grasses.

First, experiments designed to demonstrate the direct regulation of ramet initiation by the R:FR signal under field conditions have not produced consistent results. An increase in the R:FR beneath canopies of *P. dilatatum* increased ramet initiation by three times compared with control plants at the end of the growing season (Deregibus *et al.*, 1985). However, *Sporobolus indicus* plants in the same investigation were less responsive to supplemental red radiation. Similarly, a reduction in the R:FR at the bases of three morphologically distinct *Festuca rubra* ecotypes reduced ramet initiation in only two of the ecotypes during a six-month experiment (Skálová & Krahulec, 1992). A field experiment conducted with clones of *S. scoparium* demonstrated that irradiation with either red or far-red radiation at the clone bases throughout the photoperiod for 12 consecutive weeks did not significantly affect ramet initiation from existing buds on parental ramets (Murphy & Briske, 1994). The unresponsiveness of mature *S. scoparium* clones in the field is even more intriguing given that seedlings of this species produced the anticipated responses of increased leaf elongation and reduced ramet initiation following irradiation with end-of-day far-red radiation (J. S. Murphy & D. D. Briske, unpubl. manu.).

Second, the sites and locations of R:FR photoperception remain controversial and have very likely been misidentified as the axillary buds and/or sheaths at the ramet base (Casal *et al.*, 1985, 1987*a*; Deregibus *et al.*, 1985), rather than immature leaf blades located within the canopy (Skinner & Simmons, 1993). It is difficult to envisage how either sheaths or buds could efficiently function as sites of R:FR signal perception because (1) these organs are largely shielded from the ambient radiation environment by encircling older leaves (Dale, 1988) and (2) the axillary bud subtending each leaf does not develop into a new ramet until that leaf has ceased expansion

(Skinner & Nelson, 1994). If emerging leaf blades function as the predominant site of photoperception, it potentially minimizes the ecological significance of the R:FR as a density-dependent signal because the site of photoperception and low R:FR would spatially coincide only in young juvenile ramets and seedlings beneath clone canopies (Murphy & Briske, 1994).

Third, the mechanism by which low R:FR suppresses ramet initiation has not been clearly established. Although low R:FR suppression of ramet initiation has been demonstrated to occur without a reduction in the rate of leaf appearance (Casal, 1988; Skinner & Simmons, 1993), and therefore, axillary bud availability, an alternative mechanism of ramet suppression has not been identified. A plausible explanation of bud suppression following exposure to low R:FR is the temporary diversion of carbon (e.g. Yanovsky *et al.*, 1995) from axillary buds to support enhanced leaf development. Ramet suppression in response to low R:FR may be restricted to the early stages of seedling and ramet growth when competition between leaves and buds for a limited carbon supply is sufficient to affect juvenile ramet growth (J. S. Murphy & D. D. Briske, unpubl. manu.). Therefore, ramet suppression in response to low R:FR may not be a pervasive mechanism exclusively regulating axillary bud growth throughout the life of a ramet.

Summary of potential regulation mechanisms

An evaluation of the intraclonal mechanisms potentially capable of regulating ramet demography reveals that little is known about their mode of operation or relative contribution. The direct hypothesis of apical dominance is outdated from a physiological perspective and the response of ramet recruitment to apical meristem removal is less consistent than generally recognized. Physiological integration is restricted to individual ramet hierarchies and, therefore, incapable of regulating ramet recruitment within entire clones. Autonomous ramet hierarchies are not capable of equitably sampling resources, indicating that this process does not mediate interhierarchy competition to regulate ramet densities within clones. Additional information is required to define the mechanisms and ecological significance of the red:far-red ratio of solar radiation on ramet demography. Competition appears to exert the greatest influence on ramet populations, but it is uncertain whether competition exerts disproportionate effects on ramet recruitment relative to the overall influence on clonal growth.

The search for a sole physiological (e.g. apical dominance) or ecological (e.g. red:far-red ratio) mechanism of intraclonal ramet regulation may have limited progress towards understanding the regulation of ramet demography within grasses. Regulation of this important demographic process by a single mechanism would potentially constrain morphological plasticity within the wide range of habitats occupied by the caespitose growth form. Alternatively, intraclonal ramet regulation may be a multivariable process regulated by several interacting physiological and environmental variables. However, a multivariable model has received minimal research emphasis (Phillips, 1975; Murphy & Briske, 1992). In addition, the ramet hierarchy, rather than the entire clone, may be a more appropriate scale at which to investigate intraclonal ramet regulation.

Evaluation of intraclonal ramet competition

Inter-ramet competition may occur, both aboveground and belowground, at each of the following hierarchical locations: (1) within individual hierarchies, (2) among hierarchies within a clone, (3) among hierarchies within intraspecific, and (4) interspecific clones (Harper, 1985; Briske & Butler, 1989). Although the relative contributions of inter-ramet, intraclonal, and interclonal competition to clonal structure and function have not been established, the relative intensities of intraclonal and interclonal competition have been demonstrated to be comparable in several caespitose grasses (Briske & Butler, 1989; Kelley, 1989; Cheplick & Salvadori, 1991). Comparable competitive intensities within, compared with between, clones suggest that potentially intense competitive interactions may occur among autonomous ramet hierarchies within clones.

Intense competitive interactions among ramet hierarchies are established by the absence of physiological integration and non-equitable acquisition of soil resources among these genetically identical structures (Derner & Briske, 1998). However, negative competitive interactions among ramet hierarchies may increase the competitive ability of individual clones by the development of a resource depletion zone (Harper, 1985), which is consistent with the phalanx strategy of clonal plant growth (Lovett Doust, 1981). In contrast, interactions among ramets within individual hierarchies are assumed to be positive, rather than negative (Table 4.1). The benefits of physiological integration within autonomous ramet hierarchies appear to exceed the detrimental effects of competition among these connected ramet generations.

Mechanisms of competitive interactions among hierarchical levels of clonal structure are also unclear. It is uncertain whether competition

were verified in this and other regions, it would substantiate that resource accumulation in soils beneath individual clones is an important mechanism conferring ecological success to caespitose grasses.

Managerial implications

Degradation of late-seral caespitose grass populations in response to intensive long-term herbivory is well documented (Briske & Richards, 1995). Degradation is characterized by a reduction in mean basal area per clone and an increase in clone density very likely resulting from the fragmentation of individual large clones (Butler & Briske, 1988). Herbivore-induced population degradation may potentially be mediated by the reduction of SOC and N accumulation in soils beneath individual clones. Long-term intensive herbivory may reduce nutrient pools beneath clones by limiting availability of photosynthetic and/or meristematic tissues necessary for growth, and may thereby decrease organic matter input into soils in the immediate vicinity of clones (e.g. Berendse, Elberse & Geerts, 1992). A reduction in SOC and N pools beneath clones may potentially limit their ability to monopolize resources in the form of low-quality litter and maintain their competitive dominance. However, a reduction in SOC and N pools beneath clones following intensive herbivory would require decades based on the time required for organic matter turnover in temperate regions (Schimel *et al.*, 1994). Therefore, herbivore-induced population degradation of caespitose grasses must initially be mediated by the suppression of plant function because population structure of caespitose grasses can be modified by intensive herbivory within several years (Butler & Briske, 1988).

Herbivore-induced reductions of SOC and N in soils beneath clones may also have important implications for the restoration of severely degraded populations of caespitose grasses. Slow and incomplete recovery of degraded caespitose grass populations (e.g. Dyksterhuis, 1946; Riegel *et al.*, 1963) may be a consequence of the reduction in SOC and N pools beneath individual clones which minimizes clone function and competitive ability. Carbon addition to soil for purposes of N immobilization may be required to shift the competitive advantage from opportunistic mid-seral species, which often dominate grasslands following degradation, back to consolidator strategists. For example, N fertilization effectively negated the ability of the dominant caespitose grass, *S. scoparium,* to immobilize N in litter and root detritus and placed it at a competitive disadvantage with associated mid-seral perennial grasses within several years (Tilman & Wedin, 1991; Wedin, 1995). Increased N immobilization has been associated with the progression of succession in several communities (McLendon & Redente, 1992).

Summary

Although mechanisms of intraclonal ramet regulation have received more research emphasis than mechanisms conferring ecological success to caespitose grasses, a functional interpretation of intraclonal demographic regulation has not yet been developed. The search for a single regulatory mechanism, as opposed to a multivariable approach, may have retarded development of an unified ecological interpretation for this important demographic process. Organization of caespitose clones as assemblages of autonomous ramet hierarchies suggests that the ramet hierarchy may be the appropriate level at which to investigate potential mechanisms of intraclonal ramet regulation. Interactions among ramets within individual hierarchies are assumed to convey a positive growth response because the benefits of physiological integration appear to exceed the detrimental effects of competition among connected ramet generations. In contrast, the absence of physiological integration, and the inability of genetically identical ramet hierarchies to acquire soil resources equitably, potentially produces intense intraclonal competition.

The ability of caespitose clones to monopolize resources through accumulation of soil organic carbon and total nitrogen directly beneath clones represents the most plausible mechanism for the ecological success of this growth form. Additional research is required to identify the environmental constraints and trade-offs associated with this mode of resource monopolization relative to a more active foraging strategy characteristic of rhizomatous and stoloniferous grasses. Recognition that caespitose grasses are composed of autonomous ramet hierarchies suggests that future advances in clonal biology of caespitose grasses will result from investigation of intraspecific and interspecific interactions among ramet hierarchies and their potential contributions to clone, genet and population maintenance.

Acknowledgements

We wish to thank C. K. Kelly, B. Schmid, T. Wilhalm, and D. G. Williams for constructive comments on earlier versions of the manuscript. Financial support was provided by a USDA Rangeland Special Grant (92–38300–7459) to DDB.

References

Bell, A. D. (1976). The vascular pattern of Italian ryegrass (*Lolium multiflorum* Lam). 3. The leaf trace system, and tiller insertion, in the adult. *Annals of Botany*, **40**, 241–50.

Berendse, F. (1994). Litter decomposability – a neglected component of plant fitness. *Journal of Ecology*, **82**, 187–90.

Berendse, F., Elberse, W. Th. & Geerts, R. H. M. E. (1992). Competition and nitrogen loss from plants in grassland ecosystems. *Ecology*, **73**, 46–53.

Bobbink, R., den Dubbelden, K. & Willems, J. H. (1989). Seasonal dynamics of phytomass and nutrients in chalk grassland. *Oikos*, **55**, 216–24.

Briske, D. D. (1991). Developmental morphology and physiology of grasses. In *Grazing Management: An ecological perspective*, ed. R. K. Heitschmidt & J. W. Stuth, pp. 85–108. Timber Press, Portland, Oregon, USA.

Briske, D. D. & Anderson, V. J. (1990). Tiller dispersion in populations of the bunchgrass *Schizachyrium scoparium*: Implications for herbivory tolerance. *Oikos*, **59**, 50–6.

Briske, D. D. & Butler, J. L. (1989). Density-dependent regulation of ramet populations within the bunchgrass *Schizachyrium scoparium*: Interclonal versus intraclonal interference. *Journal of Ecology*, **77**, 963–74.

Briske, D. D. & Richards, J. H. (1995). Plant responses to defoliation: A physiological, morphological, and demographic evaluation. In *Wildland Plants: Physiological ecology and developmental morphology*, ed. D. J. Bedunah & R. E. Sosebee, pp. 635–710. Society for Range Management, Denver, Colorado, USA.

Burke, I. C., Lauenroth, W. K. & Coffin, D. P. (1995). Soil organic matter recovery in semiarid grasslands: Implications for the conservation reserve program. *Ecological Applications*, **5**, 793–801.

Butler, J. L. & Briske, D. D. (1988). Population structure and tiller demography of the bunchgrass *Schizachyrium scoparium* in response to herbivory. *Oikos*, **51**, 306–12.

Cain, M. L. (1990). Patterns of *Solidago altissima* growth and mortality: The role of below-ground ramet connections. *Oecologia*, **82**, 201–9.

Caldwell, M. M., Dean, T. J., Nowak, R. S., Dzurec, R. S. & Richards, J. H. (1983). Bunchgrass architecture, light interception, and water-use efficiency: Assessment by fiber optic point quadrats and gas exchange. *Oecologia*, **59**, 178–84.

Campbell, B. D. & Grime, J. P. (1989). A comparative study of plant responsiveness to the duration of episodes of mineral nutrient enrichment. *New Phytologist*, **112**, 261–7.

Caraco, T. & Kelly, C. K. (1991). On the adaptive value of physiological integration in clonal plants. *Ecology*, **72**, 81–93.

Casal, J. J. (1988). Light quality effects on the appearance of tillers of different order in wheat (*Triticum aestivum*). *Annals of Applied Biology*, **112**, 167–73.

Casal, J. J., Deregibus, V. A. & Sanchez, R. A. (1985). Variations in tiller dynamics and morphology in *Lolium multiflorum* Lam. vegetative and reproductive plants as affected by differences in red/far-red irradiation. *Annals of Botany*, **56**, 59–65.

Casal, J. J., Sanchez, R. A. & Deregibus, V. A. (1987a). The effect of light quality on shoot extension growth in three species of grasses. *Annals of Botany*, **59**, 1–7.

Casal, J. J., Sanchez, R. A. & Deregibus, V. A. (1987b). Tillering responses of

Lolium multiflorum plants to changes of red/far-red ratio typical of sparse canopies. *Journal of Experimental Botany*, 38, 1432–9.

Cheplick, G. P. (1993). Reproductive systems and sibling competition in plants. *Plant Species Biology*, 8, 131–9.

Cheplick, G. P. & Salvadori, G. M. (1991). Intra- and interclonal competition in the cleistogamous grass *Amphibromus scabrivalvis*. *American Journal of Botany*, 78, 1494–502.

Cline, M. G. (1991). Apical dominance. *Botanical Review*, 57, 318–58.

Cook, R. E. (1979). Asexual reproduction: A further consideration. *American Naturalist*, 113, 769–72.

Cook, R. E. (1985). Growth and development in clonal plant populations. In *Population Biology and Evolution of Clonal Organisms*, ed. J. B. C. Jackson, L. W. Buss & R. E. Cook, pp. 259–96. Yale University Press, New Haven, Connecticut, USA.

Crick, J. C. & Grime, J. P. (1987). Morphological plasticity and mineral nutrient capture in two herbaceous species of contrasted ecology. *New Phytologist*, 107, 403–14.

Dahl, B. E. & Hyder, D. N. (1977). Developmental morphology and management implications. In *Rangeland Plant Physiology*, ed. R. E. Sosebee, pp. 258–90. Society for Range Management, Denver, Colorado, USA.

Dale, J. E. (1988). The control of leaf expansion. *Annual Review of Plant Physiology and Plant Molecular Biology*, 39, 267–95.

Danin, A. & Orshan, G. (1995). Circular arrangement of *Stipagrostis ciliata* clumps in the Negev, Israel and near Gokaeb, Namibia. *Journal of Arid Environments*, 30, 307–13.

de Kroon, H., Hara, T. & Kwant, R. (1992). Size hierarchies of shoot and clones in clonal herb monocultures: Do clonal and non-clonal plants compete differently? *Oikos*, 63, 410–19.

de Kroon, H. & Hutchings, M. J. (1995). Morphological plasticity in clonal plants: The foraging concept reconsidered. *Journal of Ecology*, 83, 143–52.

de Kroon, H. & Kwant, R. (1991). Density-dependent growth responses in two clonal herbs: regulation of shoot density. *Oecologia*, 86, 298–304.

de Kroon, H. & Schieving, F. (1990). Resource partitioning in relation to clonal growth strategy. In *Clonal Growth in Plants: Regulation and function*, ed. J. van Groenendael & H. de Kroon, pp. 113–30. SPB Academic Publishing, The Hague, The Netherlands.

de Kroon, H. & van Groenendael, J. (1990). Regulation and function of clonal growth in plants: An evaluation. In *Clonal Growth in Plants: Regulation and function*, ed. J. van Groenendael & H. de Kroon, pp. 177–86. SPB Academic Press, The Hague, The Netherlands.

Deregibus, V. A., Sanchez, R. A. & Casal, J. J. (1983). Effects of light quality on tiller production in *Lolium* spp. *Plant Physiology*, 72, 900–2.

Deregibus, V. A., Sanchez, R. A., Casal, J. J. & Trlica, M. J. (1985). Tillering responses to enrichment of red light beneath the canopy in a humid natural grassland. *Journal of Applied Ecology*, 221, 199–206.

Derner, J. D., Briske, D. D. & Boutton, T. W. (1997). Does grazing mediate soil carbon and nitrogen accumulation beneath C_4 perennial grasses along an environmental gradient? *Plant and Soil*, 191 (in press).

Derner, J. D. & Briske, D. D. (1998). An isotopic assessment of intraclonal regulation in C_4 perenial grasses: ramet interdependence, independence or both? *Journal of Ecology* (in press).

Clonal biology of caespitose grasses 131

Dyksterhuis, E. J. (1946). The vegetation of the Fort Worth Prairie. *Ecological Monographs*, **16**, 1–29.

Ekstam, B. (1995). Ramet size equalisation in a clonal plant, *Phragmites australis. Oecologia*, **104**, 440–6.

Etter, A. G. (1951). How Kentucky bluegrass grows. *Annals of Missouri Botanical Garden*, **38**, 293–375.

Fischer Walter, L. E., Hartnett, D. C., Hetrick, B. A. D. & Schwab, A. P. (1996). Interspecific nutrient transfer in a tallgrass prairie plant community. *American Journal of Botany*, **83**, 180–4.

Gatsuk, L. E., Smirnova, O. V., Vorontzova, L. I., Zaugolnova, L. B. & Zhukova, L. A. (1980). Age states of plants of various growth forms: A review. *Journal of Ecology*, **68**, 675–96.

Geiger, D. R. (1979). Control of partitioning and export of carbon in leaves of higher plants. *Botanical Gazette*, **140**, 241–8.

Grace, J. B. (1993). The adaptive significance of clonal reproduction in angiosperms: an aquatic perspective. *Aquatic Botany*, **44**, 159–80.

Grant, S. A., Barthram, G. T. & Torvell, L. (1981). Components of regrowth in grazed and cut *Lolium perenne* swards. *Grass and Forage Science*, **36**, 155–68.

Harberd, D. J. (1961). Observations on population structure and longevity of *Festuca rubra* L. *New Phytologist*, **60**, 184–206.

Harberd, D. J. (1962). Some observations on natural clones in *Festuca ovina. New Phytologist*, **61**, 85–100.

Harberd, D. J. (1967). Observations on natural clones of *Holcus mollis. New Phytologist*, **66**, 401–8.

Hardwick, R. C. (1986). Physiological consequences of modular growth in plants. *Philosophical Transactions of Royal Society London B*, **313**, 161–73.

Harper, J. L. (1981). The concept of population in modular organisms. In *Theoretical Ecology: Principles and applications*, 2nd edn, ed. R. M. May, pp. 53–77. Backwell, Oxford.

Harper, J. L. (1985). Modules, branches and the capture of resources. In *Population Biology and Evolution of Clonal Organisms*, ed. J. B. C. Jackson, L. W. Buss & R. E. Cook, pp. 1–33. Yale University Press, New Haven, Connecticut, USA.

Hartnett, D. C. (1989). Density- and growth stage-dependent responses to defoliation in two rhizomatous grasses. *Oecologia*, **80**, 414–20.

Hartnett, D. C. (1993). Regulation of clonal growth and dynamics of *Panicum virgatum* (Poaceae) in tallgrass prairie: Effects of neighbor removal and nutrient addition. *American Journal of Botany*, **80**, 1114–20.

Heckathorn, S. A. & DeLucia, E. H. (1994). Drought-induced nitrogen retranslocation in perennial C4 grasses of tallgrass prairie. *Ecology*, **75**, 1877–86.

Hitch, P. A. & Sharman, B. C. (1968a). Initiation of procambial strands in leaf primordia of *Dactylis glomerata* L. as an example of a temperate herbage grass. *Annals of Botany*, **32**, 153–64.

Hitch, P. A. & Sharman, B. C. (1968b). Initiation of procambial strands in axillary buds of *Dactylis glomerata* L., *Secale cereale* L., and *Lolium perenne* L. *Annals of Botany*, **32**, 667–76.

Hook, P. B., Burke, I. C. & Lauenroth, W. K. (1991). Heterogeneity of soil and plant N and C associated with individual plants and openings in North American shortgrass steppe. *Plant and Soil*, **138**, 247–56.

132   *D. D. Briske and J. D. Derner*

Hutchings, M. J. (1979). Weight-density relationships in ramet populations of clonal perennial herbs, with special reference to the −3/2 power law. *Journal of Ecology*, 67, 21–33.

Hutchings, M. J. & Barkham, J. P. (1976). An investigation of shoot interactions in *Mercurialis perennis* L., a rhizomatous perennial herb. *Journal of Ecology*, 64, 723–43.

Hutchings, M. J. & Bradbury, I. K. (1986). Ecological perspectives on clonal perennial herbs. *BioScience*, 36, 178–82.

Hutchings, M. J. & de Kroon, H. (1994). Foraging in plants: The role of morphological plasticity in resource acquisition. *Advances in Ecological Research*, 25, 159–238.

Hutchings, M. J. & Mogie, M. (1990). The spatial structure of clonal plants: Control and consequences. In *Clonal Growth in Plants: Regulation and function*, ed. J. van Groenendael & H. de Kroon, pp. 57–76. SPB Academic Press, The Hague, The Netherlands.

Jackson, R. B. & Caldwell, M. M. (1989). The timing and degree of root proliferation in fertile-soil microsites for three cold-desert perennials. *Oecologia*, 81, 149–53.

Jackson, R. B. & Caldwell, M. M. (1992). Shading and the capture of localized soil nutrients: Nutrient contents, carbohydrates, and root uptake kinetics of a perennial tussock grass. *Oecologia*, 91, 457–62.

Jackson, R. B. & Caldwell, M. M. (1993). Geostatistical patterns of soil heterogeneity around individual perennial plants. *Journal of Ecology*, 81, 683–92.

Jonasson, S. & Chapin, F. S. III (1991). Seasonal uptake and allocation of phosphorus in *Eriophorum vaginatum* L. measured by labeling with ^{32}P. *New Phytologist*, 118, 349–57.

Kelley, S. E. (1989). Experimental studies of the evolutionary significance of sexual reproduction. VI. A greenhouse test of the sub-competition hypothesis. *Evolution*, 43, 1066–74.

Kelly, C. K. (1995). Thoughts on clonal integration: Facing the evolutionary context. *Evolutionary Ecology*, 9, 575–85.

Langer, R. H. M. (1972). *How Grasses Grow*. Edward Arnold, London.

Leith, H. (1978). Primary productivity in ecosystems: Comparative analysis of global patterns. In *Patterns of Primary Productivity in the Biosphere*, ed. H. F. H. Leith, pp. 300–21. Dowden, Hutchinson and Ross, Stroudberg, Pennsylvania, USA.

Lord, J. M. (1993). Does clonal fragmentation contribute to recruitment in *Festuca novae-zelandiae*? *New Zealand Journal of Botany*, 31, 133–8.

Lovett Doust, L. (1981). Population dynamics and local specialization in a clonal perennial (*Ranunculus repens*). I. The dynamics of ramets in contrasting habitats. *Journal of Ecology*, 69, 743–55.

Lovett Doust, L. & Lovett Doust, J. (1982). The battle strategies of plants. *New Scientist*, 95, 81–4.

McLendon, T. & Redente, E. F. (1992). Effects of nitrogen limitation on species replacement dynamics during early secondary succession on a semiarid sagebrush site. *Oecologia*, 91, 312–7.

Murphy, J. S. & Briske, D. D. (1992). Regulation of tillering by apical dominance: Chronology, interpretive value, and current perspectives. *Journal of Range Management*, 45, 419–29.

Murphy, J. S. & Briske, D. D. (1994). Density-dependent regulation of ramet recruitment by the red:far-red ratio of solar radiation: A field evaluation with the bunchgrass *Schizachyrium scoparium*. *Oecologia*, 97, 462–9.

Newman, E. I. (1988). Mycorrhizal links between plants: Their functioning and ecological significance. *Advances in Ecological Research*, 18, 243–70.

Olson, B. E. & Richards, J. H. (1988). Spatial arrangement of tiller replacement in *Agropyron desertorum* following grazing. *Oecologia*, 76, 7–10.

Pedersen, B. & Tuomi, J. (1995). Hierarchical selection and fitness in modular and clonal organisms. *Oikos*, 73, 167–80.

Phillips, I. D. J. (1975). Apical dominance. *Annual Review of Plant Physiology*, 26, 341–67.

Pitelka, L. F. (1984). Application of the −3/2 power law to clonal herbs. *American Naturalist*, 123, 442–9.

Pitelka, L. F. & Ashmun, J. W. (1985). Physiology and integration of ramets in clonal plants. In *The Population Biology and Evolution of Clonal Organisms*, ed. J. B. G. Jackson, L. W. Buss & R. E. Cook, pp. 399–435. Yale University Press, New Haven, Connecticut, USA.

Pugnaire, F. I. & Haase, P. (1996). Comparative physiology and growth of two perennial tussock grass species in a semi-arid environment. *Annals of Botany*, 77, 81–6.

Rice, B. L., Westoby, M., Griffin, G. F. & Friedel, M. H. (1994). Effects of supplementary soil nutrients on hummock grasses. *Australian Journal of Botany*, 42, 687–703.

Riegel, D. A., Albertson, F. W., Tomanek, G. W. & Kinsinger, F. E. (1963). Effects of grazing and protection on a twenty-year-old seeding. *Journal of Range Management*, 22, 60–3.

Robinson, D. (1996). Resource capture by localized root proliferation: Why do plants bother? *Annals of Botany*, 77, 179–85.

Rubinstein, B. & Nagao, M. A. (1976). Lateral bud outgrowth and its control by the apex. *Botanical Review*, 42, 83–113.

Ryel, R. J., Beyschlag, W. & Caldwell, M. M. (1993). Foliage orientation and carbon gain in two tussock grasses as assessed with a new whole-plant gas-exchange model. *Functional Ecology*, 7, 115–24.

Ryel, R. J., Beyschlag, W. & Caldwell, M. M. (1994). Light field heterogeneity among tussock grasses: Theoretical considerations of light harvesting and seedling establishment in tussocks and uniform tiller distributions. *Oecologia*, 98, 241–6.

Sachs, T., Novoplansky, A. & Cohen, D. (1993). Plants as competing populations of redundant organs. *Plant, Cell and Environment*, 16, 765–70.

Schimel, D. S., Braswell, B. H., Holland, E. A., McKeown, R., Ojima, D. S., Painter, T. H., Parton, W. J. & Townsend, A. R. (1994). Climatic, edaphic, and biotic controls over storage and turnover of carbon in soils. *Global Biogeochemistry Cycles*, 8, 279–93.

Schmid, B. (1990). Some ecological and evolutionary consequences of modular organization and clonal growth in plants. *Evolutionary Trends in Plants*, 4, 25–34.

Schwinning, S. (1996). Decomposition analysis of competitive symmetry and size structure dynamics. *Annals of Botany*, 77, 47–57.

Sims, P. L., Singh, J. S. & Lauenroth, W. K. (1978). The structure and function of ten western North American grasslands. I. Abiotic and vegetation characteristics. *Journal of Ecology*, 66, 251–81.

Skálová H. & Krahulec, F. (1992). The response of three *Festuca rubra* clones to changes in light quality and plant density. *Functional Ecology*, 6, 282–90.

Skinner, R. H. & Nelson, C. J. (1994). Epidermal cell division and the coordination of leaf and tiller development. *Annals of Botany*, 74, 9–15.

Skinner R. H. & Simmons, S. R. (1993). Modulation of leaf elongation, tiller appearance and tiller senescence in spring barley by far-red light. *Plant, Cell and Environment*, 16, 555–62.

Steinger, T., Körner, C. & Schmid, B. (1996). Long term perspective in a changing climate: DNA analysis suggest very old ages of clones of alpine *Carex curvula*. *Oecologia*, 105, 94–9.

Thimann, K. V. & Skoog, F. (1933). Studies on the growth hormone of plants. 3. The inhibiting action of the growth substance on bud development. *Proceedings of National Academy of Sciences of the USA*, 19, 714–6.

Tiffney, B. H. & Niklas, K. J. (1985). Clonal growth in land plants: A paleobotanical perspective. In *The population Biology and Evolution of Clonal Organisms*, ed. J. B. C. Jackson, L. W. Buss & R. E. Cook, pp. 35–66. Yale University Press, New Haven, Connecticut, USA.

Tilman, D. & Wedin, D. (1991). Plant traits and resource reduction for five grasses growing on a nitrogen gradient. *Ecology*, 72, 685–700.

Tomlinson, P. B. (1974). Vegetative morphology and meristem dependence – the foundation of productivity in seagrasses. *Aquaculture*, 4, 107–30.

Tuomi, J. & Vuorisalo, T. (1989). What are the units of selection in modular organisms? *Oikos*, 54, 227–33.

Van Auken, O. W., Manwaring, J. H. & Caldwell, M. M. (1992). Effectiveness of phosphate acquisition by juvenile cold-desert perennials from different patterns of fertile-soil microsites. *Oecologia*, 91, 1–6.

Vinton, M. A. & Burke, I. C. (1995). Interactions between individual plant species and soil nutrient status in shortgrass steppe. *Ecology*, 76, 1116–33.

Vorontzova, L. I. & Zaugolnova, L. B. (1985). Population biology of steppe plants. In *The Population Structure of Vegetation*, ed. J. White, pp. 143–78. Dr W. Junk Publishers, Dordrecht, The Netherlands.

Walter, H. (1979). *Vegetation of the Earth and Ecological Systems of the Geobiosphere*. Springer, New York, USA.

Watkinson, A. R. & White, J. (1986). Some life-history consequences of modular construction in plants. *Philosophical Transactions of Royal Society London B*, 313, 31–51.

Watson, M. A. & Casper, B. B. (1984). Morphogenetic constraints on patterns of carbon distribution in plants. *Annual Reviews of Ecology and Systematics*, 15, 233–58.

Wedin, D. A. (1995). Species, nitrogen, and grassland dynamics: The constraints of stuff. In *Linking Species and Ecosystems*, ed. C. G Jones & J. H. Lawton, pp. 253–62. Chapman and Hall, New York, USA.

Wedin, D. & Tilman, D. (1993). Competition among grasses along a nitrogen gradient: Initial conditions and mechanisms of competition. *Ecological Monographs*, 63, 199–229.

Welker, J. M. & Briske, D. D. (1992). Clonal biology of the temperate, caespitose, graminoid *Schizachyrium scoparium*: A synthesis with reference to climate change. *Oikos*, 63, 357–65.

Welker, J. M., Briske, D. D. & Weaver, R. W. (1987). Nitrogen-15 partitioning within a three generation tiller sequence of the bunchgrass *Schizachyrium scoparium*: Response to selective defoliation. *Oecologia*, 74, 330–4.

Welker, J. M., Briske, D. D. & Weaver, R. W. (1991). Intraclonal nitrogen allocation in the bunchgrass *Schizachyrium scoparium*: An assessment of the physiological individual. *Functional Ecology*, **5**, 433–40.

West, N. E., Rea, K. H. & Harniss, R. O. (1979). Plant demographic studies in sagebrush-grass communities of southeastern Idaho. *Ecology*, **60**, 376–88.

White, J. (1979). The plant as a metapopulation. *Annual Reviews of Ecology and Systematics*, **10**, 109–45.

Wilhalm, T. (1995). A comparative study of clonal fragmentation in tussock-forming grasses. In *Clonality in Plant Communities*, ed. B. Oborny & J. Podani, Proceedings of the 4th International workshop on clonal plants. *Abstracta Botanica*, **19**.

Williams, D. G. & Briske, D. D. (1991). Size and ecological significance of the physiological individual in the bunchgrass *Schizachyrium scoparium*. *Oikos*, **62**, 41–7.

Yanovsky, M. J., Casal, J. J., Salerno, G. L. & Sánchez, R. A. (1995). Are phytochrome-mediated effects on leaf growth, carbon partitioning and extractable sucrose-phosphate synthase activity the mere consequence of stem-growth responses in light-grown mustard? *Journal of Experimental Botany*, **46**, 753–7.

Zhukova, L. A. & Ermakova, I. M. (1985). Structure and dynamics of coenopopulations of some temperate grasses. In *The population Structure of Vegetation* ed. J. White, pp. 179–205. Dr W. Junk Publishers, Dordrecht, The Netherlands.

5
Ecological aspects of sex expression in grasses
JAMES A. QUINN

Starting with a brief introduction to the diversity of breeding systems in the grasses, this chapter will initially focus on the one-hormone model of sex determination and its relation to environmental effects and to effects of minor genetic changes on sex expression and the breeding system. This will be followed by a discussion of potential factors influencing the breeding system, and its phenotypic plasticity, of populations of wide-ranging species.

Diversity of breeding systems in the grasses

Grasses display an extraordinary diversity of breeding systems – hermaphroditism to most forms of monoecism, dioecism, and apomixis (Connor, 1979, 1981, 1987). Hermaphroditism is the usual situation in the grass family; however, some grasses have unisexual flowers in various numbers, combinations, or locations (Connor, 1981). Sometimes the different types of flowers occur together throughout the inflorescence as with the male and perfect flowers of the andromonoecious *Andropogon gerardi*, or the male and female flowers may be grouped but closely adjacent within the same inflorescence as in *Tripsacum dactyloides*, or other grasses may have the unisexual flowers aggregated on separate inflorescences (the ear and tassel of *Zea mays*). The ultimate separation of the male and female flowers is in dioecy where the male and female flowers are produced on separate plants, as in *Buchloe dactyloides* (Fig. 5.1). This separation of male and female flowers in space, whether on the same or separate plants, in addition to self-incompatibility and dichogamy (differential timing of male and female development) are mechanisms promoting outbreeding and reducing inbreeding. Conversely, different amounts and types of cleistogamy promote selfing and inbreeding. Fig. 5.2 illustrates the

Fig. 5.1. Female plant of *Buchloe dactyloides* above and male plant below (in this drawing leaf heights are 10 cm and male inflorescence culms reach 17 cm). A pistillate floret (4.5 mm), a coalesced female inflorescence (7 mm), and a staminate spikelet (4 mm) are inserted above the plants at 10 × the scale of the plants. From Quinn, 1991.

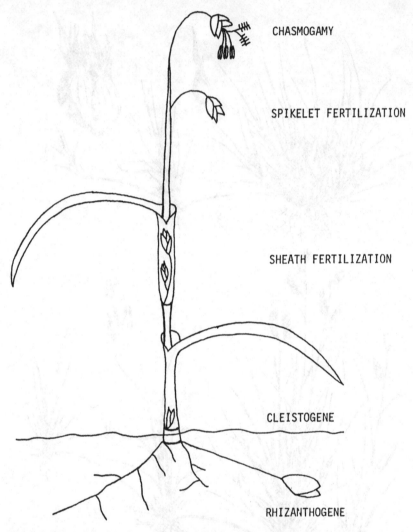

CHASMOGAMY

SPIKELET FERTILIZATION

SHEATH FERTILIZATION

CLEISTOGENE

RHIZANTHOGENE

Fig. 5.2. Schematic drawing of a grass plant showing chasmogamy and the four types of cleistogamy. Drawing by C. S. Campbell.

different types based on the enclosing structure and location, as described by Campbell *et al.* (1983). The above mechanisms, which either promote outbreeding or increase the frequency of selfing, may alter or 'fine-tune' the level of outbreeding or selfing, as they can be affected to varying degrees by the prevailing environmental conditions. Where phenotypic variation in the breeding system occurs, it is often vaguely attributed to 'environmental effects', 'hormonal effects', or 'physiological stress'.

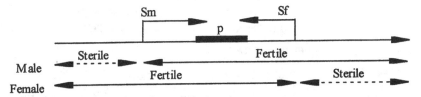

Fig. 5.3. A scheme to illustrate sex determination in plants at the individual level. The horizontal axis, hormone concentration; Bar, the normal range of the hormone concentration in the plant; Sm, the sensitivity level of the male receptor; Sf, the sensitivity level of the female receptor; p, perfect flowers. From Yin & Quinn, 1995*b*.

The one-hormone model of sex determination

The one-hormone model of sex determination (Yin & Quinn, 1992) provides a mechanistic interpretation of environmental effects and predicts that minor genetic changes can have major effects on sex expression. This model assumes that one hormone has male and female receptors to inhibit one sex, and induce the other independently. Fig. 5.3 illustrates how the range of hormone concentration (denoted by the horizontal bar) and the two sensitivity levels of the receptors (Sm, male, and Sf, female) could interact to regulate sex expression, creating male and female sterile regions. In this case, the male and female fertile regions overlap at normal hormone concentrations and perfect flowers are produced. The two sensitivity levels can produce only two types of arrangements: male and female fertile regions overlapping, and male and female sterile regions overlapping. As shown in Fig. 5.4, for each type, there are six possible ways to organize the three key components (the hormone concentration, Sm, and Sf). **p, f, m,** and **n** above the bar refer to the type of flowers produced at that hormone concentration, and the resultant sex form is also indicated below each diagram. Interestingly, every possible relative position of the three components matches a sex form in nature, and all existing sex forms in nature can be explained by these relationships (Yin & Quinn, 1992).

Genetic changes as interpreted by the one-hormone model

Tripsacum dactyloides provides an excellent example of a minor genetic change having major effects on sex expression and the breeding system. The species is normally monoecious, but Dewald *et al.* (1987) have reported that in some individuals the upper or terminal staminate florets are converted to mostly pistillate with a few perfect flowers at the tip, producing a gynomonoecious plant. They have determined that a single recessive muta-

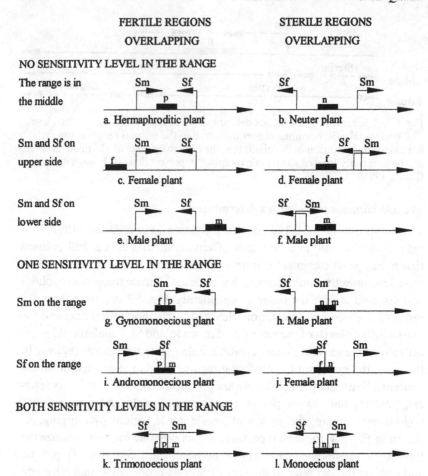

Fig. 5.4. All possible relative positions of the three key components and their corresponding sex forms. f, female flowers; m, male flowers; n, neuter; p, perfect flowers. From Yin & Quinn, 1995*b*.

tion is responsible. Because plants with this gene do not show reduced growth rates or size (Jackson & Dewald, 1994), my hypothesis is that the single gene change involved a change in the female sensitivity level (Sf) (see Fig. 5.4 and compare l. with g.). This Sf would be less sensitive and thus recessive.

Buchloe dactyloides is usually considered a 'dioecious' species because most plants are either male or female, but some populations contain a significant number of individuals that are labile or monoecious in their sex expression (Yin & Quinn, 1994). Perfect-flowered inflorescences arise rarely

a. Monoecious

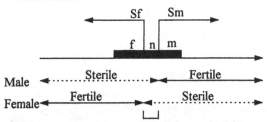

b. Trimonoecious

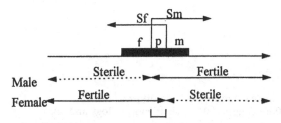

Fig. 5.5. The arrangement of the three components for monoecious (*a*) and trimonoecious (*b*) genotypes. Sf and Sm denote the sensitivity levels for female and male, respectively, f stands for female, n for neuter, p for perfect, and m for male. The brackets indicate the hormone concentration where the male and female sterile or fertile regions overlap. From Yin & Quinn, 1995*a*.

when large numbers of plants are being grown in greenhouses or gardens. They have also been noted in Mexico, postulated area of origin of *Buchloe* and its closest relatives. Thus, Fig. 5.4 can be used to illustrate the diversity of *Buchloe* sex forms that can arise by one or a few mutations from a hermaphroditic ancestor. Indeed, in my 20 years of research with *Buchloe* I have observed individuals that correspond to most of the types in Fig. 5.4.

Environmental effects as explained by the one-hormone model

For certain of the *Buchloe* genotypes, the genetically determined sex expression can be readily affected by environmental conditions or experimental manipulation of the internal hormone concentration. Because monoecious and trimonoecious plants have both sensitivity levels within the normal hormone range (Fig. 5.5), and they are fairly close together, these plants should be particularly sensitive to environmental or experimental manipulation of the hormone level. Indeed, Tongjia Yin and I found that we could

Fig. 5.6. Sex expression in a single trimonoecious genotype of *Buchloe*
dactyloides, ranging from inflorescences 100% male in morphology and sex
expression (at left) through intermediates in both morphology and sex expression
(including perfect flowers) to the normal female coalesced inflorescence (at right).
From Yin & Quinn, 1994.

dramatically shift the relative numbers of female and male flowers in
monoecious plants, and the relative numbers of female, perfect, and male
flowers in trimonoecious plants through the use of gibberellin and its
inhibitor, paclobutrazol or PAC (Yin & Quinn, 1995*a*). Fig. 5.6 shows the
range of sex expression in a single trimonoecious genotype, ranging from
inflorescences 100% male in morphology and sex expression (at left)
through intermediates in both morphology and sex expression (including
perfect flowers) to the normal female coalesced inflorescence (at right). A
high concentration of gibberellin applied over a sufficient length of time
will lead to a plant with all male inflorescences (similar to the one on the
left), and a high concentration of PAC will produce plants with all female
inflorescences (similar to the one on the right).

In contrast, the normal unisexual males and females are extremely stable
in their sex expression. In my lab, we have been able to induce the inflorescences of the opposite sex in only one genotype. This was a female that
apparently had an Sm only slightly above the normal hormone concentration (Fig. 5.7*a*). Apparently, most females have a wider separation of sensitivity levels (as indicated in Fig. 5.7*b*). It may not be physiologically

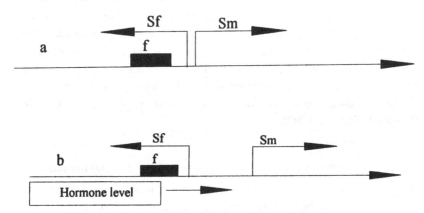

Fig. 5.7. Potential variation in separation of Sf and Sm sensitivity levels in female plants. (*a*) shows a female genotype with the Sm only slightly above the normal hormone concentration. (*b*) is a genotype with a wider separation of sensitivity levels. Modified from Yin & Quinn, 1995*a*.

possible to force unisexual plants to produce inflorescences of the other sex if the two sensitivity levels are widely separated and/or major changes in the hormone range are required. In all of our experiments to date, males lose vigour and/or die without producing female inflorescences at high levels of PAC (Yin & Quinn, 1995*a*).

Thus, the relative location of sensitivity levels and the normal hormone range (and its tight or loose regulation) should be strikingly genotype-specific, producing extreme variation among individuals in type and lability of sex expression, and indeed, we have found this in *Buchloe dactyloides* (Yin & Quinn, 1994). Many *Poa* species also show great variation in sexual/apomictic expression and its lability among individuals and populations (Clausen, Keck & Hiesey, 1946; Soreng, 1986; Kellogg, 1987). Once plastic individuals (or different sex forms) arise in a species, they may increase in a local population under locally specific conditions if they produce an increased proportion of the reproductives.

Factors influencing the breeding system and its phenotypic plasticity

Species-specific phylogenetic and/or genetic and morphological constraints
Briefly, this refers to the lack of certain options in large groups of related species, not because there is a lack of ecological conditions that would favour them, but presumably because of the lack of the necessary genes, gene sequences, and developmental morphology in those groups. Examples

would be (1) agamospermous apomixis mostly absent from the entire sub-family Bambusoideae, certain tribes, and large genera (Connor, 1979; Watson, 1990), (2) caespitose groups of species that lack the advantages of stolons and rhizomes, e.g. local vegetative spread and foraging, longevity of successful genotypes, and maintenance of genetic variation, and (3) the absence of dioecism in the Bambusoideae and strict monoecism in the Pooideae (Watson, 1990).

r- vs K-selection/open vs closed, stable communities/DI vs DD mortality

Gadgil and Solbrig (1972) stated that the central idea of r- and K-selection is that populations living in environments imposing high density-independent (DI) mortality (r-strategists) will be selectively favoured to allocate a greater proportion of resources to reproductive activities, and conversely, populations living in environments imposing high density-dependent (DD) regulation (K-strategists) will be selectively favoured to allocate a greater proportion of resources to non-reproductive activities. This hypothesis was first tested in grass populations by Roos & Quinn (1977), who compared six successional populations of Andropogon scoparius in New Jersey (USA) in reproductive allocation and plasticity. The youngest populations flow-ered earlier and had a greater reproductive effort. Most of the field differences were attributable to local habitat effects on phenotypic expres-sion. However, the possibility of some genetic differentiation paralleling successional age was raised by the consistent (though sometimes not sta-tistically significant) differences between the early (3-year) and late (40-year) successional populations in time of flowering and in reproductive effort under two light intensities and uniform greenhouse conditions. In addition, the oldest population showed the greatest phenotypic plasticity for reproductive effort. At the same time in north-west England and North Wales, Law, Bradshaw & Putwain (1977) showed that plants of Poa annua from pasture situations (said to be under DD regulation) are longer lived, make more vegetative growth, take much longer to reach a sexual phase, and invest less in sexual reproduction than those selected under open (DI) conditions. Many subsequent studies with grasses and other herbaceous plants have shown comparable trends.

Perhaps one of the more striking examples of changes in breeding system from early, temporary habitats to stable, persisting habitats is the Asian wild rice (Oryza perennis) studied by Morishima (1985) in Japan. Populations of this species show an annual to perennial continuum. The annual type occurs in shallow, temporary swamps and is characterized by early flowering, high reproductive allocation, and a high selfing rate, while the perennial type persisting in the stable habitats of deep swamps is characterized by late flowering, low reproductive effort, and a high outcrossing rate. This example, along with those cited by Richards (1990) in a review on reproductive versatility in the grasses, provides support for Richards' suggestion that in closed, stable communities there may be selection for breeding systems that maintain genetic variation, because in the closed community, almost all reproduction is vegetative from persisting individuals. As Grime (1973) has suggested for species densities, with time and a lack of defoliation and damage a relatively few successful genets should predominate, lowering the genotypic variation.

In *Buchloe dactyloides*, there is evidence that population densities may affect both the gender allocation of individual monoecious plants, and the frequency of sex forms within natural populations. Certain monoecious plants produce predominately male inflorescences after cloning and transplanting to new soil (J. A. Quinn, personal observation). There have also been reports that male inflorescences predominate at high nitrogen levels (Huff & Wu, 1987), on plants with longer stolons (Plank, 1892; Arber, 1934), in mowed fields (Wu, Harvindi & Gibeault, 1984), and in open/low density sites (Shaw, Bern & Winkler, 1987; Quinn, personal observation). Although this male inflorescence production may be mostly a response to increased light and resources, it does obviate a pollen limitation problem for *Buchloe* at low densities. A Nebraska study (Jones & Newell, 1946) documented the markedly limited dispersal of pollen from the short inflorescence culms, and my personal observations indicate that the number of seeds within diaspores is affected by the percentage cover of buffalograss and/or distance from male plants (Quinn, 1985).

Considering the frequency of sex forms within populations in relation to density, Huff & Wu (1992) reported a relationship between percentage cover of *Buchloe* and the frequency of monoecious plants in eight populations in two E–W transects across the shortgrass prairie of central North America. As can be seen in Table 5.1, the number of monoecious plants (%I) in the vegetative sample on the left ranged from 0% at Guymon, Oklahoma (in the central part of the range with 91% *Buchloe* cover) to 29% at Apache Springs, New Mexico (a population on the periphery of *Buchloe*'s range with 17% cover of *Buchloe*). Markedly more monoecious plants were found in seed samples (right side of table), and in general in peripheral populations where the *Buchloe* sod is less continuous. Huff & Wu (1992) hypothesized that in areas where *Buchloe* sod is sparse and discontinuous and potential mates are rare, monoecious genotypes may have an

Table 5.1. *Sex form distribution of vegetative plant samples and seed samples collected from eight natural populations of buffalograss, Buchloe dactyloides*

Sex forms are denoted as male (M), inconstant male (IM), inconstant female (IF), and female (F). The percentage of inconstant (or monoecious) sex forms within each sample is labelled as %I. Chi-square was used to test the independence between sex form distribution and population location.

| | Vegetative sample | | | | | | Seed sample | | | | | |
| | Sex form distribution | | | | | | Sex form distribution | | | | | |
Population	N	M	IM	IF	F	%I	N	M	IM	IF	F	%I
Hennessey	15	6	1	1	7	13	–	–				–
Chillicothe	26	9	2	5	10	27	37	6	9	16	6	68
Guymon	20	10	0	0	10	0	82	36	2	3	41	6
Wilderado	21	11	2	0	8	10	25	11	0	3	11	12
Clayton	19	9	0	3	7	16	24	10	2	2	10	17
Glenrio	18	7	0	3	8	17	45	12	4	6	23	22
Springer	20	9	1	3	7	20	–	–				–
Apache	17	6	3	2	6	29	–	–				–
Pooled	156	67	9	17	63	17	213	75	17	31	91	22
χ^2 independence			10.1						60.2***			

Notes:
*** Significant at the 0.001 probability level.
–, no data.
Source: From Huff & Wu, 1992.

advantage over unisexual individuals by possessing the ability to self-fertilize. However, at vegetatively more dense locations, they suggested that the self-fertility of monoecious genotypes might lead to inbreeding depression and less competitive and less variable progenies than those arising from cross-fertilization. They pointed out that there are alternative hypotheses, and that much additional research will be required to determine the genetic and ecological significance of such population variation in sex form frequencies.

Ecological history of frequency and intensity of disturbance

Those populations that have been historically subjected to high frequencies and intensities of disturbance can show great plasticity in their breeding systems and reproductive strategies, often switching between the generation of variable progeny and the rapid production of similar, locally adapted progeny. Such populations often have a short juvenile period, or show a 'pessimistic strategy' of producing locally adapted progeny early and larger numbers of potentially more variable progeny later, if time and resources are available (Cheplick & Quinn, 1982).

More than 30 years ago, Heslop-Harrison (1961, 1966) reported that certain species in the tribe Andropogoneae produced cleistogamous (CL) inflorescences in short days and chasmogamous (CH) inflorescences under long days, so that under natural conditions CL would prevail during the early flowering period, giving way later to CH. It was noted that those populations of *Bothriochloa decipiens* and other species showing a tendency to CL during the early part of the flowering period are short-lived, opportunistic plants of climates with hot, dry summers. The capacity for building up large seed populations in the early season by selfing may have been a selective advantage (Heslop-Harrison, 1966). Similarly, Australian populations of *Dichanthium aristatum* (also in the Andropogoneae) have been reported to show seasonal and population differences in the balance between sexual and asexual seed production (Knox, 1967). There was a relatively low (*c.* 50%) incidence of apomictic embryo sacs in the early part of the flowering season at daylengths exceeding 13 hours, while towards the end of the season under shorter days apospory reached a maximum (*c.* 90%). The longer daylengths are associated with conditions of stress during the dry season, while the high levels of apomixis occurred during conditions favourable for growth and flowering after summer rains.

In the Pinelands of New Jersey (USA), where historically fire has been a frequent but unpredictable event and the sandy soils can become very dry by mid-summer, the annual *Amphicarpum purshii* pursues a 'pessimistic strategy' by producing a few subterranean CL seeds early (the rhizanthogenes of Fig. 5.2) and larger numbers of aerial CH flowers later, if time and resources permit (Cheplick & Quinn, 1982). For the first few years following major disturbance and removal of vegetation, plants produce more aerial seeds than subterranean seeds. As succession progresses, plant cover and competitive interactions increase, and vigour of *Amphicarpum* plants declines dramatically. The ratio of subterranean to aerial seeds markedly increases to a point where more subterranean than aerial are produced and finally to the point where no aerial and only a few subterranean seeds are produced (McNamara & Quinn, 1977). Thus, the relative levels of selfing and outcrossing not only change during a season but also over a successional sequence.

In summary, all of the examples in this subsection feature a plastic reproductive system that alternates sporadically, or on a regular basis, between flowers with the potential for cross-pollination and seeds that are produced by selfing or agamospermy.

Heterogeneity of habitat in time and space

Unpredictability and heterogeneity should select for flexible breeding systems that maintain genetic variation through frequent or occasional recombinational episodes. Richards (1990) in his review of reproductive versatility in the grasses suggests that 'frequency-dependent pressures responding to levels of environmental heterogeneity will favour genetic variability, and so may indirectly select for mothers possessing breeding systems which will maximize this variability.' In addition, in those habitats where individuals are subjected to marked environmental fluctuations during their lifetimes, or where progeny are subjected to environments very different from those of their parents, we might expect selection for adaptive plasticity in the breeding system.

Dichanthelium (*Panicum*) *clandestinum* produces exserted inflorescences in late spring with exposed anthers and stigmas (CH flowers), and in late summer produces reduced inflorescences which are almost totally enclosed within the leaf sheaths (CL flowers). In a study of the effect of spatial heterogeneity on the plasticity of the breeding system, Bell & Quinn (1987) grew plants from six closely adjacent New Jersey populations along a soil moisture gradient to determine their relative allocation to CH and CL reproduction. The two most plastic populations whose plants had a relatively higher allocation to CL reproduction at low soil moisture were also the two populations which were exposed to the lowest soil moistures in the field. CL flowers are at least 10% cheaper to make and quicker to mature, and thus the plant can produce some CL offspring under conditions that may not permit production of the more costly CH offspring (Bell & Quinn, 1987). Stebbins (1957), and many others (see references in Campbell *et al.*, 1983), have regarded self-fertilization and CL as derived specialized characters that can be of value when conditions are unfavourable towards the end of the growing season.

In addition, populations of a widely distributed species ranging from Mexico to Canada may be exposed to very different heterogeneity of habitat in time. In a classic study, McMillan (1965) programmed six growth chambers to simulate a series of climates separated by 2000 miles. Genotypes from each of the simulated sites, of six grass species, were split into pieces and placed into each chamber. The northern clones responded rapidly and showed sensitivity to diverse environmental conditions, i.e. they showed different reactions to the diverse simulated climates. Many southern clones, requiring more than 20 weeks for flowering, showed developmental homeostasis under the different climates, i.e. they showed stability and continued vegetative growth under the diverse conditions. Considering the N–S contrast in climates, both types of population responses to environment could be adaptive. Because of the longer frost-free period, southern plants are exposed to a longer, climatically variable growth period. Genotypes that respond to weather fluctuations with flowering or dormancy and prematurely complete growth before the end of the growing season would be at a competitive disadvantage. In contrast, northern plants are exposed to a short growing season and weather fluctuations that necessitate a more immediate response. Those not setting seeds before the killing frosts will be at a selective disadvantage.

Biome-level differences in growing season, resources, and biotic interactions

Although there are sometimes unique aspects of the environment associated with different biomes, the breeding systems and reproductive strategies of grasses in different biomes are not unique and are often broadly overlapping. This is a result of (1) the outward spread of widely distributed grasses from centres of origin, (2) the wide range of habitats within each biome in regard to moisture, nutrients, and light availability as a result of geological substrate and disturbance followed by succession, and (3) the documented fact that differing environments often place comparable selection pressures on breeding systems and plasticity in reproductive strategies (i.e. the categories of factors previously discussed). However, if we compare grasses of tropical biomes or tundra biomes with grasses of temperate regions, we do find some pronounced trends in reproductive strategies that are linked with the unique features of those biomes. For illustration, the arctic and alpine environment and the comparative reproductive strategies of arctic and alpine grasses can be outlined as follows:

1. The growing season is short and cool in the arctic and alpine; however, the shorter but relatively predictable and stable arctic growing season contrasts with the unpredictable and relatively harsh alpine environment (featuring greater winds and temperature fluctuations) (Bliss, 1956, 1988; Billings, 1988).
2. Grasses of both environments thus show strategies for accelerated (e.g. autumn floral initiation, vegetative proliferation of spikelets or 'vivip-

ary') and conservative (e.g. agamospermy, vegetative spread) reproduction (Evans, 1964; Latting, 1972; Smith, 1984), while alpine grasses show more plasticity in type of reproduction, opportunistic flowering, and often a greater allocation to sexual reproduction (Billings, 1988; Körner, 1995).

3. In relation to grasses as a family, there are very few arctic and alpine grasses that are annuals or short-lived perennials (Callaghan, 1977; Bliss, 1988). Plants that are annual elsewhere, e.g. *Poa annua*, are only perennial in the Arctic and sub-Antarctic (Smith, 1984). In fact, some long-lived perennials in the arctic have been reported to have unusually long juvenile periods (up to 20 years) prior to the initiation of flowering (Grulke & Bliss, 1988), and some long-lived perennials flower irregularly, at intervals of up to 5 years (Latting, 1972).

4. Monoecy and dioecy are almost absent (except for *Poa*), and agamospermy and vegetative proliferation are more common, especially in the more severe arctic and alpine environments (Quinn, unpublished compilation). Some arctic poas have three asexual mechanisms within the same population, i.e. vegetative spread, vegetative proliferation, and agamospermy (Richards, 1990).

That these trends reflect more than just the predominance of C_3 species and the subfamily Pooideae is indicated by the changes in traits and strategies within a species when it occurs in the alpine and Arctic or sub-Antarctic. In a mini-review on vivipary Elmqvist & Cox (1996) noted that it may occur in specific alpine or arctic populations within otherwise non-viviparous species with a broader geographic distribution. For *Poa fendleriana* in western North America, sexually reproducing populations are more likely to be found in relatively mild and mesic environments, and asexual/apomictic populations are more likely in the alpine; in general for poas of western North America, apomicts occur in colder climates with shorter frost-free seasons, and are more common in alpine zones (Soreng, 1986; Soreng & Van Devender, 1989). *Deschampsia antarctica* shows a gradient of sexual to asexual reproduction in going south in South America to the sub-Antarctic (Edwards, 1974). And in *Festuca contracta* the relatively severe South Georgia sub-Antarctic environment appears to have favoured the selection of exclusively CL plants; also experimentally documented was the selection of a more vegetative reproduction strategy as the degree of exposure increased locally (Tallowin, 1977).

Summary and conclusions

Grasses display an extraordinary diversity of breeding systems – hermaphroditism to dioecism. In addition, many have marked environmentally governed reproductive versatility, combining a mixture of outcrossing, selfing, and asexual strategies.

The one-hormone model of sex determination provides a mechanistic interpretation of environmental effects on sex expression and predicts that minor genetic changes can have major effects on sex expression and the breeding system. The relative location of sensitivity levels and the normal hormone range (and its tight or loose regulation) should be strikingly genotype-specific, producing considerable variation among individuals in type and lability of sex expression. Once such plastic individuals (or different sex forms) arise, they may increase in a local population if the conditions are such that they produce an increased proportion of the reproductives.

Excluding phylogenetic and/or genetic and morphological constraints, the breeding system, and its phenotypic plasticity, of a population should be influenced by the levels of r- vs K-selection and DI vs DD mortality, ecological history of frequency and intensity of disturbance, heterogeneity of habitat in time and space, and biome-level differences in growing season, resources, and biotic interactions.

As a result, 'No two plant populations have exactly similar breeding systems and exactly similar patterns of variation.' (Richards, 1986).

Considering this 'individualistic' population concept (Quinn, 1987), we must avoid breeding system studies that utilize only one population of a species or only two or three individuals each from a few populations. If we are to appreciate the range of adaptive variation in sex expression in grasses, we must not only compare the differential responses of individuals and populations to a range of environments, but also equally emphasize the amounts and patterns of plasticity in reproductive strategies exhibited across this range of environments.

Although there are sometimes unique aspects of the environment associated with different biomes, and we can recognize some pronounced trends in arctic and alpine grasses, the breeding systems and reproductive strategies of grasses in different biomes are not unique and are often broadly overlapping.

Finally, the breeding system is not the sole determiner of the success or failure of a plant. A plant may be deficient or maladapted at other stages of the life cycle, or an extremely vigorous, competitive plant may appear to persist in spite of its breeding system.

References

Arber, A. (1934). *The Gramineae. A Study of Cereal, Bamboo, and Grass.*
Cambridge: Cambridge University Press.
Bell, T. J. & Quinn, J. A. (1987). Effects of soil moisture and light intensity on the
chasmogamous and cleistogamous components of reproductive effort of
Dichanthelium clandestinum populations. *Canadian Journal of Botany,* **65,**
2243–49.
Billings, W. D. (1988). Alpine vegetation. In *North American Terrestrial
Vegetation,* ed. M. G. Barbour & W. D. Billings, pp. 391–420. Cambridge:
Cambridge University Press.
Bliss, L. C. (1956). A comparison of plant development in microenvironments of
arctic and alpine tundra. *Ecological Monographs,* **26,** 303–37.
Bliss, L. C. (1988). Arctic tundra and polar desert biome. In *North American
Terrestrial Vegetation,* ed. M. G. Barbour & W. D. Billings, pp. 1–32.
Cambridge: Cambridge University Press
Callaghan, T. V. (1977). Adaptive strategies in the life cycles of South Georgian
graminoid species. In *Adaptations within Antarctic Ecosystems,* ed. G. A.
Llano, pp. 981–1002. Washington, DC: Smithsonian Institution Press.
Campbell, C. S., Quinn, J. A., Cheplick, G. P. & Bell, T. J. (1983). Cleistogamy in
grasses. *Annual Review of Ecology and Systematics,* **14,** 411–41.
Cheplick, G. P. & Quinn, J. A. (1982). *Amphicarpum purshii* and the 'pessimistic
strategy' in amphicarpic annuals. *Oecologia,* **52,** 327–32.
Clausen, J., Keck, D. D. & Hiesey, W. M. (1946). *Poa* investigations. *Carnegie
Institution of Washington Year Book,* **45,** 117–20.
Connor, H. E. (1979). Breeding systems in the grasses: a survey. *New Zealand
Journal of Botany,* **17,** 547–74.
Connor, H. E. (1981). Evolution of reproductive systems in the Gramineae.
Annals of the Missouri Botanical Garden, **68,** 48–74.
Connor, H. E. (1987). Reproductive biology in the grasses. In *Grass Systematics
and Evolution,* ed. T. R. Soderstrom, K. W. Hilu, C. S. Campbell & M. E.
Barkworth, pp. 117–32. Washington, DC: Smithsonian Institution Press.
Dewald, C. L., Burson, B. L., DeWet, J. M. J. & Harlan, J. R. (1987).
Morphology, inheritance and evolutionary significance of sex reversal in
Tripsacum dactyloides (Poaceae). *American Journal of Botany,* **74,** 1055–9.
Edwards, J. A. (1974). Studies in *Colobanthus quitensis* (Kunth) Barl. and
Deschampsia antarctica Desv. VI. Reproductive performance on Signy
Island. *British Antarctic Survey Bulletin,* **39,** 67–86.
Elmqvist, T. & Cox, P. A. (1996). The evolution of vivipary in flowering plants.
Oikos, **77,** 3–9.
Evans, L. T. (1964). Reproduction. In *Grasses and Grasslands,* ed. C. Barnard, pp.
126–53. London: Macmillan.
Gadgil, M. & Solbrig, O. T. (1972). The concept of *r*- and *K*-selection: Evidence
from wild flowers and some theoretical considerations. *American Naturalist,*
106, 14–31.
Grime, J. P. (1973). Competitive exclusion in herbaceous vegetation. *Nature,* **242,**
344–7.
Grulke, N. E. & Bliss, L. C. (1988). Comparative life history characteristics of two
high arctic grasses, Northwest Territories. *Ecology,* **69,** 484–96.
Heslop-Harrison, J. (1961). The function of the glume pit and the control of
cleistogamy in *Bothriochloa decipiens* (Hack.) C. E. Hubbard.
Phytomorphology, **11,** 378–83.

Heslop-Harrison, J. (1966). Reflections on the role of environmentally-governed reproductive versatility in the adaptation of plant populations. *Transactions of the Botanical Society of Edinburgh*, **40**, 159–68.

Huff, D. R. & Wu, L. (1987). Sex expression in buffalograss under different environments. *Crop Science*, **27**, 623–6.

Huff, D. R. & Wu, L. (1992). Distribution and inheritance of inconstant sex forms in natural populations of dioecious buffalograss (*Buchloe dactyloides*). *American Journal of Botany*, **79**, 207–15.

Jackson, L. L. & Dewald, C. L. (1994). Predicting evolutionary consequences of greater reproductive effort in *Tripsacum dactyloides*, a perennial grass. *Ecology*, **75**, 627–41.

Jones, M. D. & Newell, L. C. (1946). *Pollination cycles and pollen dispersal in relation to grass improvement*. Nebraska Agricultural Experiment Station Research Bulletin 148. Lincoln: University of Nebraska.

Kellogg, E. A. (1987). Apomixis in the *Poa secunda* complex. *American Journal of Botany*, **74**, 1431–37.

Knox, R. B. (1967). Apomixis: seasonal and population differences in a grass. *Science*, **157**, 325–6.

Körner, C. (1995). Alpine plant diversity: A global survey and functional interpretations. In *Arctic and Alpine Biodiversity: Patterns, Causes and Ecosystem Studies*, Ecological Studies, vol. 113, ed. F. S. Chapin III & C. Körner, pp. 45–62. Berlin: Springer-Verlag.

Latting, J. (1972). Differentiation in the grass inflorescence. In *The Biology and Utilization of Grasses*, ed. V. B. Youngner & C. M. McKell, pp. 365–99. New York: Academic Press.

Law, R., Bradshaw, A. D. & Putwain, P. D. (1977). Life-history variation in *Poa annua*. *Evolution*, **31**, 233–46.

McMillan, C. (1965). Grassland community fractions from central North America under simulated climates. *American Journal of Botany*, **52**, 109–16.

McNamara, J. & Quinn, J. A. (1977). Resource allocation and reproduction in populations of *Amphicarpum purshii* (Gramineae). *American Journal of Botany*, **64**, 17–23.

Morishima, H. (1985). Habitat, genetic structure and dynamics of perennial and annual populations of the Asian wild rice *Oryza perennis*. In *Genetic Differentiation and Dispersal in Plants*, ed. P. Jacquard, G. Heim & J. Antonovics, pp. 179–90. NATO ASI Series, vol. 65. Berlin: Springer-Verlag.

Plank, E. N. (1892). *Buchloe dactyloides*, Engelm., not a dioecious grass. *Bulletin of the Torrey Botanical Club*, **19**, 303–6.

Quinn, J. A. (1985). Validity of breeding for a female bias in the dioecious buffalograss (*Buchloe dactyloides*). In *Proceedings of the XV International Grassland Congress*, ed. T. Okubo & M. Shiyomi, pp. 297–9. Nishi-nasuno, Tochigi-ken: The Science Council of Japan and the Japanese Society of Grassland Science.

Quinn, J. A. (1987). Complex patterns of genetic differentiation and phenotypic plasticity versus an outmoded ecotype terminology. In *Differentiation Patterns in Higher Plants*, ed. K. M. Urbanska, pp. 95–113. London: Academic Press.

Quinn, J. A. (1991). Evolution of dioecy in *Buchloe dactyloides* (Gramineae): tests for sex-specific vegetative characters, ecological differences, and sexual niche-partitioning. *American Journal of Botany*, **78**, 481–88.

Richards, A. J. (1986). *Plant Breeding Systems*. London: George Allen & Unwin.

Richards, A. J. (1990). The implications of reproductive versatility for the

structure of grass populations. In *Reproductive Versatility in the Grasses*, ed.
G. P. Chapman, pp. 131–53. Cambridge: Cambridge University Press.
Roos, F. H. & Quinn, J. A. (1977). Phenology and reproductive allocation in
Andropogon scoparius (Gramineae) populations in communities of different
successional stages. *American Journal of Botany*, **64**, 535–40.
Shaw, R. B., Bern, C. M. & Winkler, G. L. (1987). Sex ratios of *Buchloe
dactyloides* (Nutt.) Engelm. along catenas on the shortgrass steppe. *Botanical
Gazette*, **148**, 85–9.
Smith, R. I. L. (1984). Terrestrial plant biology of the sub-Antarctic and
Antarctic. In *Antarctic Ecology*, vol. 1, ed. R. M. Laws, pp. 61–162. London:
Academic Press.
Soreng, R. J. (1986). *Distribution and Evolutionary Significance of Apomixis in
Diclinous* Poa *of Western North America*. Ph.D. Dissertation. Las Cruces:
New Mexico State University.
Soreng, R. J. & Van Devender, T. R. (1989). Late Quaternary fossils of *Poa
fendleriana* (muttongrass): Holocene expansions of apomicts. *Southwestern
Naturalist*, **34**, 35–45.
Stebbins, G. L. (1957). Self-fertilization and population variability in the higher
plants. *American Naturalist*, **91**, 337–54.
Tallowin, J. R. B. (1977). The reproductive strategies of a subantarctic grass,
Festuca contracta T. Kirk. In *Adaptations within Antarctic Ecosystems*, ed. G.
A. Llano, pp. 967–80. Washington, DC: Smithsonian Institution Press.
Watson, L. (1990). The grass family, Poaceae. In *Reproductive Versatility in the
Grasses*, ed. G. P. Chapman, pp. 1–31. Cambridge: Cambridge University
Press.
Wu, L., Harvindi, A. H. & Gibeault, V. A. (1984). Observations on buffalograss
sexual characteristics and potential for seed production improvement.
HortScience, **19**, 505–6.
Yin, T. & Quinn, J. A. (1992). A mechanistic model of a single hormone
regulating both sexes in flowering plants. *Bulletin of the Torrey Botanical
Club*, **119**, 431–41.
Yin, T. & Quinn, J. A. (1994). Effects of exogenous growth regulators and a
gibberellin inhibitor on sex expression and growth form in buffalograss
(*Buchloe dactyloides*), and their ecological significance. *Bulletin of the Torrey
Botanical Club*, **121**, 170–9.
Yin, T. & Quinn, J. A. (1995a). Tests of a mechanistic model of one hormone
regulating both sexes in *Buchloe dactyloides* (Poaceae). *American Journal of
Botany*, **82**, 745–51.
Yin, T. & Quinn, J. A. (1995b). Tests of a mechanistic model of one hormone
regulating both sexes in *Cucumis sativus* (Cucurbitaceae). *American Journal
of Botany*, **82**, 1537–46.

6

Interspecific variation in plasticity of grasses in response to nitrogen supply

ERIC GARNIER

Introduction

Over the last three decades, evidence has accumulated showing that plant species originating from nutrient-poor habitats tend to have a lower growth potential – i.e. when measured under optimal environmental conditions in the laboratory – than species from nutrient-rich habitats (e.g. Bradshaw *et al.*, 1964; Grime & Hunt, 1975; reviews in Chapin, 1980, 1988 and Lambers & Poorter, 1992). Whereas the adaptive value of a high growth potential is relatively clear, that of a low one has been the subject of much debate (Chapin, 1980, 1988; Lambers & Dijkstra, 1987; Berendse & Elberse, 1990; Aerts & van der Peijl, 1993). One of the most convincing hypotheses currently put forward is that rather than the growth rate itself, it is one of its underlying components (namely the specific leaf area, the ratio of leaf area to leaf dry mass) which is the target of selection (Lambers & Dijkstra, 1987; Lambers & Poorter, 1992).

Species from habitats differing in fertility (or differing in growth potential) also exhibit differences in plasticity, defined as a variation in morphology and/or physiology in response to changes in environmental conditions (Schlichting, 1986). It has commonly been observed that the growth rate of species from nutrient-poor habitats (and slow-growing species) is less sensitive to a variation in external nutrient supply than that of species from nutrient-rich habitats (and fast-growing species: reviews in Chapin, 1980, 1988 and Lambers & Poorter, 1992). Furthermore, plants respond to nutrient stress by altering simultaneously a number of other traits: for example, when nitrogen availability decreases, internal plant nitrogen concentration decreases and biomass allocation to roots increases (both reviewed in Larsson, 1994). The increase in proportional allocation to roots when nutrient availability decreases has long been observed and

155

interpreted as an adaptive feature allowing plants to increase the acquisition of the resource that most limits growth (Brouwer, 1963; Davidson, 1969; Bloom, Chapin & Mooney, 1985). Based on considerations related to the periodicity of nutrient flushes in their natural habitats and the costs of adjusting morphology, both Grime (1979) and Chapin (1980, 1988) predict that the plasticity of this trait should be higher in species from nutrient-rich habitats than in those from nutrient-poor habitats. Chapin (1980, 1988) further argues that tissue nutrient concentration should exhibit more plasticity in the latter species, because in conditions of high nutrient supply, slow-growing species from nutrient-poor habitats tend to accumulate nutrients in excess ('luxury consumption') to a greater extent than species from nutrient-rich habitats.

The aim of the present paper is to test these predictions related to the plasticity of growth rate, biomass allocation and plant nitrogen within the grass family, using nitrogen as a model element. In temperate climates, grasses have colonized virtually all types of habitats from very poor to very rich or disturbed (e.g. Grime, Hodgson & Hunt, 1988), and members of the family have been extensively used as crops worldwide (e.g. Harlan, 1982). This, combined with the fact that the analysis of covariation between growth and its underlying components may be confused when comparisons are carried out with taxonomically remote species (e.g. Garnier, 1991), makes grasses a good family to address these questions.

Both plant growth (e.g. Ingestad, 1979; Ågren, 1985; Ingestad & Ågren, 1992) and biomass allocation (e.g. Ågren & Ingestad, 1987; Levin, Mooney & Field, 1989; van der Werf *et al.*, 1993*a*) are tightly related to internal nitrogen concentration. There is therefore a 'syndrome' of plant response to nitrogen availability, presenting an interesting case of phenotypic integration (Schlichting, 1986), to which a sound physiological interpretation can be given. Here, I will analyse differences in plasticity among grass species within this framework, using data from the literature complemented by some as yet unpublished results.

Methods

Data

In all the studies retained for this meta-analysis (i.e. a synthesis of independent experiments), at least two grass species (or genotypes) compared at a minimum of two nitrogen (or nutrient) treatments were examined. Those in which plant nitrogen concentration was not measured are commented on and referred to directly in the text. The others (17 studies briefly described in Table 6.1) have been used to form a data set called NGRASS hereafter. It concerns 26 wild species (11 annuals and 15 perennials), 2 annual grain crops (*Avena sativa* and *Hordeum vulgare* [4 cultivars]), and 3 perennial pasture species (*Bromus inermis, Lolium perenne* [3 cultivars] and *Phalaris aquatica*); all are C_3 and herbaceous. These studies were selected because it was possible to extract data on – at least – whole plant mean nitrogen concentration (PNC), relative growth rate (RGR, the increase in total plant biomass per unit time and biomass already present in the plant) and root mass ratio (RMR, the ratio of root mass to total plant mass) from either tables or figures. In one of them (van de Vijver *et al.*, 1993), RGR values were not available, but this study was still retained because the two species under investigation were grown under eight nitrogen levels, which permits a thorough analysis of the relationship between plant internal nitrogen and biomass allocation.

In all studies, plants were grown isolated (except in Ryser & Lambers, 1995, where there were four plants per pot), and remained in the vegetative stage of growth. They were grown either hydroponically or on sand, allowing the experimenter to monitor carefully the concentration and/or amount of nitrogen supplied to the plants (the data selected are for treatments where nitrogen alone was purposely varied). Three methods were used to control the supply of nitrogen in these studies (Table 6.1): control of nitrogen concentration in the nutrient solution, periodic supply of nutrient solutions differing in nitrogen concentrations, and control of relative addition rate of nitrogen (cf. Ingestad & Ågren, 1992). There was usually no indication regarding the changes in nitrogen concentration that occurred during the course of the experiments, except for that by Robinson & Rorison (1988), who estimated the drop in nitrogen concentration to approximately 50% of the initial concentration in two of the nitrogen treatments. There were 2 nitrogen levels applied in 10 of the studies, 3 levels in 3 studies, and more than 3 levels (4–8) in the remaining 4 studies (Table 6.1). These data will be used for two purposes: (a) a description and analysis of the relationships between PNC and RGR on the one hand, and PNC and RMR on the other hand, which will be done with all data from these four latter studies (Mattsson *et al.*, 1991; van der Werf *et al.* 1993*d*; van de Vijver *et al.*, 1993; E. Garnier, M.-L. Navas & R. M. Gifford, unpublished), and (b) an analysis of interspecific differences in plasticity, which will be done using the whole data set, retaining only data for the lowest and highest non-supra-optimal nitrogen supply when more than two levels were applied.

In this chapter, I will refer to a 'fast-growing (or slow-growing) species'

Table 6.1. *List of studies used to assess the plasticity of plant nitrogen concentration, relative growth rate and root mass ratio in response to nitrogen supply in grass species (NGRASS data set)*

Superscript A or P following the species name is for annual and perennial, respectively; for cultivated species, the cultivar name is given between brackets. Information on growing conditions ('cultivation' column) concerns the overall environment of the experiment (GD, GH and GR are for experimental garden, glasshouse and growth room respectively), the growing medium (hydroponics [hydro] or sand), the way nutrient supply was controlled (CC, control of nutrient concentration in solution culture; PS, periodic supply of nutrient solution; RA, control of relative addition rate of nitrogen), and the number of nutrient levels in the experiment (2 to 8). RGR_{HN} is the relative growth rate of the species under the highest nutrient supply of the experiment. S_N, S_G and S_R are respectively the ratios of plant nitrogen concentration, relative growth rate and root mass ratio between the lowest and highest nutrient supply of the experiment, averaged across species (see text). Blank entries indicate a repetition of the entry above.

Reference	Species	Cultivation	RGR_{HN} $(g\ g^{-1}\ d^{-1})$	S_N, S_G, S_R
Boot & den Dubbelden (1990)	*Agrostis vinealis*[P]	GH, sand, PS, 2	0.051	0.82, 0.71, 1.18
	Corynephorus canescens[P]		0.041	
Boot & Mensink (1991)	*Deschampsia flexuosa*[P]	GH, sand, PS, 2	0.036	0.79, 0.80, 1.23
	Festuca ovina[P]		0.055	
	Festuca rubra[P]		0.063	
	Holcus lanatus[P]		0.065	
	Molinia caerulea[P]		0.062	
Garnier & Gaslonde (unp.)	*Bromus erectus*[P]	GR, hydro, CC, 2	0.146	0.89, 0.79, 1.08
	Bromus hordeaceus[A]		0.239	
Garnier et al. (1989)	*Bromus hordeaceus*[A]	GR, hydro, CC, 2	0.168	0.92, 0.91, 1.23
	Bromus intermedius[A]		0.175	
Garnier et al. (unp.)	*Danthonia richardsonii*[P]	GH, sand, PS, 6	0.117	0.52, 0.35, 2.00
	Phalaris aquatica[P] (Holdfast)		0.174	
Gobin et al. (unp.)	*Avena barbata*[A]	GR, hydro, CC, 2	0.219	0.60, 0.68, 1.54
	Avena sativa[A] (Manoire)		0.179	
	Bromus erectus[P]		0.127	
	Bromus inermis[P] (Barton)		0.192	
	Bromus madritensis[A]		0.202	

Reference	Species	Conditions		
	Hordeum murinum[A]		0.198	
	Hordeum vulgare[A] (Maïté)		0.152	
	Lolium perenne[P]		0.228	
	L. perenne[P] (Préférence)		0.226	
	Lolium rigidum[A]		0.259	
Hull & Mooney (1990)	Avena fatua[A]	GR, sand, PS, 2	0.123	0.50, 0.54, 1.57
	Bromus diandrus[A]		0.114	
	Bromus hordeaceus[A]		0.167	
	Lolium multiflorum[A]		0.147	
Kachi & Rorison (1989, 1991)	Festuca ovina[P]	GR, hydro, RA, 3	0.171	0.29, 0.36, 1.90
	Holcus lanatus[P]		0.239	
Mattsson et al. (1991)	Hordeum vulgare[A] (Golf)	GR, hydro, RA, 7	0.148	0.24, 0.18, 1.79
	H. vulgare[A] (Laevigatum)		0.147	
	H. vulgare[A] (Mette)		0.150	
Muller & Garnier (1990)	Bromus erectus[P]	GR, hydro, CC, 2	0.121	0.74, 0.72, 1.17
	Bromus sterilis[A]		0.176	
Pavlik (1983)	Ammophila arenaria[P]	GH, sand, PS, 2	0.055	0.49, 0.93, 1.43
	Elymus mollis[P]		0.056	
Poorter et al. (1995)	Deschampsia flexuosa[P]	GR, sand, PS, 2	0.087	0.57, 0.47, 1.39
	Holcus lanatus[P]		0.163	
Robinson & Rorison (1988)	Deschampsia flexuosa[P]	GR, hydro, PS, 3	0.149	0.64, 0.58, 1.90
	Holcus lanatus[P]		0.242	
	Lolium perenne[P] (Norlea)		0.214	
	L. perenne[P] (S23)		0.210	
	Poa annua[A]		0.229	
Ryser & Lambers (1995)	Brachypodium pinnatum[P]	GD, sand, PS, 3	0.073	0.54, 0.79, 1.27
	Dactylis glomerata[P]		0.137	
van der Werf et al. (1993b)	Brachypodium pinnatum[P]	GR, hydro, RA, 2	0.149	0.25, 0.20, 1.71
	Holcus lanatus[P]		0.238	
van der Werf et al. (1993d)	Briza media[P]	GR, hydro, RA, 4	0.113	0.32, 0.76, 1.93
	Dactylis glomerata[P]	GR, hydro, RA, 5	0.213	
van de Vijver et al. (1993)	Deschampsia flexuosa[P]	GR, sand, PS, 8	—	0.46, −, 1.73
	Holcus lanatus[P]		—	

if that particular species has a high (or low) relative growth rate *when measured under non-limiting nutrient conditions* (RGR_{max}).

Comparisons of NGRASS experiments

There are a number of problems to be overcome when comparing and synthesizing results obtained from studies conducted on various species grown under very different environmental conditions (see e.g. Gurevitch & Hedges, 1993; Reynolds & D'Antonio, 1996), as is the case for NGRASS. Below, I will point to three of them, and explain how the data have been analysed to try to overcome them in the present review.

Firstly, because we are interested in comparing species originating from habitats differing in fertility, an index of this fertility is necessary. One possibility is to use the nitrogen figure (N-figure) of a species, as defined by Ellenberg. In an attempt to characterize the ecological behaviour of plant species of Central Europe, Ellenberg (1979, 1988) devised a system of classification whereby roughly 2000 species were attributed a series of six 'indicator values', each corresponding to a particular habitat factor, either climatic or edaphic. One of these values – the N-figure – classifies the different species with respect to their occurrence in sites of different nitrogen availabilities. It varies from one (species found in nitrogen-poor habitats) to nine (species found in nitrogen-rich habitats). In this chapter, N-figures will be used to characterize the different species in some instances, but they present two major difficulties: (a) they have been defined essentially for ecotypes of species within Central Europe, and may thus not apply for ecotypes of the same species found elsewhere (although some generalizations seem to be possible: Thompson *et al.*, 1993), and (b) species from other continents are not classified. I will therefore use the finding that species from nutrient-rich habitats tend to have a high potential relative growth rate (RGR_{max}) when grown under optimum conditions (Grime & Hunt, 1975; Chapin, 1980, 1988), which translates into a positive relationship between the N-figure of a species and its RGR_{max} (Poorter & Remkes, 1990). Thus, the assumption is that a perfect equivalence exists between the RGR_{max} of a species and its occurrence in a site of particular fertility. Doing so merges the competitive and the ruderal strategies into fast-growing plants, which are then compared with stress-tolerators, according to Grime's (1979) terminology. This grouping has already been done to contrast physiological properties of fast- and slow-growing species (Grime *et al.*, 1988; p. 36).

Secondly, the 17 NGRASS experiments have been conducted under very different environmental conditions. This is illustrated by data on *Holcus lanatus*, which was studied on five occasions, and whose RGR and PNC under the 'high' nitrogen treatments varied between 0.065 g g^{-1} d^{-1} (PNC=1.05 mmol g^{-1}: Boot & Mensink, 1991) and 0.242 g g^{-1} d^{-1} (PNC=3.00 mmol g^{-1}: Robinson & Rorison, 1988). RGR of each species in the high nutrient treatment (RGR_{HN}, given in Table 6.1) was corrected within each experiment, so that it could be considered as the reference RGR_{max} for that species. This was done by dividing the RGR_{HN} of that particular species by the average RGR_{HN} calculated for all the species of that experiment. Table 6.2 shows the absolute and corrected RGR_{HN} values for a slow-growing (*Deschampsia flexuosa*) and a fast-growing (*Holcus lanatus*) species studied in three different experiments with very contrasting environments. The four-fold difference in absolute RGR_{HN} among experiments for the two species translates into a 10–12% difference in corrected RGR_{HN} values.

Thirdly, the estimation of the plasticity of a given species in different experiments will depend on the difference between the highest and lowest nutrient treatment applied in each experiment. For example, Table 6.2 shows that the drop in plant nitrogen concentration between the highest and lowest nitrogen treatments in *D. flexuosa* and *H. lanatus* is substantially less in the study by Boot & Mensink (1991) than in that by Poorter *et al.* (1995). To account for that effect, the ratio between the PNC in the lowest (PNC_{LN}) and the highest (PNC_{HN}) nutrient treatment was calculated for each species in a given experiment, and these values were averaged across the species of that experiment:

$$S_{Nj} = \left[\sum_{i=1}^{n} (PNC_{LN}/PNC_{HN})_i \right] / n \tag{1}$$

where S_{Nj} may be defined as the mean degree of stress imposed in experiment *j*, PNC_{LN} and PNC_{HN} are the plant nitrogen concentration of species *i* in the lowest and highest nutrient treatments respectively, and *n* is the number of species in the experiment. The lower this ratio, the higher the difference between the nutrient treatments within an experiment. The corrected plasticity (subscript c) of species *i* for PNC in that particular experiment was then estimated as:

$$(PNC_{LN}/PNC_{HN})_c = (PNC_{LN}/PNC_{HN})_i / S_{Nj} \tag{2}$$

Table 6.2 shows these corrected values for *D. flexuosa* and *H. lanatus* in three experiments. Similar procedures were applied for RGR and RMR. Each experiment may thus be characterized by three coefficients (given in

Table 6.2. *Examples of absolute and corrected values in three experiments where* Holcus lanatus (*a fast-growing species*) *and* Deschampsia flexuosa (*a slow-growing species*) *have both been studied*

RGR_{HN} (g g^{-1} d^{-1}) *and* $cRGR_{HN}$ *are the absolute and corrected values of relative growth rate measured under high nitrogen supply;* PNC_{LN}/PNC_{HN} *is the ratio of plant nitrogen concentration under the lowest and highest nutrient levels respectively, and* $(PNC_{LN}/PNC_{HN})c$ *is the same ratio corrected by the mean nitrogen stress imposed in the experiment (see text for details).*

Reference	Holcus lanatus				Deschampsia flexuosa			
	RGR_{HN}	$cRGR_{HN}$	$\dfrac{PNC_{LN}}{PNC_{HN}}$	$\left(\dfrac{PNC_{LN}}{PNC_{HN}}\right)c$	RGR_{HN}	$cRGR_{HN}$	$\dfrac{PNC_{LN}}{PNC_{HN}}$	$\left(\dfrac{PNC_{LN}}{PNC_{HN}}\right)c$
Boot & Mensink (1991)	0.065	1.157	0.816	1.036	0.036	0.641	0.781	0.991
Poorter et al. (1995)	0.163	1.304	0.597	1.044	0.087	0.696	0.547	0.957
Robinson & Rorison (1988)	0.242	1.158	0.570	0.893	0.149	0.713	0.894	1.402

Table 6.1) representing the average response of each trait (PNC, RGR and RMR) of all species in that experiment.

These calculations actually imply that the relationships between external nitrogen supply and each variable is linear. This is probably an acceptable assumption for RGR and PNC, provided that the highest nutrient supply remains below supra-optimal levels (e.g. Mattsson *et al.*, 1991, and see Fig. 6.3*A*). In the case of RMR, non-linearity is further discussed below. Unfortunately, these problems cannot be thoroughly analysed in the case of NGRASS, since in most studies only two levels of nitrogen were used (Table 6.1).

Assessing plasticity

There does not seem to be a single means of assessing the plasticity of a particular trait (see e.g. Schlichting, 1986). One way of doing so is to express the value taken by a particular trait under a given non-optimum environment, relative to its value under optimum conditions. For example, to study the impact of nutrient availability on the plasticity of growth and development, Robinson & Rorison (1988), Shipley & Keddy (1988), Kik, Jongman & van Andel (1991) and Cheplick (1995) used the following index (*PI*):

$$PI=(X_F - X_{NF})/X_F=1-(X_{NF}/X_F) \tag{3}$$

where X_F and X_{NF} are the mean values of the trait considered, under fertile and non-fertile conditions respectively, whereas Antonovics, Lovett & Bradshaw (1967) and Cheplick (1991) used the X_{NF}/X_F ratio. From a conceptual point of view, both expressions are strictly equivalent, and in the present review, the latter will be used. As stated by Cheplick (1991), such ratios can be interpreted as the disparity between the realized phenotype under a given non-optimum condition, and a fundamental phenotype, representing the state of the phenotype in an ideal, non-limiting environment.

Expressing plasticity in such a way is of interest not only from an ecological point of view, but also from a physiological point of view. In the body of theory developed by Ingestad and co-workers, the relationship between internal plant nitrogen and relative growth rate is best described when both parameters are expressed in relative terms (e.g. Ingestad & Ågren, 1992, and see below), taking both references under optimum conditions. Expressing results in such a way therefore helps in setting a bridge between the two fields of research.

Plasticity of biomass production and relative growth rate

As a function of the (true) fertility of the habitat of the species

In a classical experiment, Bradshaw *et al.* (1964) studied the response to nitrogen of seven grass species originating from communities differing in overall productivity. They grew plants in pots filled with sand at six nitrogen levels, covering an almost 250-fold range in concentration, under glasshouse conditions, and harvested the plants (above and belowground parts) after eight weeks of growth. Species from nutrient-rich sites produced much more biomass than those from nutrient-poor sites. For two species originating from fertile sites (*Lolium perenne* and *Agrostis stolonifera*), there was more than an 11-fold variation in plant biomass between the highest and the lowest nitrogen level; for two species coming from acidic, nutrient-poor sites (*Festuca ovina* and *Nardus stricta*) the highest nitrogen levels appeared to be in the toxic range, and the variation was less than five-fold between the highest and lowest biomass reached along the gradient. The response was intermediate with no detectable toxicity for the remaining three species (*Agrostis tenuis, Cynosurus cristatus* and *Agrostis canina*).

Such results have been observed at the intraspecific level as well in a study similar in design to that by Bradshaw *et al.* (1964), conducted on six populations of *Lolium perenne* originating from habitats differing in nitrogen availability (Antonovics *et al.*, 1967): individuals selected from a cultivated strain or coming from a population growing in a bird sanctuary (high fertility due to droppings) had much stronger biomass responses to nitrogen availability than those coming from poorer pastures.

More recently, Elberse & Berendse (1993) have grown eight perennial grass species coming from sites of different nutrient availabilities at two levels of total nutrient availability (unfertilized and fertilized), and measured a number of plant parameters on four occasions during a 16-week period. In this study, they could attribute an N-figure to the eight species, which was between one (one species from nutrient-poor habitat) and seven (three species from relatively rich habitats). As in the experiment by Bradshaw *et al.* (1964), species from nutrient-rich habitats produced substantially more biomass than those from nutrient-poor habitats. Fig. 6.1*A* shows the plasticity of the eight species (measured with the ratios described above) to the shortage of nutrients, in term of biomass produced. The decrease in biomass was tightly related to the N-figure of the species, species from nutrient rich-habitats being more sensitive (ratio closer to 0) to a mineral stress than those from nutrient-poor habitats. Such quantita-

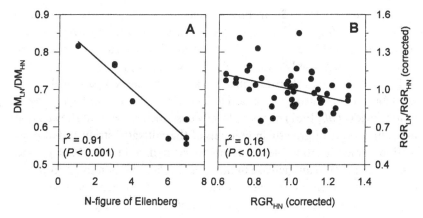

Fig. 6.1. Relationships between (*A*) the N-figure of Ellenberg and the plasticity of dry mass produced (DM_{LN}/DM_{HN}) in response to nutrient availability for eight grass species (data from Elberse & Berendse, 1993); (*B*) the relative growth rate measured under high nutrient supply (RGR_{HN}) and the plasticity of relative growth rate (RGR_{LN}/RGR_{HN}) for the pooled data of NGRASS (both variables are corrected: see Methods section). Solid lines indicate significant relationships whose r^2 and significance levels are given.

tive relationships between N-figure and sensitivity to mineral stress have also been found for species other than grasses, and when nitrogen availability alone was varied (Rorison, 1991; Fichtner & Schulze, 1992).

As a function of the RGR_{max} of the species

A number of studies comparing a small number (i.e. two to five) of grass species have shown that the higher the RGR_{HN} of a species, the larger the difference between RGR_{HN} and the RGR measured under low nutrient availability (RGR_{LN}: e.g. Robinson & Rorison, 1988; Hull & Mooney, 1990; Muller & Garnier, 1990; Poorter *et al.*, 1995). Does this still hold when plasticity is expressed in relative terms (see section on 'Assessing plasticity' above)? This question was addressed using NGRASS, for which the corrected RGR_{LN}/RGR_{HN} ratios for all the species in the data set were plotted against their corrected RGR_{HN} (Fig. 6.1*B*). There is a significant ($P<0.01$) negative trend between the two variables, showing that species with a high RGR_{HN} are more sensitive to nutrient stress than those with a low RGR_{HN}.

A similar relationship between the relative sensitivity to stress and RGR_{max} has also been found in a study comparing the growth of 28 species of emergent macrophytes grown at two levels of nutrient availability (range in RGR_{max}: 0.06–0.26 g g^{-1} d^{-1}), both monocots and dicots (Shipley &

Keddy, 1988), but for 34 tropical species of trees differing five-fold in RGR_{max} (0.03–0.15 g g^{-1} d^{-1}), this trend was significant only when plasticity was expressed as a *difference* between the RGRs measured under high and low nitrogen supply (Huante, Rincón & Acosta, 1995).

To our knowledge, there are only two studies where comparable data are available at the intraspecific level. The first one on *Lolium perenne* (Antonovics *et al.*, 1967) has been briefly described above, and showed that in the first stages of growth, individuals of the cultivated strain and of populations coming from the bird sanctuary had a higher RGR_{HN} and responded more to an increase in nitrogen availability than those from the poorer pastures. In the second one (Kik *et al.*, 1991), genotypes of three populations of the perennial *Agrostis stolonifera* coming from sites of different fertilities were grown under two nitrogen levels in a glasshouse. It was found that (a) RGR_{HN} was higher in genotypes coming from a meadow (nutrient-rich habitat) than in those coming from sand dunes (nutrient-poor habitat), with intermediate values in genotypes coming from a moderately rich polder, and (b) the ratio of RGR_{LN} to RGR_{HN} was lower (higher plasticity) for the meadow population than for the sand dune one, with intermediate values for the polder population.

From the data reviewed in these two sections, it can thus be concluded that grass species – or populations of a species – coming from nutrient-rich sites (and/or fast-growing) are more plastic in terms of biomass production and relative growth rate than species or populations coming from nutrient-poor sites (and/or slow-growing), in response to a decrease in nitrogen availability. The following section will examine the plasticity of mean plant nitrogen concentration in such different species.

Plasticity of internal plant nitrogen concentration

Plant nitrogen under high nutrient supply

First of all, is there evidence that tissue nitrogen concentration is higher in species from nutrient-poor habitats than in those from nutrient-rich habitats when plants are grown under high nutrient supply? The relationship between RGR_{max} and PNC_{max} has been drawn using data from two studies that have examined interspecific differences in these characters for 11 grass species coming from habitats differing in fertility (Poorter, Remkes & Lambers, 1990) and for 14 grass species differing in maximum relative growth rate (Garnier & Vancaeyzeele, 1994). Each of these two studies has been conducted under non-limiting nitrogen supply, and both

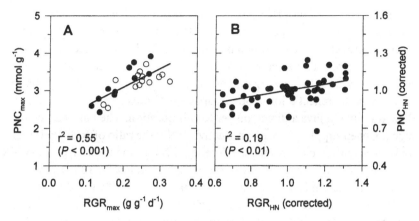

Fig. 6.2. Relationship between (*A*) the relative growth rate and plant nitrogen concentration measured for plants grown under non-limiting nutrient conditions (RGR_{max} and PNC_{max}, respectively), for the 11 grass species studied by Poorter *et al.* (1990: ●) and the 14 grass species studied by Garnier & Vancaeyzeele (1994: ○). Both regressions are significant at $P < 0.01$, but only statistics for the pooled data are given; (*B*) relative growth rate and plant nitrogen concentration both measured under high nutrient supply (RGR_{HN} and PNC_{HN} respectively) for the pooled data of the NGRASS data set (both variables are corrected: see Methods section). Solid lines indicate significant relationships whose r^2 and significance levels are given.

show a significant positive relationship between RGR_{max} and PNC_{max} (Fig. 6.2*A*).

Three points should be made concerning these PNC_{max} values. Firstly, nitrogen concentration is not uniform among the different organs of the plant, aboveground organs having a higher nitrogen concentration than belowground ones (e.g. Poorter *et al.*, 1990; van de Vijver *et al.*, 1993; Garnier & Vancaeyzeele, 1994). Thus, a high PNC_{max} may simply reflect a high allocation to aboveground organs, but not a difference in nitrogen concentration *per se*. This is not likely to be the case, since in the two screening experiments on grasses used to draw Fig. 6.2*A*, there was no relationship between RGR_{max} and any biomass allocation parameter, while there was a positive relationship between RGR_{max} and *leaf* nitrogen concentration (the relationship with root nitrogen concentration was positive and significant only for the 11 grasses of the experiment by Poorter *et al.* 1990). Secondly, when grown under high nitrogen supply, fast-growing plants tend to have higher inorganic (nitrate) nitrogen concentrations than slow-growing ones (Poorter & Bergkotte, 1992; Gobin *et al.*, submitted). Therefore, different PNC_{max} could be reached with different nitrate concentrations but with similar organic (reduced) nitrogen concentrations. Again, this does not seem to be the case, since a positive relationship between RGR_{max} and *organic* nitrogen concentration has been found in the two screening experiments described above (Poorter & Bergkotte, 1992; Gobin *et al.*, submitted). Thirdly, does the low organic-nitrogen concentration of slow-growing plants actually reflect a low storage capacity? There are presently very few data available to give a direct answer to this question. The fact that, under high nitrogen supply, nitrogen productivity (NP: the ratio of RGR to PNC) and photosynthetic nitrogen use efficiency (PNUE: the ratio of photosynthetic rate to leaf nitrogen concentration) are both lower in slow-growing grasses than that in fast-growing ones (Poorter *et al.*, 1990; van der Werf *et al.*, 1993*b*; Garnier & Vancaeyzeele, 1994 for NP; Poorter *et al.*, 1990; Pons, van der Werf & Lambers, 1994; Garnier, Gobin & Poorter, 1995 for PNUE) suggest that slow-growing species may invest more nitrogen in compounds not associated with immediate growth of carbon fixation. Whether this fraction is involved in storage or other compounds (e.g. nucleic acids, molecules involved in protection against herbivores) is currently unknown.

Studies selected in the NGRASS data set mainly provide us only with PNC values, which makes a detailed analysis of nitrogen allocation among organs or compounds not possible. As far as PNC is concerned, NGRASS studies show a similar trend to that presented in Fig. 6.2*A*. Using corrected values of RGR and PNC from the highest nitrogen supply of each experiment, it can actually be shown that under high nitrogen supply, fast-growing species tend to have a significantly higher plant nitrogen concentration than slow-growing species (Fig. 6.2*B*).

We can thus conclude that data from NGRASS and recent available evidence suggest that slow-growing species have a *lower* (organic) nitrogen concentration than fast-growing species when grown under high nutrient supply. Although this appears contrary to Chapin's (1980, 1988) statements, the discussion presented above shows that there is currently no evidence that a low nitrogen concentration actually reflects a low storage capacity. Whatever the case may be, one of the premises to infer that the plasticity of plant nutrient concentration is higher in slow-growing species than in fast-growing ones is not verified. What then happens to this latter prediction?

Plasticity of internal concentration

Fig. 6.3*A* shows how the mean internal plant concentration varies as a function of external nitrogen availability for a slow-growing (*Danthonia*

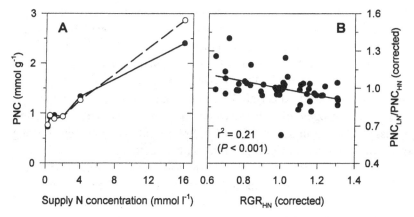

Fig. 6.3. (*A*) influence of external nitrogen concentration in the supply solution on mean plant nitrogen concentration (PNC) for *Danthonia richardsonii* (●) and *Phalaris aquatica* (○) in the experiment by E. Garnier *et al.* (unpublished); (*B*) relationship between the relative growth rate measured under high nutrient supply (RGR_{HN}) and the plasticity of plant nitrogen concentration (PNC_{LN}/PNC_{HN}) for the pooled data of NGRASS (both variables are corrected: see Methods section). A solid line indicates a significant relationship whose r^2 and significance level are given.

richardsonii) and a fast-growing (*Phalaris aquatica*) perennial species (E. Garnier *et al.*, unpublished: Table 6.1). Starting at low levels of nitrogen supply at which there is not much difference between the two species (0.73 vs 0.77 mmol g^{-1}), PNC progressively increases in a similar way in both species up to an external supply of 4 mmol N l^{-1}, but a further increase to 16 mmol N l^{-1} leads to a larger augmentation in PNC in the fast-growing than in the slow-growing species. The former thus ends up with a higher PNC (2.87 vs 2.40 mmol g^{-1}), in accordance with the conclusion of the preceding paragraph. The ratio of PNC_{LN} to PNC_{HN} is slightly lower (higher plasticity) in the fast-growing (0.27) than in the slow-growing (0.30) species. Such an increase in PNC with increasing external nitrogen availability is well documented (Larsson, 1994) and is the consequence of both a decrease in the proportional biomass allocation to roots (see below) and an increase in nitrogen concentration of all organs, particularly of leaves (e.g. Hull & Mooney, 1990; Mattsson *et al.*, 1991; van de Vijver *et al.*, 1993). Does this trend also show when the whole NGRASS data set is analysed? Fig. 6.3*B* shows that when the corrected PNC_{LN}/PNC_{HN} ratio is plotted as a function of the corrected RGR_{HN}, there is a significant ($P<0.001$) negative trend between the two variables. This means that the plasticity in PNC is higher in fast-growing than in slow-growing species, again at variance with the conclusions of Chapin's reviews. However, which fraction of plant nitrogen is affected when there is a change in external nitrogen availability is currently not well known. Nitrogen productivity and/or photosynthetic nitrogen use efficiency both tend to increase as internal nitrogen decreases (Boot, Schildwacht & Lambers, 1992; van der Werf et al., 1993b, c; Pons et al., 1994) at least down to a certain limit, pointing to an increase in the proportion of nitrogen directly involved in growth and carbon fixation processes. But to our knowledge, there are no data conclusively showing which metabolic form of nitrogen is actually affected.

Finally, if we combine the findings from this section with those from the previous one, we end up with the following conclusion: when external nitrogen availability is varied, species from nutrient-poor habitats (and/or slow growing species) exhibit *less* plasticity than those from nutrient-rich habitats (and/or fast-growing species), both in terms of relative growth rate and internal plant nitrogen concentration. I will show below that these observations are actually closely related.

Relationship between plant growth and internal nitrogen

Over the last two decades, Ingestad and co-workers have developed a body of theory formalizing the influence of nutrient supply, and in particular nitrogen, on plant growth (e.g. Ingestad, 1979; Ågren, 1985; Ingestad & Ågren; 1992; Wikström & Ågren, 1995). The analysis of a large number of experiments where external nitrogen supply was varied yielded the following conclusions: (a) there is a minimum nitrogen concentration in the plant required for growth (PNC_{min}); (b) there is an optimum nutrient concentration (PNC_{opt}) above which no further increase in growth rate occurs; (c) between these two concentrations, relative growth rate increases linearly with PNC. I examine below how this framework can be used to compare the plasticity of response to nitrogen among species described in the previous paragraphs.

To establish the response curves to nitrogen described above, a given species (or genotype) has to be grown at several levels of external nitrogen availabilities over a large enough range, which was actually done in only three studies in NGRASS (Mattsson et al., 1991; van der Werf et al., 1993b; E. Garnier, M.-L. Navas & R. Gifford, unpublished). The last of these three experiments will be used to illustrate how RGR varies with PNC in a slow-growing (*Danthonia richardsonii*) and a fast-growing (*Phalaris aquatica*) species (Fig. 6.4A). For both species, PNC_{min} is about 0.5 mmol g^{-1} and it appears that at the highest nitrogen supply, PNC values went beyond

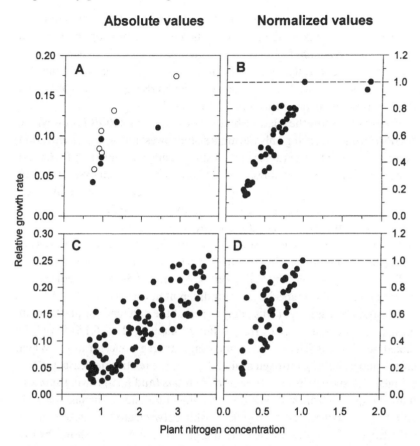

Fig. 6.4. Relationships between plant nitrogen concentration and relative growth rate drawn with (*A*) absolute values of both variables for *Danthonia richardsonii* (●) and *Phalaris aquatica* (○) in the experiment by E. Garnier *et al.* (unpublished); (*B*) normalized values of both variables (see text) pooling data from 3 of the NGRASS experiments (Mattsson *et al.*, 1991; van der Werf *et al.*, 1993*c*, and E. Garnier *et al.*, unpublished); (*C*) absolute values and (*D*) normalized values for the pooled data of NGRASS, using data from the highest and lowest nutrient treatment in each experiment. In (*A*) and (*C*), plant nitrogen concentration is in mmol g^{-1} and relative growth rate is in g g^{-1} d^{-1}.

PNC$_{opt}$. This was particularly obvious for *D. richardsonii*, for which RGR reached a maximum of 0.117 g g^{-1} d^{-1} at PNC$_{opt}$=1.3 mmol g^{-1}, with a slight decrease for higher values of PNC. For *P. aquatica*, the transition is less sharp, and PNC$_{opt}$ was estimated to be 1.6 mmol g^{-1}, for a maximum RGR of 0.174 g g^{-1} d^{-1}.

Ingestad & Ågren (1992) and Wikström & Ågren (1995) argued that the best way to express the relationship between RGR and PNC was to do so in relative values, taking the reference when nutrition was optimum (i.e. at PNC_{opt} and maximum RGR, both parameters are set to 1). When this normalization is done, differences between species and growing conditions disappear, which is indeed verified for the three NGRASS studies conducted at more than three nitrogen supplies (Fig. 6.4B).

Fig. 6.4C, D show the relationships between PNC and RGR for the whole NGRASS data set, using only the highest and lowest nitrogen concentration and relative growth rate values in each study, using absolute (Fig. 6.4C) and normalized (Fig. 6.4D) values. A major problem in the analysis is that in more than 75% of the studies, only two or three levels of external nitrogen supply were applied, which makes it impossible to define PNC_{opt}. Therefore, the normalization was done with reference to the internal nitrogen concentration measured at the highest external nitrogen supply. Given this uncertainty, the quality of the relationship shown in Fig. 6.4D compares relatively well with that published by Wikström & Ågren (1995) for 18 studies concerning mainly woody species (and for different nutrients).

The main conclusion of this analysis is that the difference of plasticity in growth rate between species from different habitats (Fig. 6.1B) is tightly linked to the plasticity of their internal nitrogen concentration. When grown under limiting nitrogen availability: (a) the relative decrease in PNC is higher for species from nutrient-rich habitats (and fast-growing species) than for species from nutrient-poor habitats (and slow-growing species), and (b) because there is a strong relationship between growth rate and internal nitrogen both at the interspecific and phenotypic levels (as shown by the relationship between normalized values of PNC and RGR: Fig. 6.4B, D), this higher plasticity in PNC leads to a higher plasticity in RGR.

In the next section, I will use a similar framework to analyse the differences in plasticity of biomass allocation to roots between species.

Plasticity in proportional allocation of biomass to roots

In the experiment by Bradshaw et al. (1964) described earlier, the extent to which the root:shoot ratio varies in response to the nitrogen gradient differs greatly among species: the response for five of the species coming from contrasting habitats is roughly similar, and more interestingly, the plasticity of this character for *Agrostis stolonifera* (coming from a fertile site) is very similar to that of *Nardus stricta* (coming from an infertile site) and is much higher than for the other five species.

In the experiment by Kik et al. (1991) on *Agrostis stolonifera*, no difference in plasticity of the root:shoot ratio in response to nitrogen supply was observed among the three populations coming from sites of differing nutrient availabilities. Comparison of results among NGRASS experiments yields a somewhat confused picture, which can be clarified if the dependency of biomass allocation pattern on plant nitrogen concentration is taken into account.

Fig. 6.5*A* shows the relationship between PNC and the proportional biomass allocation to roots (RMR) for *D. richardsonii* and *P. aquatica* in the same experiment as above (E. Garnier *et al.*, unpublished). For both species, the relationship between the two variables is negative and clearly curvilinear, with a steep, almost linear, decline in root mass ratio with increasing plant nitrogen at low values of PNC and a smoother slope at high values of PNC. This curvilinear shape has been found in several experiments, and is also predicted from theoretical models (Ågren & Ingestad, 1987; Levin *et al.*, 1989; Ingestad & Ågren, 1991; van der Werf *et al.*, 1993*a*). The plot for the whole NGRASS data set – although somewhat noisy – gives the same overall picture (Fig. 6.5*C*), with RMR values decreasing with increasing PNC up to roughly 2.5 mmol g^{-1}, and no further change for higher values of PNC.

Following the same normalization procedure as for relative growth rate cannot produce a similar homogenization of data (Fig. 6.5*B*, *D*), because of this curvilinear trend. This means that a given decrease in PNC will have a lower impact on RMR for species (and experiments!) with a high PNC$_{HN}$ (i.e. fast-growing ones) than for those with a low PNC$_{HN}$ (i.e. slow-growing ones). This is shown in Fig. 6.6*A*, where the sensitivity of the root mass ratio on internal plant nitrogen has been plotted against the RGR$_{HN}$ of the species. As a consequence, and although the plasticity in PNC is higher in fast-growing species, the analysis of NGRASS shows that there is no significant difference in RMR plasticity between fast- and slow-growing species (Fig. 6.6*B*).

This corroborates the conclusions of a recent review based on 46 studies comparing the plasticity of root:shoot ratio in approximately 100 species (both herbaceous and woody), which did not reveal any difference in plasticity among species originating from habitats of different fertility or differing in relative growth rate (Reynolds & D'Antonio, 1996).

Synthesis and discussion

Based on the present review, a scheme synthesizing how a variation in external nitrogen supply affects plant nitrogen concentration, relative growth

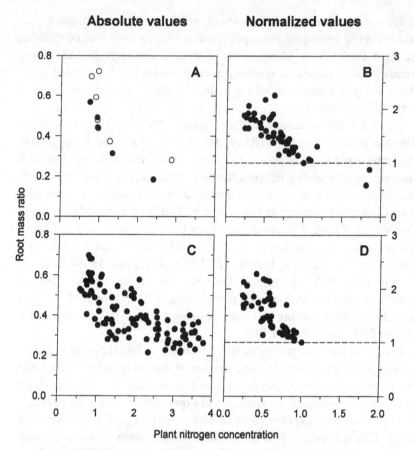

Fig. 6.5. Relationships between plant nitrogen concentration and root mass ratio
(the ratio of root mass to total plant mass) drawn with (*A*) absolute values of
both variables for *Danthonia richardsonii* (●) and *Phalaris aquatica* (○) in the
experiment by E. Garnier *et al.* (unpublished); (*B*) normalized values of both
variables (see text) pooling data from four of the NGRASS experiments
(Mattsson *et al.*, 1991; van der Werf *et al.*, 1993c; van de Vijver *et al.*, 1993, and
E. Garnier *et al.*, unpublished); (*C*) absolute values and (*D*) normalized values for
the pooled data of NGRASS, using data from the highest and lowest nutrient
treatment in each experiment. In (*A*) and (*C*), plant nitrogen concentration is in
mmol g^{-1} and root mass ratio is in g g^{-1}.

rate and biomass allocation to roots in fast- and slow-growing species is
presented (Fig. 6.7): data suggest that at high nitrogen supply, plant nitrogen concentration is higher in fast- than in slow-growing species, and that
the drop in PNC in response to a decrease in external supply is larger in the
former (Fig. 6.7*A*). Therefore, because any drop in internal nitrogen concentration produces a decrease in relative growth rate which is identical for

Responses of grasses to nitrogen 175

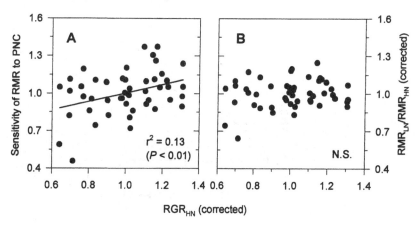

Fig. 6.6. – Relationships between the relative growth rate measured under high nutrient supply (RGR$_{HN}$) and (*A*) the sensitivity of root mass ratio (RMR) in response to a decrease in plant nitrogen concentration (PNC), measured as the RMR$_{LN}$/RMR$_{HN}$ to PNC$_{LN}$/PNC$_{HN}$ ratio; (*B*) the plasticity of root mass ratio (RMR$_{LN}$/RMR$_{HN}$). Both figures are for the pooled data of NGRASS (all variables are corrected: see Methods section). A solid line indicates a significant relationship whose r^2 and significance level are given.

fast- and slow-growing species (Fig. 6.7*B*), RGR is also more plastic in fast-growing plants (Fig. 6.7*C*). The effect of a decrease in PNC on RMR differs between species because: (a) the relationship between the two variables is curvilinear, RMR increasing slowly with decreasing PNC at high nitrogen concentrations, and more sharply at lower concentrations (Fig. 6.7*D*), and (b) PNC$_{HN}$ is higher in fast-growing species, which suggests that their mean sensitivity to a decrease in PNC is lower. As a consequence, the larger drop in PNC in fast-growing species does not induce a larger increase in RMR. Fast- and slow-growing species thus display the same plasticity in biomass allocation in response to nitrogen supply (Fig. 6.7*E*).

The general reservation concerning meta-analysis put forward by Gurevitch & Hedges (1993) undoubtedly applies here. Indeed the approach used will inevitably uncover only general mechanisms having their origin in general physiological processes depending on internal plant nitrogen. However, plants may respond to a change in nitrogen supply by some other, more specific mechanisms. For example, it may be argued that plasticity in the morphology and/or physiology of the root system itself rather than its size relative to that of the whole plant is of importance to modify the uptake capacity of a plant. This has been examined in a number of studies, where the change in specific root length (the root length to root biomass ratio) in response to a decrease in nitrogen supply has been monitored in

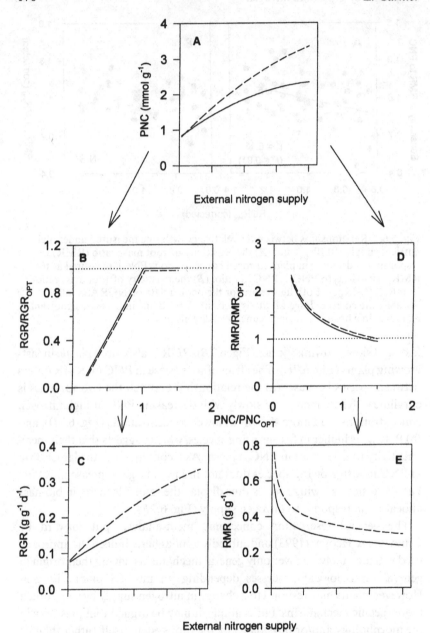

Fig. 6.7. Summary of the effects of nitrogen supply on internal plant nitrogen (PNC), relative growth rate (RGR) and proportional biomass allocation to roots (RMR) for a slow-growing (——) and a fast-growing (– – –) species. Subscript OPT refers to the value a trait recorded for plants grown under non-limiting nutrient conditions (see text for further explanation).

grasses. With the same five species as those described in Table 6.1, Boot & Mensink (1990) did not find any species × nitrogen interaction for this trait; in the study by Robinson & Rorison (1988, Table 6.1), the least plastic species was found to be the fastest growing one, while the reverse was true in that by E. Garnier & S. Gaslonde (unpublished results, Table 6.1). Only in the experiment by Ryser & Lambers (1995) was it found that the fast-growing species was the more plastic. Too few data are available yet to know whether this trait can be amenable to the same analysis as that presented above for the root mass ratio, but some data suggest that specific root length may depend on internal nitrogen concentration as well (e.g. Ryser & Lambers, 1995 for *Dactylis glomerata*). Modification of other traits (e.g. root branching or proliferation of root hairs) in response to nutrient supply have been touched upon in recent reviews by Lambers & Poorter (1992), Hutchings & de Kroon (1994) and Reynolds & D'Antonio (1996).

Another point worth noting is that the analysis conducted here has been done for studies assuming steady-state plant functioning, while some of the interspecific differences in plasticity may only become apparent when nutrients are delivered as brief pulses of enrichment (Crick & Grime, 1987).

Conclusion

Of the predictions presented in the introduction, only that concerning the plasticity of growth is verified: fast-growing grasses are more plastic than slow-growing ones in terms of growth, but contrary to what was stated, they have a higher internal nitrogen concentration under high nitrogen supply, and they are also more plastic for this trait; no difference in plasticity of the root mass ratio was found among species. The analysis presented shows that the plasticity of growth and biomass allocation can be predicted from the variation in internal nitrogen induced by the change in external nitrogen supply. Ultimately, slow-growing species are less plastic in internal nitrogen because their concentration is lower at high nitrogen supply. What sets the limit for this upper nitrogen concentration a given species can reach? Preliminary studies on grasses show that this may well be linked to anatomical features, particularly of leaves (Garnier & Laurent, 1994; van Arendonk & Poorter, 1994).

Acknowledgements

I thank David Robinson and Adrie van der Werf for giving me access to the raw data from their experiments, and Heather Reynolds for sending me a

preprint of her paper. Mike Austin, Hendrik Poorter, David Robinson, Bill
Shipley and Adrie van der Werf are warmly thanked for their comments on
the manuscript. Unpublished work on *Danthonia* and *Phalaris* was carried
out in collaboration with Marie-Laure Navas and Julianne Lilley during a
sabbatical leave in Canberra (C.S.I.R.O., Division of Plant Industry) and
was partially funded by grant n. 93/6564 from NGAC to Roger Gifford.

References

Aerts, R. & van der Peijl, M. J. (1993). A simple model to explain the dominance
    of low-productive perennials in nutrient-poor habitats. *Oikos*, **66**, 144–7.
Ågren, G. I. (1985). Theory for growth of plants derived from the nitrogen
    productivity concept. *Physiologia Plantarum*, **64**, 17–28.
Ågren, G. I. & Ingestad, T. (1987). Root:shoot ratio as a balance between
    nitrogen productivity and photosynthesis. *Plant, Cell and Environment*, **10**,
    579–86.
Antonovics, J., Lovett, J. & Bradshaw, A. D. (1967). The evolution of adaptation
    to nutritional factors in populations of herbage plants. In *Isotopes in Plant
    Nutrition and Physiology*, pp. 549–67. Vienna: International Atomic Energy
    Agency.
Berendse, F. & Elberse, W. T. (1990). Competition and nutrient availability in
    heathland and grassland ecosystems. In *Perspectives on Plant Competition*,
    ed. J. B. Grace & D. Tilman, pp. 93–116. San Diego: Academic Press.
Bloom, A. J., Chapin, F. S., III & Mooney, H. A. (1985). Resource limitation in
    plants. An economic analogy. *Annual Review of Ecology and Systematics*, **16**,
    363–92.
Boot, R. G. A. & den Dubbelden, K. C. (1990). Effects of nitrogen supply on
    growth, allocation and gas exchange characteristics of two perennial grasses
    from inland dunes. *Oecologia (Berlin)*, **85**, 115–21.
Boot, R. G. A. & Mensink, M. (1990). Size and morphology of root systems of
    perennial grasses from contrasting habitats as affected by nitrogen supply.
    Plant and Soil, **129**, 291–9.
Boot, R. G. A. & Mensink, M. (1991). The influence of nitrogen availability on
    growth parameters of fast- and slow-growing perennial species. In *Plant Root
    Growth. An Ecological Perspective*, ed. D. Atkinson, pp. 161–8. Oxford:
    Blackwell Scientific Publications.
Boot, R. G. A., Schildwacht, P. M. & Lambers, H. (1992). Partitioning of
    nitrogen and biomass at a range of N-addition rates and their consequences
    for growth and gas exchange in two perennial grasses from inland dunes.
    Physiologia Plantarum, **86**, 152–60.
Bradshaw, A. D., Chadwick, M. J., Jowett, D. & Snaydon, R. W. (1964).
    Experimental investigations into the mineral nutrition of several grass
    species. IV. Nitrogen level. *Journal of Ecology*, **52**, 665–77.
Brouwer, R. (1963). Some aspects of the equilibrium between overground and
    underground plant parts. *Jaarboek Instituut voor Biologisch en Scheikundis
    Onderzoek van Landbouwgewassen, Wageningen*, **213**, 31–9.
Chapin, F. S., III (1980). The mineral nutrition of wild plants. *Annual Review of
    Ecology and Systematics*, **11**, 233–60.
Chapin, F. S., III (1988). Ecological aspects of plant mineral nutrition. In

Advances in Plant Nutrition, vol. 3, ed. B. Tinker & A. Läuchli, pp. 161–91. New-York: Praeger.

Cheplick, G. P. (1991). A conceptual framework for the analysis of phenotypic plasticity and genetic constraints in plants. *Oikos*, **62**, 283–91.

Cheplick, G. P. (1995). Genotypic variation and plasticity of clonal growth in relation to nutrient availability in *Amphibromus scabrivalvis*. *Journal of Ecology*, **83**, 459–68.

Crick, J. C. & Grime, J. P. (1987). Morphological plasticity and mineral nutrient capture in two herbaceous species of contrasted ecology. *New Phytologist*, **107**, 403–14.

Davidson, R. L. (1969). Effect of root/leaf temperature differentials on root/shoot ratios in some pasture grasses and clover. *Annals of Botany*, **33**, 561–9.

Elberse, W. T. H. & Berendse, F. (1993). A comparative study of the growth and morphology of eight grass species from habitats with different nutrient availabilities. *Functional Ecology*, **7**, 223–9.

Ellenberg, H. (1979). Zeigerwerte der Gefäßpflanzen Mitteleuropas. *Scripta Geobotanica*, IX (2nd edition).

Ellenberg, H. (1988). *Vegetation Ecology of Central Europe*. Cambridge: Cambridge University Press.

Fichtner, K. & Schulze, E.-D. (1992). The effect of nitrogen nutrition on growth and biomass partitioning of annual plants originating from habitats of different nitrogen availability. *Oecologia (Berlin)*, **92**, 236–41.

Garnier, E. (1991). Resource capture, biomass allocation and growth in herbaceous plants. *Trends in Ecology and Evolution*, **6**, 126–31.

Garnier, E., Gobin, O. & Poorter, H. (1995). Nitrogen productivity depends on photosynthetic nitrogen use efficiency and on nitrogen allocation within the plant. *Annals of Botany*, **76**, 667–72.

Garnier, E., Koch, G. W., Roy, J. & Mooney, H. A. (1989). Responses of wild plants to nitrate availability. Relationships between growth rate and nitrate uptake parameters, a case study with two *Bromus* species and a survey. *Oecologia (Berlin)*, **79**, 542–50.

Garnier, E. & Laurent, G. (1994). Leaf anatomy, specific mass and water content in congeneric annual and perennial grass species. *New Phytologist*, **128**, 725–36.

Garnier, E. & Vancaeyzeele, S. (1994). Carbon and nitrogen content of congeneric annual and perennial grass species: relationships with growth. *Plant, Cell and Environment*, **17**, 399–407.

Gobin, O., Garnier, E., Fabreguettes, J. & Jardon, F. Influence of internal nitrogen on relative growth rate, components of carbon balance and nitrate uptake in 10 grass species. *Environmental and Experimental Botany*, submitted.

Grime, J. P. (1979). *Plant Strategies and Vegetation Processes*. Chichester: John Wiley & Sons.

Grime, J. P., Hodgson, J. G. & Hunt, R. (1988). *Comparative Plant Ecology*. London: Unwin Hyman.

Grime, J. P. & Hunt, R. (1975). Relative growth-rate: its range and adaptive significance in a local flora. *Journal of Ecology*, **63**, 393–422.

Gurevitch, J. & Hedges, L. V. (1993). Meta-analysis: combining the results of independent experiments. In *Design and Analysis of Ecological Experiments*, ed. S. M. Scheiner & J. Gurevitch, pp. 378–98. New-York: Chapman & Hall.

Harlan, J. R. (1982). Human interference with grass systematics. In *Grasses and Grasslands. Systematics and Ecology*, ed. J. R. Estes, R. J. Tyrl & J. N. Brunken, pp. 37–50. Norman: University of Oklahoma Press.

Huante, P., Rincón, E. & Acosta, I. (1995). Nutrient availability and growth rate of 34 woody species from a tropical deciduous forest in Mexico. *Functional Ecology*, **9**, 849–58.

Hull, J. C. & Mooney, H. A. (1990). Effects of nitrogen on photosynthesis and growth rates of four California annual species. *Acta Oecologica*, **11**, 453–68.

Hutchings, M. J. & de Kroon, H. (1994). Foraging in plants: the role of morphological plasticity in resource acquisition. *Advances in Ecological Research*, **25**, 159–238.

Ingestad, T. (1979). Nitrogen stress in birch seedlings. II. N, K, P, Ca and Mg nutrition. *Physiologia Plantarum*, **45**, 149–57.

Ingestad, T. & Ågren, G. I. (1991). The influence of plant nutrition on biomass allocation. *Ecological Applications*, **1**, 168–74.

Ingestad, T. & Ågren, G. I. (1992). Theories and methods on plant nutrition and growth. *Physiologia Plantarum*, **84**, 177–84.

Kachi, N. & Rorison, I. H. (1989). Optimal partitioning between root and shoot in plants with contrasted growth rates in response to nitrogen availability and temperature. *Functional Ecology*, **3**, 549–59.

Kachi, N. & Rorison, I. H. (1991). Root and shoot activity of two grasses with contrasted growth rates in response to low nutrient availability and temperature. In *Plant Root Growth. An Ecological Perspective*, ed. D. Atkinson, pp. 147–59. Oxford: Blackwell Scientific Publications.

Kik, C., Jongman, M. & van Andel, J. (1991). Variation in relative growth rate and survival in ecologically contrasting populations of *Agrostis stolonifera*. *Plant Species Biology*, **6**, 47–54.

Lambers, H. & Dijkstra, P. (1987). A physiological analysis of genotypic variation in relative growth rate: Can growth rate confer ecological advantage? In *Disturbance in Grasslands. Causes, Effects and Processes*, ed. J. van Andel, J. P. Bakker & R. W. Snaydon, pp. 237–52. Dordrecht: Dr W. Junk Publishers.

Lambers, H. & Poorter, H. (1992). Inherent variation in growth rate between higher plants: A search for physiological causes and ecological consequences. *Advances in Ecological Research*, **23**, 187–261.

Larsson, C.-M. (1994). Responses of the nitrate uptake system to external nitrate availability: a whole plant perspective. In *A Whole Plant Perspective on Carbon–Nitrogen Interactions*, ed. J. Roy & E. Garnier, pp. 31–45. The Hague: SPB Academic Publishing bv.

Levin, S. A., Mooney, H. A. & Field, C. (1989). The dependence of plant root:shoot ratios on internal nitrogen concentration. *Annals of Botany*, **64**, 71–5.

Mattsson, M., Johansson, E., Lundborg, T., Larsson, M. & Larsson, C.-M. (1991). Nitrogen utilization in N-limited barley during vegetative and generative growth. I. Growth and nitrate uptake kinetics in vegetative cultures grown at different relative addition rates on nitrate-N. *Journal of Experimental Botany*, **42**, 197–205.

Muller, B. & Garnier, E. (1990). Components of relative growth rate and sensitivity to nitrogen availability in annual and perennial species of *Bromus*. *Oecologia (Berlin)*, **84**, 513–8.

Pavlik, B. M. (1983). Nutrient and productivity relations of the dune grasses *Ammophila arenaria* and *Elymus mollis*. II. Growth and patterns of dry matter and nitrogen allocation as influenced by nitrogen supply. *Oecologia (Berlin)*, **57**, 233–8.

Pons, T. L., van der Werf, A. & Lambers, H. (1994). Photosynthetic nitrogen use efficiency of inherently slow- and fast-growing species: possible explanations for observed differences. In *A Whole Plant Perspective on Carbon–Nitrogen Interactions*, ed. J. Roy & E. Garnier, pp. 61–77. The Hague: SPB Academic Publishing bv.

Poorter, H. & Bergkotte, M. (1992). Chemical composition of 24 wild species differing in relative growth rate. *Plant, Cell and Environment*, **15**, 221–9.

Poorter, H. & Remkes, C. (1990). Leaf area ratio and net assimilation rate of 24 wild species differing in relative growth rate. *Oecologia (Berlin)*, **83**, 553–9.

Poorter, H., Remkes, C. & Lambers, H. (1990). Carbon and nitrogen economy of 24 wild species differing in relative growth rate. *Plant Physiology*, **94**, 621–7.

Poorter, H., van de Vijver, C. A. D. M., Boot, R. G. A. & Lambers, H. (1995). Growth and carbon economy of a fast-growing and a slow-growing grass species as dependent on nitrate supply. *Plant and Soil*, **171**, 217–27.

Reynolds, H. L. & D'Antonio, C. (1996). The ecological significance of plasticity in root weight ratio in response to nitrogen. *Plant and Soil* **185**, 75–97.

Robinson, D. & Rorison, I. H. (1988). Plasticity in grass species in relation to nitrogen supply. *Functional Ecology*, **2**, 249–57.

Rorison, I. H. (1991). Ecophysiological aspects of nutrition. In *Plant Growth. Interactions with Nutrition and Environment*, ed. J. R. Porter & D. W. Lawlor, pp. 157–76. Cambridge: Cambridge University Press.

Ryser, P. & Lambers, H. (1995). Root and leaf attributes accounting for the performance of fast- and slow-growing grasses at different nutrient supply. *Plant and Soil*, **170**, 251–65.

Schlichting, C. D. (1986). The evolution of phenotypic plasticity in plants. *Annual Review of Ecology and Systematics*, **17**, 667–93.

Shipley, B. & Keddy, P. A. (1988). The relationship between relative growth rate and sensitivity to nutrient stress in twenty-eight species of emergent macrophytes. *Journal of Ecology*, **76**, 1101–10.

Thompson, K., Hodgson, J. G., Grime, J. P., Rorison, I. H., Band, S. R. & Spencer, R. E. (1993). Ellenberg numbers revisited. *Phytocoenologia*, **23**, 277–89.

van Arendonk, J. J. C. M. & Poorter, H. (1994). The chemical composition and anatomical structure of leaves of grass species differing in relative growth rate. *Plant, Cell and Environment*, **17**, 963–70.

van der Werf, A., Enserink, T., Smit, B. & Booij, R. (1993*a*). Allocation of carbon and nitrogen as a function of the internal nitrogen status of a plant: modelling allocation under non-steady-state situations. *Plant and Soil*, **155/156**, 183–6.

van der Werf, A., van Nuenen, M., Visser, A. J. & Lambers, H. (1993*b*). Contribution of physiological and morphological plant traits to a species' competitive ability at high or low nitrogen supply. A hypothesis for inherently fast- and slow-growing species. *Oecologia (Berlin)*, **94**, 434–40.

van der Werf, A., Van Nuenen, M., Visser, A. J. & Lambers, H. (1993c). Effects of N-supply on the rates of photosynthesis and shoot and root respiration of inherently fast- and slow-growing monocotyledonous species. *Physiologia Plantarum*, **89**, 563–9.

van der Werf, A., Visser, A. J., Schieving, F. & Lambers, H. (1993*d*). Evidence for optimal partitioning of biomass and nitrogen at a range of nitrogen availablities for a fast- and slow-growing species. *Functional Ecology*, **7**, 63–74.

van de Vijver, C. A. D. M., Boot, R. G. A., Poorter, H. & Lambers, H. (1993).

Phenotypic plasticity in response to nitrate supply of an inherently fast-growing species from fertile habitat and an inherently slow-growing species from an infertile habitat. *Oecologia (Berlin)*, **96**, 548–54.

Wikström, F. & Ågren, G. I. (1995). The relationship between the growth rate of young plants and their total-N concentration is unique and simple: a comment. *Annals of Botany*, **75**, 541–4.

7
Population biology of intraspecific polyploidy in grasses

KATHLEEN H. KEELER

Polyploidy is the duplication of an entire nuclear genome, whether diploid or higher level (Stebbins, 1971; Thompson & Lumaret, 1992) and a frequent occurrence in plants. Stebbins (1971) estimated that 30–35% of flowering plant species are polyploid, and that many more had a polyploid event in their evolutionary history, including all members of such important families as the Magnoliaceae, Salicaceae, and Ericaceae. Goldblatt (1980) estimated 55%, but probably up to 75%, of monocotyledons had at least one polyploid event in their history, using the criterion that if the species has a base number higher than $n=13$ it is derived from a polyploid. Using the same criterion, Grant (1981) estimated that 52% of angiosperms, 49% of dicotyledon species and 60% of monocotyledons are polyploid. Masterson (1994) supports high frequencies of ancestral polyploidy using fossil evidence. Clearly, polyploids have been fixed in many lineages.

Within many genera of higher plants, individual species often have different, but uniform, ploidy levels (e.g. *Draba*, Brassicaceae, Brockman & Elven, 1992), the grasses being no exception, e.g. *Bromus*, *Elymus* (Seberg & von Bothmer, 1991; Ainouche, Misset & Huon, 1995). Intrageneric polyploid series provide another indicator of frequent polyploid events. For example, of a miscellaneous collection of 87 grass genera for which I had chromosome numbers for two or more species, 65 (75%) formed a polyploid series in relation to other members of the genus (Table 7.1).

Stebbins (1947) distinguished the forms of polyploidy based on whether the duplicated genomes are derived from one species (autopolyploidy) or two (allopolyploidy) or both (segmental allopolyploidy). Recent detailed genetic analysis has made it possible to distinguish these based on homology of the genomes: the same genomes in multiple copies (autopolyploidy), or several different genomes in the same individual (allopolyploidy) (Jackson, 1982).

183

Table 7.1. *Polyploid series within some grass genera*

Plants entered only if cytotype is available for two or more species. Taxonomy is that of the author and not necessarily modern.

Genus	Cytotypes of selected members	
Aegilops	28, 56	Hickman, 1993
Agropyron	14, 28, 42, 56	Bowden, 1965
Agrostis	14, 28, 42, 56	Bowden, 1965; Hickman, 1993
Aira	14, 28	Myers, 1947
Alopecurus	14, 28, 56, 70, 98, 112–116	Myers, 1947
Ammophila	28	Myers, 1947
Andropogon	20, 30, 40, 45, 50, 60, 80, 120, 180	Myers, 1947; Norrmann & Quarín, 1987
Anthoxanthum	10, 20, 80	Myers, 1947; Hedberg, 1967
Aristida	22, 44	Hickman, 1993
Arrhenatherum	28, 40	Myers, 1947
Arthraxon	36	Myers, 1947
Arundinaria	48, 54	Myers, 1947
Avena	14, 28, 42	Hickman, 1993
Bambusa	68, 72	Myers, 1947
Bouteloua	20, 21, 22, 35, 40, 42	Myers, 1947
Brachypodium	14, 18	Myers, 1947
Briza	10, 14	Myers, 1947
Bromus	14, 28, 42, 56, 70	Myers, 1947; Sutherland, 1986; Ainouche *et al.*, 1995
Calamagrostis	28, 42, 56, 79, 84	Myers, 1947; Sutherland, 1986
Calamovilfa	40	Sutherland, 1986
Catabrossa	20	Sutherland, 1986
Cenchrus	34, 70	Myers, 1947; Sutherland, 1986
Chimonobambusa	48	Myers, 1947
Chloris	20, 30, 40, 50, 80	Myers, 1947
Cinna	28	Myers, 1947
Coix	10, 20	Myers, 1947
Cymbopogon	20, 40	Myers, 1947
Cynodon	18, 36, 54	Sutherland, 1986
Cynosurus	14	Myers, 1947
Dactylis	14, 28, 42	Lumaret, 1988*b*
Dactylocteniaum	34, 48	Myers, 1947
Danthonia	18, 24, 36, 42, 48	Myers, 1947; Hickman, 1993
Deschampsia	14, 26, 28, 42	Myers, 1947; Rothera & Davy, 1986
Dichanthelium	18	Sutherland, 1988
Digitaria	16, 24, 28, 30, 36, 54, 72	Myers, 1947; Sutherland, 1986
Distichlis	40	Myers, 1947
Echinochloa	36, 54, 130	Myers, 1947; Sutherland, 1986
Ehrharta	24, 48	Myers, 1947
Eleusine	18, 36, 45	Myers, 1947
Elymus	14, 28, 42, 56	Myers, 1947
Eragrostis	20, 40, 42, 50, 60, 80, 100, 120	Myers, 1947; Sutherland, 1986
Erianthus	20, 60	Myers, 1947
Euchlaena	20, 40	Myers, 1947
Festuca	14, 28, 42, 56, 63, 70	Seal, 1983
Gastridium	14, 28	Myers, 1947

Table 7.1. (*cont.*)

Plants entered only if cytotype is available for two or more species. Taxonomy is that of the author and not necessarily modern.

Genus	Cytotypes of selected members	
Glyceria	10, 14, 20, 28, 40, 56	Myers, 1947; Hickman, 1993
Hilaria	18, 36	Sutherland, 1986
Holcus	14, 28	Jones, 1958; Richard *et al.*, 1995
Hordeum	14, 28, 42	von Bothmer & Jacobsen, 1986; Kankanpää *et al*, 1996
Hystrix	28, 56	Myers, 1947
Koeleria	14, 28	Myers, 1947
Leersia	48, 96	Myers, 1947
Leptochloa	20, 40	Sutherland, 1986
Lepturus	14, 26, 36	Myers, 1947
Lolium	14, 28	Sutherland, 1986
Melica	18	Myers, 1947
Milium	18, 28	Myers, 1947
Miscanthus	36, 42, 64	Myers, 1947
Muhlenbergia	18, 20, 40, 42, 60, 80	Myers, 1947; Hickman, 1993
Munroa	16	Sutherland, 1947
Oplismenus	54, 72	Myers, 1947
Oryza	24, 48	Myers, 1947
Oryzopsis	22, 24, 46, 48	Sutherland, 1988
Panicum	18, 20, 36, 40, 54, 72, 90, 108	Myers, 1947; Sutherland, 1986
Paspalum	20, 25, 40, 45, 55, 60, 80, 120, 160	Burton, 1942; Myers, 1947; Quarín *et al.*, 1982
Pennisetum	14, 27, 28, 36, 45, 54	Burton, 1942
Phalaris	12, 14, 28, 42	Myers, 1947
Phippsia	28	Myers, 1947
Phleioblastus	48	Myers, 1947
Phleum	14, 28, 42	Myers, 1947
Phyllostachys	48, 54	Myers, 1947
Poa	14, 28, 35, 42, 54, 56, 62, 64, 70, 76, 84, 106	Myers, 1947; Sutherland, 1986
Polypogon	14, 28, 42	Hickman, 1993
Puccinellia	14, 28, 42, 56	Myers, 1947; Sutherland, 1986
Saccharum	40, 80, 112	Myers, 1947
Sasa	48	Myers, 1947
Setaria	18, 36, 54, 72	Sutherland, 1986
Sitanion	28	Myers, 1947
Sorghastrum	20, 40	Myers, 1947
Sorghum	10, 20, 40	Gu *et al.*, 1984
Spartina	28, 40, 42, 56, 80, 84, 112, 128	Myers, 1947; Hickman, 1993
Sporobolus	18, 24, 36, 45, 46, 54, 72, 82, 108	Sutherland, 1986
Stipa	24, 28, 34, 36, 40, 42, 44, 46, 48, 64, 68, 70, 82	Myers, 1947
Tridens	16, 32, 40, 60	Sutherland, 1986
Triodia	28, 48	Myers, 1947
Trisetum	14, 24, 26, 28, 42	Myers, 1947
Zizania	30	Myers, 1947

In many taxa the current genome is a complex product of multiple occurrences of both allo- and autopolyploidy. Allopolyploids begin with a
hybrid. The doubling of the two genomes solves pairing problems between
its component genomes. Sometimes doubling occurs after the production
of a relatively sterile hybrid (e.g. *Spartina anglica*, Guenegou, Citharel &
Levasseur, 1988), sometimes allopolyploidy is the result of direct combination of unreduced gametes (Bretagnolle & Thompson, 1995).
Allopolyploidy can occur multiple times in the history of a species as
in *Triticum aestivum* (Sears, 1969). Autopolyploidy occurs when, within a
single lineage, the genome duplicates, usually by production of an unreduced gamete that successfully forms an embryo, either combined with a
normal (reduced) gamete or another unreduced gamete (Bretagnolle &
Thompson, 1995). Historically, more attention has been paid to allopolyploidy than autopolyploidy and it has been considered to be by far the most
important form of polyploidy (Stebbins, 1971; Grant, 1981). Recent work,
however, finds autopolyploidy to be relatively common (Thompson &
Lumaret, 1992; Bretagnolle & Thompson, 1995).

Stebbins (1947, 1971) proposed a widely accepted sequence for the development of polyploidy within lineages. First tetraploidy occurs, then the
tetraploids spread and replace the diploids. The diploids become geographically restricted, rare, and then extinct, and the process repeats as hexaploids
and octoploids are formed from the tetraploids and expand at the expense of
the tetraploids. Although Stebbins considered autopolyploidy rare, the
model needs little modification to incorporate autopolyploidy.

In an allopolyploid complex, when the ploidy levels are incompatible,
there is a tight relationship between ploidy level and taxonomy, i.e. different
ploidy levels belong in different species. Such polyploid series occur in many
grass genera (Table 7.1). Although it has not been studied, polyploid series
within genera should also result from autopolyploids becoming fixed in one
derived species, but not in another. During this process, populations containing a mixture of ploidy levels might persist for long periods of time.
Indeed, it appears that derived species may carry intraspecific polyploid
variation with them, as in the case of *Andropogon hallii*, which is clearly
derived from *A. gerardii*, and has similar intraspecific polyploidy
(Sutherland, 1986; Table 7.3). In these cases ploidy level need not correlate
with taxonomic divisions (see below). Little is known about the relative frequency of these processes.

This paper discusses polyploidy within species of grasses. For more
general reviews of polyploidy, see Stebbins (1971), Lewis (1980), Grant
(1981), Lumaret (1988a), and Thompson & Lumaret (1992).

Distribution of intraspecific polyploidy

Species that are composed of individuals and populations with differing ploidy levels are known from many families, including the Chenopodiaceae (Dunford, 1985; Freeman & McArthur, 1989), Fabaceae (Grant, Brown & Grace, 1984; Hymowitz, Parker & Singh, 1991), Rosaceae (Campbell, Greene & Bergquist, 1987), reviewed in Lewis (1980).

Variation in ploidy level occurs within many grass species (Federov, 1974; Lewis, 1980; Keeler & Kwankin, 1989; Table 7.2). Of the grass species listed in the floras of the Great Plains (Sutherland, 1986) and California (Hickman, 1993) about 21% of the species were reported to have intraspecific polyploidy (Table 7.3). It is difficult to evaluate the quality of this value: (a) Some old counts produced with untrustworthy methods are probably still being cited (see discussion in Church, 1936). (b) With higher numbers of chromosomes, accurate counting is more difficult and so less accurate. (c) Some reports of intraspecific polyploidy may be the result of taxonomic confusion: certainly taxonomic revision can greatly simplify the cytogenetics of some genera. (d) On the other hand, for some species, the cytotype is known from a single count, therefore it is not known whether the species is chromosomally variable. (e) Other species are only very sketchily sampled or from only part of an extensive range, likewise making undetected intraspecific polyploidy possible. I think it is premature to analyse patterns of intraspecific polyploidy within the Poaceae, although it is obvious that a substantial number of species have been reported to possess intraspecific polyploidy.

Genetics of polyploids

Polyploidy usually produces profound changes in the genetics of the species. These changes can include change in Mendelian inheritance patterns and modification of dominance relationships, declines in fertility, loss of incompatibility, greater retention of genetic diversity under selfing and loss of interfertility with other members of the (former) species (Haldane, 1930; Mather, 1936; Fisher, 1949; Levin, 1983; Fowler & Levin, 1984; reviewed in Bever & Felber, 1993). There is a rich literature on the genetics of polyploids of agricultural importance, such as *Solanum* and *Triticum* (e.g. Simmonds, 1976; Tsuchiye & Gupta, 1991). Theoretical studies of polyploid genetics go back 60 years (Haldane, 1930; Mather, 1936), and there is a growing body of more recent population genetics theory (Ehlke & Hill, 1988; Bever & Felber, 1993, Rodríguez, 1996a,b). Only a brief review relevant to intraspecific polyploid variation will be given here.

Table 7.2. *Grasses reported to have within-species polyploidy*

Genus and species	Ploidy	Chromosome numbers	Reference
Distribution studied			
Agropyron dasystachyum	2n, 6n	14, 42	Sadasivaiah & Weijer, 1981
Agrostis stolonifera	6n, 9n	28, 35, 42	Björkman, 1984; Kik et al., 1992
Andropogon gerardii		60, 90	Norrmann et al., 1997
Anthoxanthum odoratum	2n, 3n, 4n	10, 15, 20	Hedberg, 1967
Bouteloua curtipendula	3n, 4n, 5n, 6n, 8n, 10n, 14n	21, 28, 35, 40, 42, 45, 50, 52, 56, 70, 98	Harlan, 1949; Gould & Kapadia, 1962; Kapadia & Gould, 1964
Dactylis glomerata	2n, 4n, 6n	14, 28, 42	Müntzing, 1937; Stebbins & Zohary, 1959; Zohary & Nur, 1959; Lumaret, 1988b
Deschampsia caespitosa	2n, 4n	26,52	Rothera and Davy, 1986
Holcus mollis	4n, 5n, 6n, 7n	28, 35, 42, 49	Jones, 1958; Jones & Carroll, 1962
Panicum virgatum	2n, 4n, 6n, 8n, 10n, 12n	18, 21, 25, 30, 32, 36, 54, 56–65, 70, 72, 90, 108	Nielsen, 1944; McMillan & Weiler, 1959
Paspalum brunneum	2n, 4n	20, 40	Norrmann et al., 1989
P. cromyorrhizon	2n, 4n	20, 40	Quarín et al., 1982
P. haumanii	2n, 4n	20, 40	Norrmann et al., 1989
P. hexastachyum	2n, 4n, 6n	20, 40, 60	Quarín & Hanna, 1980
P. intermedium	2n, 4n	20, 40	Norrmann et al., 1989
P. maculosum	2n, 4n	20, 40	Norrmann et al., 1989
P. quadrifarium	2n, 3n, 4n	20, 30, 40	Norrmann et al., 1989
P. rufum	2n, 4n	20, 40	Norrmann et al., 1989

Intraspecific polyploidy reported

Species		Chromosome numbers	Reference
Andropogon hallii	6n, 7n, 10n	60, 70, 100	Nielsen, 1939; Brown, 1950
Aristida purpurea	2n, 4n, 6n, 8n	22, 44, 66, 88	Hickman, 1993
Arrhenatherum elatius	2n, 4n, 6n	14, 28, 42	Hickman, 1993
Bouteloua gracilis	2n, 4n, 6n	20, 29, 35, 40, 42, 60, 61, 77, 84	Fults, 1942; Snyder & Harlan, 1953
B. hirsuta		12, 20, 21, 28, 37, 42, 46	Fults, 1942; Gould, 1958
Buchloe dactyloides	2n, 4n, 6n	20, 40, 60	Sutherland, 1986
Elymus canadensis	2n, 4n	28, 42	Nielsen & Humphrey, 1937
Festuca elatior	2n, 4n, 6n,	14, 28, 42	Myers, 1947; Myers & Hill, 1947
Koeleria pyramidata	2n, 4n, 8n, 10n, 12n	14, 28, 56, 70, 84	Hickman, 1993
Phalaris arundinacea	2n, 4n, 5n, 6n	14, 27, 28, 29, 30, 31, 35, 42, 48	Hansen & Hill, 1953
Phleum alpinum	2n, 4n	14, 28	Hickman, 1993
P. pratense	2n, 3n, 4n	14, 21, 28	Hickman, 1993
Phragmites australis	4n, 6n	36, 44, 46, 48, 49–52, 72, 84, 96	Hickman, 1993
Saccharum spontaneum	not certain 2n (=8x=64), 4n ?	40, 48, 54, 56, 60, 64, 72, 80, 96, 104, 112, 128	Panje & Babu, 1960; Al-Janabi *et al.*, 1993
Spartina pectinata	4n, 6n, 12n	28, 40, 42, 80, 84	Myers, 1947; Sutherland, 1986
Sporobolus cryptandrus	2n, 4n, 6n, 8n	18, 36, 38, 72	Hickman, 1993
Trisetum canescens	2n, 4n	14, 28	Hickman, 1993
T. flavescens	2n, 4n	14 28	Hickman, 1993
T. spicatum	2n, 4n, 6n	14 28 42	Hickman, 1993

Table 7.3. *Frequency of grass species with intraspecific polyploidy in two recent floras*

Flora	Intraspecific polyploid series reported	One chromosome number in the species	Variation but not a polyploid series	No value reported	Total grass species
Great Plains	84 (33.5%)	127 (50.6%)	19 (7.6 %)	21 (8.4%)	251
California	82 (16.6%)	307 (62.0%)	34 (6.9 %)	72 (14.5%)	495
Shared species	27	45	6	0	78
Combined	138 (20.7%)	390 (58.4%)	47 (7.0%)	93 (13.9%)	668

Note:
Taken from Sutherland (1986); Hickman (1993).

For ease of description I will outline the differences using diploids and tetraploids. Diploids are the basis on normal Mendelian genetics: each individual has two copies of each (non-duplicated) locus, and so can be heterozygous $(a_1 a_2)$ or homozygous $(a_1 a_1,$ or $a_2 a_2)$. If there is dominance, the heterozygote shows some phenotype other than one intermediate between the homozygotes. Under selfing, diploids approach two pure lines rapidly: half of the progeny are expected to be homozygotes and so frequency of heterozygotes drops by 50% each generation. Progeny receive just one of the two alleles of each parent.

In contrast, tetraploids have four copies of each locus, and normal terminology for heterozygosity immediately breaks down. While homozygous tetraploids are clear: $a_1 a_1 a_1 a_1$ and $a_2 a_2 a_2 a_2$, heterozygotes could be $a_1 a_1 a_1 a_2$, $a_1 a_1 a_2 a_2$ and $a_1 a_2 a_2 a_2$. But of course they are not limited to just two alleles per individual the way diploids are, so additional forms of heterozygosity exist, to the extreme of $a_1 a_2 a_3 a_4$. Dominance between the alleles is not *per se* affected by ploidy levels, but since so many more alleles can occur in the same individual, that dominant allele may mask a greater variety of different genetic combinations. In addition, the phenotypes produced by interactions between co-occurring alleles are far more complicated ($a_1 a_1 a_2 a_3$ might be functionally distinguishable from $a_1 a_2 a_2 a_3$ and $a_1 a_2 a_3 a_3$). At the enzyme level, being tetraploid opens up the possibility for a wide array of dimeric enzymes produced by combinations of two allele products. Under selfing, a much lower percentage of the progeny will be homozygotes. Assume for simplicity we start with $a_1 a_1 a_2 a_2$, selfed: 17% of the progeny will be homozygotes, so that 83% of the genetic variation is retained. After five generations, when 97% of the diploid population would be expected to be homozygous under selfing (0.5^5), 60.6% of the tetraploid population will be homozygous (0.83^5), a much slower loss of genetic variation (Haldane, 1930). Finally, progeny receive two copies of the genome from each parent. This is more complex than it seems, because at meiosis, crossing over can recombine the alleles with respect to the centromere. Genetic exchange among the two chromosomes of diploids breaks up linkage groups. In tetraploids, since crossing over and segregation can result in double reduction, the same allele being twice included in a gamete, it is therefore possible for $a_1a_2a_2a_2$, selfed, to produce among its progeny $a_1a_1a_2a_2$. In some cases the frequency of double reduction could be 50% (Mather, 1936). Thus the genetic recombination of tetraploids is more than just a simple doubling of all the values of the diploid: much more variation is possible. For more detail see the review by Bever & Felber (1993).

The comments made for tetraploids relative to diploids apply likewise to hexaploids and octoploids and other higher multiples (summarized by Bever & Felber, 1993). The variety possible is a geometric, not linear, expansion.

One of the key observations about polyploidy is that where there is not strict control of meiotic segregation, there is significant loss of fitness due to production of gametes with incomplete or partially duplicated genomes. In addition, segregation in polyploids that are odd multiples of the basic number, i.e. triploids ($3n$), pentaploids ($5n$), etc., often results in gametes receiving partial genomes and, therefore, chromosome complements that function poorly or not at all. There is widespread documentation of reduced fertility of these odd-numbered ploidy levels although the situation is complex and species differ greatly (Stebbins, 1971; Grant 1981).

Chromosome combinations which produce few viable gametes are important in understanding the evolution of polyploidy because chromosome complements and crosses that result in unbalanced gametes are usually at a fitness disadvantage compared with conspecifics with even multiples of the genome. Where polyploidy confers an obvious decrease in fitness, it is clear that doubling the genome is not a neutral trait, and an explanation for maintenance or recurrent origin of the polyploids is required.

As emphasized by Bever & Felber (1993) much more work remains to be done on the genetics and population genetics of polyploids.

Implications of intraspecific polyploidy

Intraspecific polyploidy is not necessarily a result of poor taxonomy. Disparate ploidy levels are frequently both interfertile and morphologically indistinguishable: the evolutionary unit in many grass species is a population of plants of diverse ploidy levels. Although historically plants with different ploidy levels have been held to be reproductively isolated from each other, this has proven to be an oversimplification.

First, gene flow can occur between ploidy levels. *Dactylis glomerata* and *Andropogon gerardii*, both autopolyploid complexes, show interploidy level fertility (i.e. gene flow) (Zohary & Nur, 1959; Lumaret, 1988b; Norrmann, Quarín & Keeler, 1997). In *Holcus*, allopentaploids produce fertile diploid progeny (Richard et al., 1995).

In general, fertility is greater in even multiples of the genome, and intraploidy level hybrids such as triploids and pentaploids are of significantly lowered fertility (Zohary & Nur, 1959; Stebbins, 1971; Grant, 1981; Richard et al., 1995; Norrmann et al., 1997). However, in a surprising number of species, some individuals with odd-numbered ploidy levels, e.g. pentaploids in *Holcus mollis* (Jones, 1958; Jones & Carroll, 1962), triploids in *Dactylis glomerata* (Zohary & Nur, 1959; Borrill, 1978), enneaploids ($9x$) in *Andropogon gerardii* (Norrmann et al., 1997) have good fertility, linking the ploidy series into an evolutionary whole (Jackson, 1976; Lumaret, 1988a; Thompson & Lumaret, 1992).

Secondly, from a pragmatic identification standpoint, many ploidy levels are not recognizable as distinguishable separate taxa. So long as polyploids are recurrently created as multiple events, whether allo- or autopolyploid, there is going to be a range of characters that do not lend themselves to distinguishing ploidy levels easily. In some cases this has been carefully studied, e.g. *Dactylis* (Stebbins & Zohary, 1959), *Anthoxanthum odoratum*, (Hedberg, 1967), and *Deschampsia* (Rothera & Davy, 1986).

Stebbins & Zohary (1959) summarize the problem succinctly:

. . . the evolutionary relationships within *Dactylis* would be reasonably well expressed by only two ways of recognizing species. One would be to recognize a single tetraploid and one or two diploid species . . . [but] . . . the only absolute criterion which separates all diploids from all tetraploids is the chromosome number. The separation of *Dactylis* into two such variable and similar species is not only impractical from a taxonomic point of view, but is also not altogether compatible with a species concept based upon reproductive isolation.

Thus, they reduce the diploid and tetraploid forms to subspecific status. It is a clear reflection on the reality of gene flow and morphological variation in the complex that the many subsequent workers have largely accepted the merging of eleven diploid taxa, two to five tetraploid taxa and one hexaploid taxon into one named species, *Dactylis glomerata* (e.g. references cited below).

Given that some polyploid complexes are considered intraspecific because the ploidy levels are neither morphologically nor genetically discrete, it is important to consider what this might tell us about the population ecology of grasses. I will describe two well-studied examples.

Dactylis glomerata L., cocksfoot, orchard grass, is a rhizomatous grass naturally distributed throughout Europe to northern Africa and Asia that has been widely introduced elsewhere in the world (Müntzing, 1937; Myers, 1941; Stebbins & Zohary, 1959; Zohary & Nur, 1959; Borrill, 1961, 1978; Parker & Borrill, 1968; Lumaret *et al.*, 1987; Lumaret, 1988*b*; Bretagnolle & Thompson, 1996). Diploid, tetraploid, and hexaploid plants and populations are known. The diploids are narrowly distributed and often readily distinguished from each other morphologically (Parker & Borrill, 1968; Lumaret, 1988*b*). They hybridize to some degree with each other and the diploid hybrids are fertile, although less so than the parents (Borrill, 1961, 1978). Tetraploids in this complex are more abundant and more widespread than the diploids. In many cases (Stebbins & Zohary, 1959; Borrill, 1961) they cannot be readily distinguished from diploids occurring in the same area. Furthermore, tetraploids are fully interfertile (Stebbins & Zohary, 1959). Where diploids and tetraploids occur together, natural triploids form. Triploids have about 1% fertility and are responsible for gene flow between the ploidy levels (Zohary & Nur, 1959). As noted above, the complexity of the group led Stebbins & Zohary (1959) to classify all the recognizable types as simply subspecies of *Dactylis glomerata*.

A detailed study looking for microhabitat differences in northern Spain found local tetraploids to be distributed more broadly than diploids, but particularly they inhabited open and disturbed habitats, whereas the corresponding diploids were confined to shaded areas (Lumaret *et al.*, 1987). Experimental studies indicated that the correlation with shade resulted from the moister conditions under the trees, rather than shade tolerance (Lumaret *et al.*, 1987). The picture was complicated by the presence of tetraploids of subspecies *Dactylis glomerata glomerata* which escapes from cultivation and had hybridized with local (Galician) tetraploids. Plants of subspecies *glomerata* and Galacian tetraploids were concentrated in different habitats and their hybrid was found in between them (Lumaret *et al.* 1987; Fig. 7.1*A*). The outcome of the extensive study of *Dactylis* is a picture of a taxon in which local plants form autopolyploids fairly frequently, but diploids can form interspecific hybrids occasionally, the tetraploids are interfertile, and triploids hybridize with tetraploids and other triploids. Thus gene flow links the complex at many points. The variation in the tetraploids matches that of diploids in their areas of origin, but

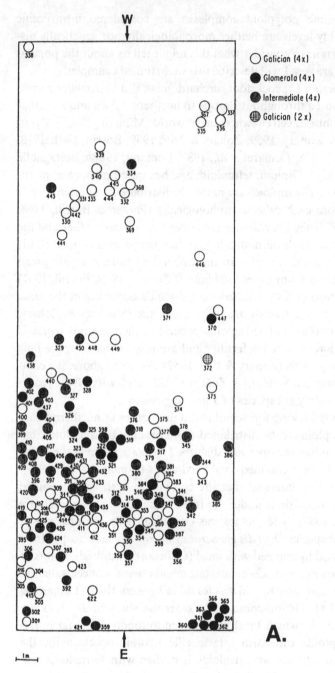

Fig. 7.1. Spatial distribution of polyploids within populations. (*A*) *Dactylis glomerata* (fig 5 from Lumaret *et al.*, 1987). The diploid and tetraploid morphs are mapped. (Reproduced with permission of Springer-Verlag.)

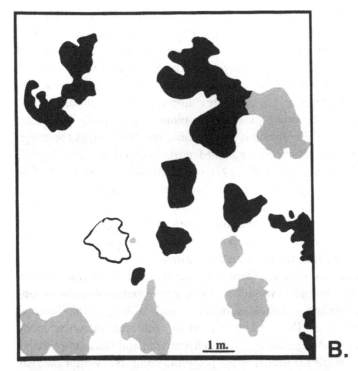

Fig. 7.1. (*cont.*)
(*B*) *Andropogon gerardii* (K. H. Keeler unpublished) population in Boulder Co.,
CO. Shaded are *A. gerardii* individuals: dark, 60 chromosomes; light, 90
chromosomes; outline intermediate value (probable aneuploid).

overall exceeds that of diploids, because although there is some increase in
variation with chromosomal increase, the greater genetic variation is also a
result of the multiple origins of tetraploids and their hybridization with
each other (Lumaret, 1988*b*).

Panicum virgatum, switchgrass, is a tall rhizomatous perennial of central
North America. A polyploid series was recognized by Nielsen (1944):
$2n=18$, 36, 54, 72, 90, 108. He found no geographic patterns and in one
population the complete range of cytotypes were present. Comparing the
morphological characteristics of agronomic potential between the cytotypes, Nielsen (1944) found no consistent differences or any reason to prefer
higher polyploids as forage grasses. McMillan & Weiler (1959) conducted
a study of the central USA and again found multiple cytotypes within populations (although none with more than three cytotypes). They too found
no consistent pattern for characters they compared between cytotypes (e.g.
date of first flowering, height). They reported that clones from one region

were more similar to clones of the other ploidy levels found in that area, than to the same ploidy level found elsewhere. In Oklahoma and Kansas, *Panicum virgatum* has two distinct morphotypes that are described as 'lowland' and 'upland' forms. A series of studies (Porter, 1966; Barnett & Carver, 1967; Brunken & Estes,1975) found the lowland form, which is conspicuously taller, more robust and more clumped, to be tetraploid, while the upland form was both hexaploid or octoploid, often equally frequently. Recent studies indicate that this pattern does not hold across the range: upland forms from Nebraska included tetraploids (Hultquist, Vogel & Kaeppler, 1996; Hultquist *et al.*, 1997), which led Hultquist *et al.* (1997) to suggest separate origins of the two forms and an autopolyploid series within the upland race, something suggested by Brunken & Estes (1975) from much less detailed information. Hybridization between the upland and lowland races has not been reported in the literature but can occur (K. P. Vogel & J. Martinez Reyna, unpublished data).

Thus, in *Panicum virgatum* there is widespread and significant chromosomal and morphological variation, with a strong geographic component to both cytotype and morphology, but apparently with no simple relationship between them. Despite extensive study, however, there are many regions in which *P. virgatum* has not been analysed and which may clarify the situation. While the details of the polyploid complexes within *Dactylis glomerata* and *Panicum virgatum* differ greatly, one could conclude that both species show important and widespread cytotypic and morphological variation which are not related in any simple manner. Both taxa suggest that, whether or not intraspecific polyploidy is adaptive, it is evolving independently of morphology.

Population ecology of intraspecific polyploidy

As discussed above, there is usually gene flow between the cytotypes in a grass population with several ploidy levels. In *Dactylis glomerata*, for example, diploids and tetraploids produced viable hybrids at a frequency of at least 3 triploids per 2000 plants. While the triploids were largely pollen-sterile, they produced viable seeds, including tetraploids and pentaploids (Zohary & Nur, 1959; Lumaret & Barrientos, 1990). Thus, genes from the diploids could move via triploids into tetraploid populations. In the case of Zohary & Nur's study, this logical argument was supported by the presence of B chromosomes in the progeny of the triploids, because in the Israeli populations studied, B chromosomes were found only in diploid plants. Similarly inter-cytotype ($6x \times 9x$) hybrids of *Andropogon gerardii* form a potential bridge between the ploidy levels of that species because they have some pollen and seed fertility, despite being largely aneuploid (Norrmann *et al.*, 1997).

The genetic complexity of some populations can hardly be overemphasized. Nielsen (1944) found five ploidy levels within a single population of *Panicum virgatum*, and populations with three cytotypes were common (McMillan & Weiler, 1959). Lumaret *et al.* (1987) mapped *Dactylis glomerata* plants with a plot of 10 × 80 m that had a local diploid and tetraploid cytotype and an agronomically introduced tetraploid cytotype and the tetraploid hybrids of the two tetraploid races (Fig. 7.1*A*). *Andropogon gerardii* also has populations with plants of different ploidy levels intermingling (Keeler 1992, Fig. 7.1*B*). Because these cytotypes are interfertile, all contribute to the evolutionary population.

When the species-wide variation is considered, many grass species are very complex indeed. Over its range, *D. glomerata* has a very wide variety of cytotypes, genotypes and phenotypes, occurring as single or multiple cytotype populations (Lumaret 1988*b*; Lumaret & Barrientos, 1990). The diploid:tetraploid:pentaploid complex of *Holcus mollis* × *H. lanatus* likewise has regional differences superimposed on the local populations that contain varying combinations of cytotypes (Jones, 1958; Richard *et al.*, 1995). Others that have not been as intensely studied are likely to be very complex as well, e.g. *Bouteloua curtipendula* (Gould & Kapadia, 1962; Kapadia & Gould, 1964), *Poa pratensis, Phragmites australis* (Table 7.2).

For most intraspecific polyploid complexes studied, ploidy levels cannot be distinguished at the individual level with sufficient accuracy for taxonomic distinctions (Hedberg, 1967; Rothera & Davy 1986; Norrmann *et al.*, 1997). Populations of different ploidy levels often differ, but there is such extensive overlap, especially by the higher ploidy levels, that individuals are difficult to categorize. Generally this is a result of the broader variation in the higher ploidy levels (Stebbins 1947, 1971; Hedberg, 1967; Rothera & Davy, 1986). To the degree that selection acts on the phenotype, not the genotype, such cytotypic variation is cryptic variation, invisible to selection.

Ecological and geographic differentiation

Except for their impact on speciation, the evolutionary and population biology consequences of polyploidy and intraspecific polyploidy are largely unexplored. Intraspecific polyploid complexes could, like any form of genetic variation, be, for example, adaptive, deleterious, neutral or transient.

Some differences between ploidy levels seem as if they should be subject to selection. Hedberg (1967) found that tetraploid plants of *Anthoxanthum odoratum* had hairier leaves, leaf sheaths and glumes than diploids, but the occasional glabrous tetraploid eliminated that for use as a distinguishing character. Diploids on the whole also had smaller spikelets and pollen, but the overlap again precluded using these characters diagnostically. Rothera & Davy (1986) found that tetraploid *Deschampsia caespitosa* often, but not always, had larger florets than diploids. This and other characters formed a general syndrome that distinguished diploids and tetraploids, but it broke down for individual plants. Enneaploid ($9x$) *Andropogon gerardii* are usually taller than hexaploids ($6x$), but there is so much phenotypic plasticity that the difference is detectable only statistically within populations, not for individuals (K. Keeler, unpublished). Differences in winter-hardiness and growth rate to flowering have been reported in *Dactylis* (e.g. Bretagnolle & Thompson, 1996). Other differences have been correlated with ploidy level (e.g. Stebbins, 1971; Lewis, 1980; Grant, 1981; Roy & Lumaret, 1987; Warner, Ku & Edwards, 1987, Masterson, 1994). Without a series of direct experiments, it is difficult to judge whether the ploidy levels respond differently enough to stresses in the environment to show different relative fitnesses: although the types of differences suggest they do, the lack of consistent responses within these polyploid complexes argue they do not.

Small-scale ecological differences occur between ploidy levels in *Dactylis glomerata* (Lumaret *et al.*, 1987; Bretagnolle & Thompson, 1996); the *Holcus lanatus – Holcus molis* complex (Richard *et al.*, 1995), *Paspalum hexastachyum* (Quarín & Hanna, 1980), *Agrostis stolonifera* (Kik, Linder & Bijlsma, 1992), and *Anthoxanthum odoratum* (Hedberg, 1967). Small-scale ecological differentiation was not found among cytotypes of *Deschampsia cespitosa* (Rothera & Davy, 1986) or *Andropogon gerardii* (Keeler, 1990, 1992). For *Panicum virgatum*, the situation appears to vary across its range (Nielsen, 1944; McMillan & Weiler, 1959; Porter, 1966; Barnett & Carver, 1967; Brunken & Estes, 1975; Hultquist *et al.*, 1996, 1997).

If the cytotypes cannot be morphologically distinguished in any reliable way, yet can be shown to have distributions correlated with environmental variables, the ecological patterns presumably stem from the fine differences resulting from doubling of the genome, although experiments are needed to eliminate the possibility that the pattern is a stochastic artifact. In *Dactylis glomerata* there is sufficient data to suggest that patterns are deterministic not random (Stebbins & Zohary, 1959; Lumaret *et al.*, 1987; Roy & Lumaret, 1987; Bretagnolle & Thompson, 1996), while for *Andropogon gerardii*, the lack of match between patterns at different scales (Keeler *et al.*, 1987; Keeler, 1990, 1992) could reasonably result from stochastic effects. Small-scale maps often reveal intimate mixing of cytotypes (Fig. 7.1*A,B*), whether or not the species shows ecological differentiation. For *Dactylis*, levels of mixing vary greatly across its range (Lumaret, 1988*b*). *Andropogon gerardii* is mainly hexaploid in the eastern part of its extensive range, but the populations in the west have roughly equal frequencies of 60 and 90 chromosome plants, thoroughly intermingled (Keeler *et al.*, 1987; Keeler, 1990, 1992, unpublished data, Fig. 7.1*B*). Mixing of the cytotypes of *Anthoxanthum odoratum* has been enhanced by human activities (Hedberg, 1967). *Panicum virgatum* populations are usually a mix of cytotypes (Nielsen, 1944; McMillan & Weiler, 1959), but there is ecological separation as well (e.g. Porter, 1966). *Holcus lanatus* in France shows both ecological and geographic differences among cytotypes (Richard *et al.*, 1995).

The simplest of the consequences of multiple cytotypes within populations is that a wide array of morphologies and ecological adaptations are available within the complex. Lumaret *et al.* (1987) demonstrate the power of cytotypic variation (Fig. 7.1*A*), in the sense that one evolutionary unit (*Dactylis glomerata sensu lato*) occupies three microhabitats. Given the ecological importance of grasses (they are after all the only plant family with a major ecosystem named for them, and that ecosystem occupies every continent except Antarctica), the ecological differentiation afforded by polyploid complexes deserves to be looked at as a potential adaptive strategy. Much work needs to be done to understand the implication of these consequences of polyploidy to mixed populations and to adaptive evolution of grasses. *Andropogon gerardii*, for example, is an ecosystem dominant despite cytotypic variation that should lower fitness (Keeler, 1990; Norrmann *et al.*, 1997) (Fig. 7.2). A frequently burned prairie in what is now the 'corn belt' was a virtual monoculture of *A. gerardii*, with more than 80% of the biomass from this single species (Weaver, 1954). In the face of arguments that 'nature abhors a monoculture' one is moved to ask whether ploidal variation may have helped ameliorate the disadvantages of a monoculture (e.g. lack of genetic variation to resist diseases), facilitating ecological dominance by a single lineage.

Other important points in the population ecology of intraspecific polyploidy are illustrated in *Holcus* (Jones, 1958; Richard *et al.*, 1995), *Anthoxanthum* (Hedberg, 1967), *Dactylis* (Lumaret 1988*b*), and *Panicum* (Hultquist *et al.*, 1996, 1997), where in all cases there are inconsistencies in the breeding relationships and behaviour of cytotypes (such as hybridization

Fig. 7.2. Photo of *A. gerardii* -dominated prairie (Konza Prairie, Manhattan, KS).
Photo by K. H. Keeler

and genetic composition of populations) in different parts of the species' range. This is understandable given the recurrent formation of auto- and allopolyploids, and backcrossing within the complex. It does, however, mean that caution must be used in extrapolating experimental results across the geographic range of a polyploid complex.

The taxon cycle of Stebbins (1947) provides a description of reality but begs the question of causation. What are the forces that increase ploidy level and allow higher ploidy levels to survive (cf. Levin, 1983; Fowler & Levin, 1984; Rodríguez, 1996*a*, *b*)? Are taxa found at different places in Stebbins' progression simply because of different periods since they originated, or is polyploidy adaptive in some contexts but not in others?

Polyploids have been described as being more variable than diploids (see above). Their differences result from the hybrid origin of allopolyploids or the multiple origin of autopolyploids in different local populations, which must produce selective advantage under some conditions. Other differences ascribed to polyploidy include loss of the incompatibility systems of diploids or other means of greater reproductive promiscuity (Lumaret, 1988*a*,*b*; Bretagnolle & Thompson, 1995). In several *Paspalum* species, diploids are obligate outcrossers but autotetraploids are largely apomictic (Quarín & Hanna, 1980; Quarín, Hanna & Fernández, 1982).

It seems unlikely that all or even most grass species with intrapopulation polyploidy will have the same fitness relationships among the cytotypes, but it is reasonable to expect that there are a finite number of combinations of genetics and ecology which select for (or permit) the realized distribution of cytotypes. Given the importance of grasses, both ecologically and economically, and the frequency of intraspecific polyploidy (Tables 7.2, 7.3), the evolutionary forces underlying intraspecific polyploidy present a rich area for future research.

Summary

Polyploid series within local populations occur in many grass species. The evidence suggests recurring autopolyploidy is often the cause of the polyploid variation, but recurrent allopolyploidy is indicated in some cases. Often the different ploidy levels cannot be distinguished morphologically and they exchange genes at least occasionally; thus the evolutionary unit is a cytologically complex population. The implications of the presence of multiple cytotypes for local adaptation have scarcely begun to be investigated.

Acknowledgements

This work was supported in part by NSF 95-09139 and a University of Nebraska faculty research leave. It is dedicated to G. L. Stebbins.

References

Ainouche, M., Misset, M-T, & Huon, A. (1995). Genetic diversity in Mediterranean diploid and tetraploid *Bromus* L. (section *Bromus* Sm.) populations. *Genome*, **38**, 879–88.

Al-Janabi, S. M., Honeycutt, R. J., McClelland, M., and Sobral, B. W. S. (1993). A genetic linkage map of *Saccharum spontaneum* L. 'SES 208'. *Genetics*. **14**, 1249–60.

Barnett, F. L. & Carver, R. F. (1967). Meiosis and pollen stainability in switchgrass, *Panicum virgatum* L. *Crop Science*, **7**, 301–4.

Bever, J. D. & Felber, F. (1993). The theoretical population genetics of autopolyploidy. *Oxford Surveys in Evolutionary Biology*, **8**: 185–217.

Björkman, S. O. (1984). Chromosome studies in *Agrostis*. II. *Hereditas*, **40**, 254–68.

Borrill, M. (1961). The pattern of morphological variation in diploid and tetraploid *Dactylis*. *Journal of the Linnean Society (Botany)*, **56**, 441–52.

Borrill, M. (1978). Evolution and genetic resources in cocksfoot. *Report of the Welsh Plant Breeding Station (Aberystwyth, Wales)* pp. 190–209.

Bowden, W. M. (1965). Chromosome numbers and taxonomic notes on some northern grasses. III. Twenty five genera. *Canadian Journal of Botany*. **38**, 541–57.

Bretagnolle, F. & Thompson, J. D. (1995). Gametes with the somatic chromosome number: mechanisms for their formation and role in the evolution of autopolyploid plants. *New Phytologist.*, **129**, 1–22.

Bretagnolle, F. & Thompson, J. D. (1996). An experimental study of ecological differences in winter growth between sympatric diploid and autotetraploid *Dactylis glomerata*. *Journal of Ecology*, **84**, 343–51.

Brockman, C. & Elven, R. (1992). Ecological and genetic consequences of polyploidy in arctic *Draba* (Brassicaceae). *Evolutionary Trends in Plants*, **6**, 111–24.

Brown, W. V. (1950). A cytological study of some Texas Gramineae. *Bulletin of the Torrey Botanical Club*. **77**, 63–76.

Brunken, J. N. & Estes, J. R. (1975) Cytological and morphological variation in *Panicum virgatum* L. *The Southwestern Naturalist*, **19**, 379–85.

Burton, G. W. (1942). A cytological study of some species in the tribe Panicaceae. *American Journal of Botany*, **29**, 355–9.

Campbell, C. S., Greene, G. W. & Bergquist, S. E. (1987). Apomixis and sexuality in three species of amelanchier, shadbush (Rosaceae, Maloideae). *American Journal of Botany*, **74**, 321–8.

Church, G. L. (1936). Cytological studies in the Gramineae. *American Journal of Botany*, **23**, 12–16.

Dunford, M. P. (1985). A statistical analysis of morphological variation in cytotypes of *Atriplex canescens* (Chenopodiaceae) *The Southwestern Naturalist*, **30**, 377–81.

Ehlke, N. J. & Hill, R. R., Jr. (1988). Quantitative genetics of allotetraploid and allopolyploid populations. *Genome*, **30**: 63–69.

Federov, V. (1974). *Chromosome Numbers of Flowering Plants.* West Germany: O. Koeltz Science Publishers.

Fisher, R. A. (1949). *The Theory of Inbreeding.* New York: Academic Press.

Fowler, N. L. & Levin, D. A. (1984). Ecological constraints on the establishment of a novel polyploid in competition with its diploid progenitor. *American Naturalist,* **124,** 703–11.

Freeman, D. C. & McArthur, E. D. (1989). *Atriplex canescens. CRC Handbook of Flowering Plants.* VI. pp. 75–86. Boca Raton, FL: CRC Press.

Fults, J. L. (1942) Somatic chromosome numbers in *Bouteloua. American Journal of Botany,* **29,** 45–56.

Goldblatt, P. (1980). Polyploidy in Angiosperms: Monocotyledons. In *Polyploidy: Biological relevance,* ed. W. H. Lewis. New York: Plenum. pp. 219–41.

Gould, F. W. (1958). Chromosome numbers in southwest grasses. *American Journal of Botany,* **10:** 757–86.

Gould, F. W. & Kapadia, Z. J. (1962). Biosystematic studies in the *Bouteloua curtipendula* complex. I. The aneuploid rhizomatous *B. curtipendula* of Texas. *American Journal of Botany,* **49,** 887–92.

Grant, J. E., Brown, A. D. H. & Grace, J. P. (1984). Cytogenetic and isozyme diversity in *Glycine tomentella* Hayata (Leguminosae). *Australian Journal of Botany,* **32,** 655–63.

Grant, V. (1981). *Plant speciation.* 2nd edition. New York: Columbia University Press.

Gu, M.-H., Ma, H.-T. & Liang, G. H. (1984). Karyotype analysis of seven species in the genus *Sorghum. The Journal of Heredity,* **75,** 196–202.

Guenegou, M. C., Citharel, J. & Levasseur, J. E. (1988).The hybrid status of *Spartina anglica* (Poaceae). Enzymatic analysis of the species and of the presumed parents. *Canadian Journal of Botany,* **66,** 1830–3.

Haldane, J. B. S. (1930). Theoretical genetics of autopolyploids. *Journal of Genetics.* **22,** 359–72.

Harlan, J. R. (1949). Apomixis in side-oats grama. *American Journal of Botany,* **36,** 495–9.

Hansen, A. A. & Hill, H. D. (1953). The occurrence of aneuploidy in *Phalaris* sp. *Bulletin of the Torrey Botanical Club,* **80,** 172–6.

Hedberg, I. (1967). Cytotaxonomic studies on *Anthoxanthum odoratum* L s. lat. II/ Investigations of some Swedish and a few Swiss population samples. *Symbolae Botanicae Upsalienses,* **18,** 1–88.

Hickman, J. C. (ed.) (1993). *The Jepson Manual of the Higher Plants of California.* Berkeley: University of California Press.

Hultquist, S. J., Vogel, K. P., & Kaeppler, S. (1996). Chloroplast DNA and nuclear DNA content variation among cultivars of switchgrasses *Panicum virgatum* L. *Crop Science,* **36,** 1049–52.

Hultquist, S. J., Vogel, K. P., Lee, D. J, Arumuganathan, K. & Kaeppler, S. (1997). DNA content and chloroplast DNA polymorphisms among switchgrasses from remnant midwestern prairies. *Crop Science,* 37, in press.

Hymowitz, T., Parker, R. G. & Singh, R. J. (1991). Cytogenetics of the genus *Glycine.* In *Chromosome Engineering in Plants,* ed. T. Tsuchiye & P. K. Gupta, pp. 53– 81. New York: Elsevier.

Jackson, R. C. (1976). Evolutionary and systematic significance of polyploidy. *Annual Review of Ecology and Systematics,* 7, 209–34.

Jackson, R. C. (1982). Polyploidy and diploidy: new perspectives on chromosome pairing and its evolutionary implications. *American Journal of Botany,* **69:** 1512–1523.

Jones, K. (1958). Cytotaxonomic studies in *Holcus* I. The chromosome complex of *Holcus mollis* L. *New Phytologist*, **57**, 191–210.

Jones, K. & Carroll, C. P. (1962). Cytotaxonomic studies in *Holcus*. II. Morphological relationships in *Holcus mollis* L. *New Phytologist*, **61**, 63–71.

Kankanpää, J., Mannonen, L. & Schulman, A. L. (1996). The genome sizes of *Hordeum* species show considerable variation. *Genome*, **39**, 730–5.

Kapadia, Z. J. & Gould, F. W. (1964). Biosystematic studies in the *Bouteloua curtipendula* complex. III. Pollen size as related to chromosome numbers. *American Journal of Botany*, **51**, 166–72.

Keeler, K. H. (1990). Distribution of polyploid polymorphism in big bluestem, *Andropogon gerardii* in the tallgrass prairie region. *Genome*, **33**, 95–100.

Keeler, K. H. (1992). Local polyploid variation in the native prairie grass *Andropogon gerardii*. *American Journal of Botany*, **79**, 1229–32.

Keeler, K. H. & Kwankin, B. (1989). Polyploid polymorphism in prairie grasses. In *Plant population biology*, ed. J. H. Bock & Y. B. Linhart, pp. 99–128. Boulder CO: Westview Press.

Keeler, K. H., Kwankin, B., Barnes, P. W. & Galbraith, D. W. (1987). Polyploid polymorphism in *Andropogon gerardii* Vitman (Poaceae). *Genome*, **29**, 374–9.

Kik, C., Linder, Th. E., & Bijlsma, R. (1992). The distribution of cytotypes in ecologically contrasting populations of the clonal perennial *Agrostis stolonifera*. *Evolutionary Trends in Plants*, **6**, 93–8.

Lewis, W. H. (ed.) (1980). *Polyploidy: Biological Relevance*. New York: Plenum.

Levin, D. A. (1983). Polyploidy and novelty in flowering plants. *The American Naturalist*, **122**, 1–25.

Lumaret, R. (1988a). Adaptive strategies and ploidy levels. *Acta Oecologia/Oecologia Plantarum*, **9**, 83–93.

Lumaret, R. (1988b). Cytology, genetics and evolution in the genus *Dactylis*. *CRC Critical Reviews of Plant Science*, **7**, 55–91.

Lumaret, R. & Barrientos, E. (1990). Phylogenetic relationships and gene flow between sympatric diploid and tetraploid plants of *Dactylis glomerata* (Gramineae). *Plant Systematics and Evolution*, **169**, 81–96.

Lumaret R., Guillerm, J-L, Delay, J., Loutfi, A. L., Izco, J. & Jay, J. (1987). Polyploidy and habitat differentiation in *Dactylis glomerata* L. from Galicia (Spain). *Oecologia (Berlin)*, **73**, 436–46.

Masterson, J. (1994). Stomatal size in fossil plants: evidence for polyploidy in majority of Angiosperms. *Science*, **264**, 421–4.

Mather, K. (1936). Segregation and linkage in autotetraploids. *Journal of Genetics*, **32**, 287–314.

McMillan, C. & Weiler, J. (1959). Cytogeography of *Panicum virgatum* in central North America. *American Journal of Botany*, **46**, 590–3.

Müntzing, A. (1937). The effects of chromosomal variation in *Dactylis*. *Hereditas*, **29**, 113–235.

Myers, W. M. (1941). Genetic consequences of chromosomal behavior in orchard grass *Dactylis glomerata* L. *Journal of the American Society of Agronomists*, **33**, 893–900.

Myers, W. M. (1947). Cytogenetics and genetics of forage grasses. *The Botanical Gazette*, **6**, 319–421.

Myers, W. M. & Hill, H. D. (1947). Distribution and nature of polyploidy in *Festuca elatior* L. *Bulletin of the Torrey Botanical Club*, **2**, 99–111.

Nielsen, E. L. (1939). Grass studies III. Additional somatic chromosome complements. *American Journal of Botany*, **26**, 366–72.

Nielsen, E. L. (1944). Analysis of variation in *Panicum virgatum*. *Journal of Agricultural Research*, **69**, 327–53.

Nielsen, E. L. & Humphrey, L. M. (1937) Grass studies I. Chromosome numbers in certain members of the tribes Festuceae, Hordeae, Aveneae, Agrostideae, Chlorideae, Phalaridaceae and Tripsaceae. *American Journal of Botany*, **24**, 276–9.

Norrmann, G. A. & Quarín, C. L. (1987). Permanent odd polyploidy in a grass *(Andropogon ternatus)*. *Genome*, **29**, 340–4.

Norrmann, G. A, Quarín, C. L., & Burson, B. L. (1989). Cytogenetics and reproductive behavior of different chromosome races in six *Paspalum* species. *Journal of Heredity*, **80**, 24–8.

Norrmann, G. A, Quarín, C. L., & Keeler, K. H. (1997) Evolutionary implications of meiotic chromosome behavior, reproductive biology and hybridization in 6*x* and 9*x* cytotypes of *Andropogon gerardii* (Poaceae), *American Journal of Botany*, **84**, 201–7.

Panje, R. R. & Babu, C. N. (1960). Studies in *Saccharum spontaneum* distribution and geographic association of chromosome numbers. *Cytologia*, **25**, 152–72.

Parker, P. F. & Borrill, M. (1968). Studies on *Dactylis* I. Fertility relationships in some diploid subspecies. *New Phytologist*, **67**, 649–62.

Porter, C. L. Jr. (1966). An analysis of variance between upland and lowland switchgrass, *Panicum virgatum* L. *Ecology*, **47**, 980–92.

Quarín, C. L., & Hanna, W. W. (1980). The effect of three ploidy levels on meiosis and mode of reproduction in *Paspalum hexastachyum*. *Crop Science*, **20**, 69–75.

Quarín, C.L., Hanna, W.W. & Fernández, A. (1982). Genetic studies in diploid and tetraploid *Paspalum* species. *The Journal of Heredity*, **76**, 254–6.

Richard, M., Jubier, M-F., Bajon, R., Gouyon, P-H. & Lejeune, B. (1995). A new hypothesis for the origin of pentaploid *Holcus* from diploid *Holcus lanatus* L. and tetraploid *Holcus mollis* L. in France. *Molecular Evolution*, **4**, 29–38.

Rodríguez, D. J. (1996*a*). A model for establishment of polyploidy in plants. *American Naturalist*, **147**, 33–46.

Rodríguez, D. J. (1996*b*). A model for the establishment of polyploidy in plants: viable but infertile hybrids, iteroparity, and demographic stochasticity. *Journal of Theoretical Biology*, **180**, 189–96.

Rothera S. L. & Davy, A. J. (1986). Polyploidy and habitat differentiation in *Deschampsia caespitosa*. *New Phytologist*, **102**, 449–67.

Roy, J. & Lumaret, R. (1987). Associated clinal variation in leaf tissue water relations and allozyme polymorphism in *Dactylis glomerata* L. populations. *Evolutionary Trends in Plants*, **1**, 9–19.

Sadasivaiah, R. S. & Weijer, J. (1981). The origin and meiotic behaviour of hexaploid northern wheatgrass *(Agropyron dasystachyum)*. *Chromosoma (Berl.)*, **82**, 121–32.

Seal, A. G. (1983). DNA variation in *Festuca*. *Heredity*, **50**, 225–36.

Sears, E. R. (1969). Wheat cytogenetics. *Annual Review of Genetics*, **3**, 451–68.

Seberg, O. & von Bothmer, R. (1991). Genome analysis of *Elymus angulatus* and *E. patagonicus* (Poaceae) and their hybrids in North and South American *Hordeum* spp. *Plant Systematics and Evolution*, **174**, 75–82.

Simmonds, N. W. (1976). *Evolution of Crop Plants*. London: Longman.

Snyder, L. A. & Harlan, J. R. (1953). A cytological study of *Bouteloua gracilis* from western Texas and eastern New Mexico. *American Journal of Botany*, **40**, 702–7.

Stebbins, G. L. (1947). Types of polyploidy I. Their classification and significance. *Advances in Genetics*, **1**, 403–29.

Stebbins, G. L. (1971). *Chromosomal Evolution in Higher Plants*, Reading, MA: Addison-Wesley.

Stebbins, G. L. & Zohary, D. (1959). Cytogenetic and evolutionary studies in the genus *Dactylis* I. Morphology, distribution, and interrelationships of the diploid subspecies. *University of California Publications in Botany*, **31**, 1–40.

Sutherland, D. (1986). Poaceae, Barnh., the grass family. In *Flora of the Great Plains*. Great Plains Flora Association. pp. 1113–235. Lawrence KS: University Presses of Kansas.

Thompson, J. D. & Lumaret, R. (1992). The evolutionary dynamics of polyploid plants: origin, establishment and persistence. *Trends in Ecology and Evolution*, **7**, 302–7.

Tsuchiye, T. & Gupta, P. K. (eds.) (1991). *Chromosome Engineering in Plants*. New York: Elsevier.

von Bothmer, R. & Jacobsen, N. (1986). Interspecific crosses in *Hordeum* (Poaceae). *Plant Systematics and Evolution*, **153**, 49–64.

Warner, D. A., Ku, M. S. B., & Edwards, G. E. (1987). Photosynthesis, leaf anatomy, and cellular constituents in the polyploid C4 grass *Panicum virgatum*. *Plant Physiology*, **84**, 461–6.

Weaver, J. E. (1954). *North American Prairie*. Lincoln NE: Johnsen.

Zohary, D. & Nur, U. (1959). Natural triploids in the orchard grass, *Dactylis glomerata* L. polyploid complex and their significance for gene flow from diploid to tetraploid levels. *Evolution*, **13**, 311–17.

Part two

Ecological interactions

8

Plant–plant interactions in grasses and grasslands

W. K. LAUENROTH AND M. O. AGUILERA

Plants do not experience average population densities or overall levels of resource supply. Spatial structure of plant populations and communities is an important consequence of the sessile nature of plants and of the discrete quality of individual organisms (Silvertown & Lovett Doust, 1993; Tilman, 1994). Close neighbours exploit the same resources, endure the same macro- and microenvironmental controls, and interact more or less independently with herbivores, pathogens and mutualistic organisms. Although plant–plant interactions are affected by all these factors, in this chapter we will focus on direct effects between neighbours and individual interactions mediated by resource exploitation. The remaining influences are covered elsewhere in this volume (see Chapters 9, 10 and 11).

The reciprocal effects between plants and their local environment are modulated by their life cycle. Establishment, resource acquisition, growth, recruitment, and death all affect and are affected by attributes of the local environment or neighbourhood. These continuous interactions among plants and their local environment are experienced by close neighbours over a range of temporal scales. The results of these interactions ramify upwards through the system to affect population, community, and ecosystem level properties.

In our analysis of plant–plant interactions, we will first deal with the spatial pattern of individuals, focusing on neighbourhood characteristics and their significance for individual performance and the recruitment of new individuals. We will then turn our attention to competitive ability. Here we will discuss both the effects of grasses on resource availability as well as the response of grasses to resource availability. We will end with an evaluation of the significance of plant–plant interactions for community and ecosystem processes.

Space partitioning and temporal dynamics of resource capture

Individual plants compete for the space they require after establishment (Ross & Harper, 1972). Performance of individual plants of *Festuca rubra* became increasingly correlated with available space as growth proceeded (Liddle, Butt & Hutchings, 1982). Space occupation by individual plants can be characterized in terms of neighbourhoods. The temporal dynamics of space capture after colonization most often results in overlapping neighbourhoods or domains of individual plants. In these areas of contest, the portions of space or volume can be under intense competition. However, when establishment events are rare, such as in many perennial grasslands, new space made available by the mortality of an individual is often pre-empted by established plants. Consequently, although roots and shoots can overlap, the area occupied by the individual, its neighbourhood, will be associated with its performance (Mead, 1966; Mithen, Harper & Weiner, 1984; Matlack & Harper, 1986; Aguilera & Lauenroth, 1993b). Despite the overlap in root and shoot systems, neighbourhoods are often operationally represented as though there was no overlap (Mead, 1966).

In the shortgrass steppe, where *Bouteloua gracilis* accounts for a large fraction of the total basal cover of plants (Lauenroth & Milchunas, 1992), a bunchgrass neighbourhood can be defined as the basal area of the plant plus the surrounding bare area closer to the edges of the plant than to any other individual (Aguilera & Lauenroth, 1993a). Performance of individual plants was significantly and positively related to the size of the neighbourhood (Aguilera & Lauenroth, 1993a). Neighbourhood quality is related to the proportion of bare-soil area of the neighbourhood. The more bare soil a neighbourhood has, the more potential for the plant to grow and reproduce. Aboveground net primary production and reproductive output of individual bunchgrasses were significantly associated with the fraction of bare soil of the neighbourhood (Fig. 8.1). Differences in the availability of bare soil in local neighbourhoods is probably related to past disturbances (Coffin & Lauenroth, 1988; Aguilera & Lauenroth, 1993a; Hook *et al.*, 1994). Small-scale disturbances similar to the sizes of individuals are the most common events causing mortality of existing individuals in the shortgrass steppe (Coffin & Lauenroth, 1988; Aguilera & Lauenroth, 1995). The subsequent temporal dynamics of resource capture are related to interactions between seedling establishment and the ability of the dominant grass to pre-empt the belowground space. Most of the area of openings caused by small disturbances (>99%) is within the range of the root systems of the surrounding individual of *Bouteloua gracilis* (Hook, Lauenroth & Burke, 1994). Thórhallsdóttir (1990) characterized the

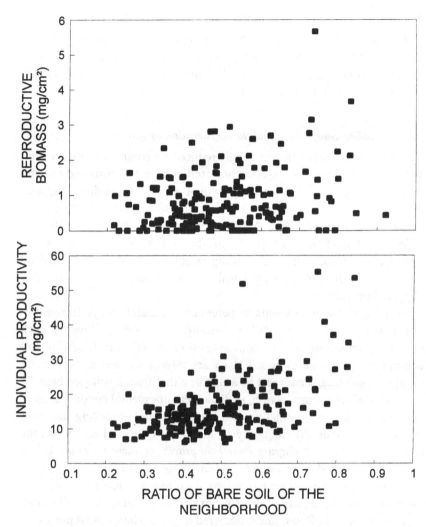

Fig. 8.1. Relationships between the proportion of bare ground in a neighbourhood and the reproductive output and productivity of individuals of *Bouteloua gracilis* (Aguilera & Lauenroth, submitted)

dynamics of a grassland community by deterministic competitive interactions within growing seasons and by disturbance and colonization obscuring the deterministic pattern over longer temporal scales. In a simulation study of patterns and processes in an annual grassland, patch disturbances were found to promote coexistence by preventing competitive exclusion (Wu & Levin, 1994).

Seedling recruitment and the spatial pattern of established grasses

Established grasses provide a heterogeneous environment for seedling establishment. Individual plant–plant interactions include asymmetric interactions between the established plants and seedlings as well as symmetric interactions between seedlings.

Adult–seedling competition and the regeneration of grasses

Seedling recruitment can be strongly reduced by competitive interactions with established grasses. Experimental studies have demonstrated negative effects on emergence, survival and growth of grass seedlings due to the presence of established grasses (Cook & Ratcliff, 1984; Gurevitch, 1986; Howe & Snaydon, 1986; Aguilera & Lauenroth, 1993b, 1995 [but see Fowler, 1988]). Potvin (1993) working in the Sand Hills of Nebraska, USA, reported complete failure of seedling establishment in native grassland unless the existing plants were killed or their aboveground material was clipped frequently.

Seedling recruitment events in perennial grasslands range from infrequent to rare (Erikkson, 1989; Lauenroth et al., 1994). However, such events are important for the conservation of population variability and community composition even if they are infrequent. For a recruitment event to take place, seeds must be available either from a soil seed bank or as a result of recent production, and microenvironmental conditions must be appropriate for germination and establishment. Modelling has been applied to estimate the frequency of microenvironmental conditions for recruitment of the bunchgrass *Bouteloua gracilis* (Lauenroth et al., 1994). Assuming that seeds were available, the analysis asked how frequently soil water and soil temperature conditions were met for successful seedling establishment. Precipitation and soil texture were found to be the most important controls. Recruitment occurred in years when annual precipitation was both above and below the long-term mean because of the critical role played by the seasonal distribution. Recruitment was predicted to occur on average from once in 50 years for soils with high silt contents to once in 5000 years for very sandy soils (>90% sand). The fact that seed production is variable among years (Coffin & Lauenroth, 1989, 1992) emphasizes the low frequency of recruitment by seedlings in some perennial grasslands. On the other hand, the importance of recruitment events is not always reflected by the frequency with which they occur. Grass seedlings have the ability to resist desiccation for relatively long periods (Chippindale, 1948; Fenner, 1978) and may, as has been found with other species, exhibit a large capacity to survive under shade (Mahmoud & Grime, 1974). Both of these may reduce the significance of the low frequency of recruitment events.

Sexual recruitment assures the conservation of genetic variability, although ramet expansion sustains the cover of grasses in many grasslands. The effectiveness of ramet expansion as a process to restore grass cover is constrained by the size of the disturbance and the speed of clonal expansion (Coffin & Lauenroth, 1988, 1989, 1993). Slow rates of ramet expansion may not be effective in recolonizing large disturbances. Establishment of individuals as a result of fragmentation of existing grasses has been studied using isozyme electrophoresis (McNeilly & Roose, 1984; Belsky, 1986; Lord, 1993). Lord (1993) found that within a 3 × 3 m plot of *Festuca novae-zelandiae*, 58% of the plants had shared isozyme profiles and the maximum distance between pairs of isozymically identical plants was 2.68 m. Fair (1996) investigated clonal fragmentation in the bunchgrass *Bouteloua gracilis* using a 35-year chart quadrat data set from a central North American grassland (Albertson & Tomanek, 1965). She found that in some years fragments contributed little to the population (i.e. each genet consisted of a single individual) and in other years there were twice as many individuals as genets (Fig. 8.2). On average there were 13 genets per square metre and 19 individual plants (1.46 individuals per genet). Both the results from *Festuca novae-zelandiae* and from *Bouteloua gracilis* emphasize the importance of clonal fragmentation as a process contributing to the size and the spatial structure of grass populations.

Seedling–seedling competition and the location of surviving seedlings

Following an establishment event, cohorts of seedlings compete for the available resources and changes in density over time often follow the 3/2 thinning law where the rate of elimination of genets is related to the rate of growth of the survivors (Kays & Harper, 1974). By contrast, seedling–seedling competition has been found to be of little significance in many studies in native plant communities (Silvertown, 1981; Gartner, Chapin & Shaver, 1983; McConnaughay & Bazzaz, 1990). In addition, seedling density has been associated with seedling growth indicating microsite favourability (Fowler, 1988). However, seedling–seedling interactions may be important for annual grasses with limited dispersal (Cheplick, this volume).

Intense sibling competition was highly significant among seedlings of *Sporobolus vaginiflorus*, an annual cleistogamous grass. Due to restricted

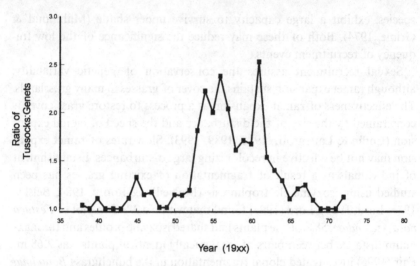

Fig. 8.2. The relationship between the number of genets and the number of tussocks of *Bouteloua gracilis* from 1938 to 1972 in a southern mixed prairie in central North America (Fair, 1996).

dispersal, competition was different in a high-density zone around the parent plant compared with an outer low-density zone. Survival and reproductive and vegetative performance were highest in the relaxed density zone (Cheplick, 1993). Differences in space availability determine local density variation for individual seedlings. Consequently, location of the mother plant determined the quality of the competitive environment for seedlings of the next generation.

Facilitation, competition and spatial patterns of other growth forms

The spatial distribution of patches of grasses is often important for the distribution of other growth forms in the community. In particular, influences are of ecological significance when the other growth forms are important for ecosystem function. Ecosystem properties can be influenced by the spatial patterns of colonizing species which in turn are determined by the microenvironment provided by established grasses.

In the Sonoran Desert, the succulent *Agave deserti* establishes in sheltered microhabitats under the canopy of the bunchgrass *Hilaria rigida* (Jordan & Nobel, 1979). The bunchgrass facilitates seedling establishment by reducing soil-surface temperatures and increasing nutrient availability, although the potential growth of the seedling is reduced by competition

(Franco & Nobel, 1988). Densely vegetated patches dominated by the tussock grass *Hilaria mutica* alternate with almost bare areas in the Chihuahuan Desert in Mexico (Montaña, 1992). Montaña (1992) found that a reduction in the abundance of tussock grasses was associated with a decrease in overall species richness, suggesting a link between the presence of *Hilaria* and the recruitment and survival of other species. In a *Chionochloa rigida* tussock grassland in New Zealand, seedling establishment of woody species was most frequent on the southern side of the tussocks, indicating the value of shelter in high elevation grasslands (Allen & Lee, 1989). On the other hand, strong competitive interactions with grasses for belowground resources can limit establishment of woody seedlings in montane grasslands (Wilson, 1993).

The opposite situation occurs when the dominant non-grass species determines the spatial distribution of grasses. In the Patagonian steppe of Argentina, shrubs provide favourable microsites under their canopies for colonization by grasses. After grasses establish, negative competitive interactions reduce the probability of further establishment (Aguiar, Soriano & Sala, 1992; Aguiar & Sala, 1994). The patch structure of the steppe consists of shrubs encircled by a ring of grass tussocks and scattered tussock grasses remaining following the death of a shrub. The spatial distribution of grasses has a large influence on the pattern of resource use because the rings of grasses constitute only 18% of cover but contribute 44% of total primary productivity (Soriano, Sala & Perelman, 1994).

Interactions due to the presence of roots or litter

Seedling recruitment is affected by microenvironmental processes that are the result of the presence of the root systems and/or litter deposition of established plants. Established perennial grasses provide a continuous source of roots and dead leaves as localized inputs in the space occupied by the plant. Annual grasses leave both their above- and belowground structure after completing their life cycle. In both cases, clumped deposition of litter influences microsite favourability. Direct effects of litter on seedling establishment include delay of emergence, reduction of competitive effects of neighbours, and increase in herbivore activity (Facelli, 1994; Facelli & Pickett, 1991).

Bergelson (1990) studied the fates of seedlings of *Capsella* and *Senecio*, both rapidly germinating annual dicots, under a background provided by the spatial distribution of dead plants of the annual grass *Poa annua*. Field experiments showed higher emergence of seedlings in a patchy distribution

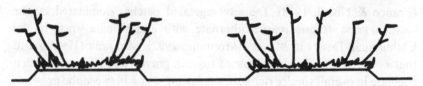

Fig. 8.3. Illustration of the relationship of microtopographic variability to the location of tussocks.

of remains of *Poa* than in a random distribution. The improved success of seedling emergence when the distribution was patchy was due to the presence of areas of relative low density of dead *Poa* plants. Furthermore, seedling survival was negatively affected by aboveground dead biomass of *Poa annua* (Bergelson, 1990).

Negative effects of microsites associated with the presence of a live or dead established plant were observed during emergence and early survival of *Bouteloua gracilis* seedlings (Aguilera & Lauenroth, 1995). In a two-year experiment, the number of seedlings per microsite in disturbed and control plots showed a consistently higher emergence in bare soil than within live or dead *Bouteloua* plants. Interactions between disturbance and microsite type were non-significant. Differences were related to the microsite environment beyond water extraction by living roots. Bare-soil microsites showed higher superficial (0–5 cm) soil water contents than plant microsites in an experimental period that included frequent watering, although disturbed and control plots showed similar soil water conditions (Aguilera & Lauenroth, 1995). Differences are probably associated with soil microtopography. Bare-soil microsites are in small depressions among hummocks of established plants; consequently plant microsites are, in absolute terms, shallower than bare-soil microsites (Fig. 8.3). Standing litter often does not support a favourable environment for seedlings to establish either in perennial or in annual grasses, affecting recruitment of conspecific dominants and subordinate species.

Competitive ability of grasses

Having established the local nature of plant–plant interactions in grasslands we turn now to the issue of competitive ability of grasses. Goldberg (1990) recognized that most plant–plant interactions occurred through intermediaries such as resources. Competitive ability can be evaluated in terms of the ability of an individual to suppress other individuals by depleting resources (competitive effects) and the ability of an individual to avoid being suppressed by responding to shifts in resource availability (competitive responses) (Goldberg & Landa, 1991). Species attributes are often linked to competitive ability encompassing both competitive effects and competitive responses. Species responses to environmental signals determined by plant–plant interactions include changes in morphology and physiology that allow individuals to exploit resources better and to avoid future competition by neighbours (Aphalo & Ballaré, 1995). We will discuss interactions between grasses by dealing with competitive effects and competitive responses separately and dividing each into aboveground and belowground resources.

The aboveground and belowground structure of grasses provide them with a particular set of characteristics with respect to resource use and competitive ability. Aboveground the laminate leaves of grasses tend to be displayed in an upright fashion which influences light interception and limits their ability to shade shorter plants compared with dicots. Furthermore, they have limited aboveground structural tissue which constrains the height over which they can display their leaves before shelf-shading becomes a detriment. Belowground, the adventitious root system of grasses allows them to have a very large root surface area in the soil close to their tillers both horizontally and vertically. Walter (1979) described grass root systems as 'intensive', meaning that they have a very high density of absorption capacity in the portion of the soil they occupy. Therefore, grasses can be very competitive for soil resources that occur near the plant and less competitive for aboveground resources. Under conditions that will support woody plants, grasses can be rapidly displaced by competition for light.

Competitive effects of grasses

Effects on aboveground resources: light

Considerable information exists on the influence of grasses on the quality and quantity of the light microenvironment. Photosynthetic photon flux density (PPFD), infrared radiation (IR), and shortwave radiation all decrease with increases in leaf area in patterns that differ according to species and type of stand (Baldocchi & Collineau, 1994). Species architecture linked to tiller arrangement determines the pattern of depletion of PPFD. Plant density controls the proportion of incident radiation intercepted per plant and the ratio of red to far red radiation (R/FR) at the plant base (Casal, Sanchez & Deregibus, 1986). However, there is scant literature on the differential competitive effects of canopy attributes of grass species in relation to light quality. Canopies of different species of grasses differentially change the R/FR. *Holcus lanatus* caused a greater reduction in the R/FR than canopies of *Lolium perenne* or *Agrostis tenuis*, which increased the negative effect on branching of *Trifolium* target plants (Thompson & Harper, 1988).

Both conopy density and architecture influence the effect a plant will have on its neighbours via light interception (Baldocchi & Collineau, 1994). The leaf area index that results in interception of 95% of the incident light is greater for grass canopies than for herbaceous dicots (Brougham, 1958). Tremmel & Bazzaz (1993) compared the effects of *Setaria faberi*, a bunchgrass with an open canopy, with three herbaceous dicots in terms of their effects on the distribution of light within their canopies. The bunchgrass had the greatest canopy depth (85 cm), the smallest leaf area, and at 20 cm above the soil surface allowed an order of magnitude more light to penetrate compared with the dicots. *Setaria* was found to be the weakest competitor of the four species evaluated.

Effects on belowground resources: soil nutrients

Depletion of soil nutrients is a key component of the successional and competitive model proposed by Tilman (1988). After three years, monoculture plots of five perennial grasses differed in the levels to which they reduced extractable nitrate and ammonium concentration in a nitrogen gradient experiment (Tilman & Wedin, 1991). The ability of each species to deplete belowground nutrients was associated with plant traits. Root mass explained 73% of the observed variance in nitrate. The late successional species *Schizachyrium scoparium* and *Andropogon gerardii* reduced soil solution nitrogen to the lowest levels on infertile soils, had lower vegetative growth rates, higher root allocation, lower reproductive allocation, and lower tissue nitrogen contents than the early successional species. The association of the resource-ratio hypothesis with the successional process is appropriate because of the overwhelming importance of competitive interactions among co-occurring grasses in the system studied by Tilman and his associates.

Depletion of soil nutrients is not the only way plants affect mineral nutrient resources. Locations of plants further influence distribution of nutrients in the landscape. Soils under grass plants have been found to have higher organic carbon, organic nitrogen, net nitrogen mineralization, and respirable carbon compared with adjacent bare soil microsites (Hook,

Lauenroth & Burke, 1991; Burke, Lauenroth & Coffin, 1995; Vinton & Burke, 1995). The mechanism involved is associated with the ability of grasses to remove nutrients from the interspaces and deposit them underneath the plant (Gibson, 1988; Nobel, 1989; Hook, *et al.*, 1991). Contour maps for extractable phosphate and potassium were consistently coupled with the locations of tussocks of *Pseudoroegneria spicata* in a sagebrush steppe (Jackson & Caldwell, 1993). However, plant cover was reported to have a larger effect on soil properties than the attributes of a particular species in a semi-arid grassland where water is the ultimate constraint (Vinton & Burke, 1995). In addition, nutrient distributions can influence the processes of species establishment and growth. In the shortgrass steppe, the death of individual bunchgrasses initially provided areas of enhanced nutrient supply under dead plants but the effect did not persist beyond several months (Kelly & Burke, 1997). Plants by acquiring nutrients and converting them into organic compounds increase the amount of nutrients in an ecosystem and are responsible for the patchiness in the distribution of nutrients and organic matter in the soil.

Effects on belowground resources: soil water

Grasses differ in their ability to deplete soil water. Harris (1967) reported that the advantage that the annual grass *Bromus tectorum* had over seedlings of the native bunchgrass steppe species *Agropyron spicatum* was its ability to extend the depth of its roots rapidly and deplete soil water before *Agropyron* seedlings became established. *Agropyron desertorum* extracted water faster than did *Agropyron spicatum,* resulting in the two species having different effects on the target *Artemisia* transplants (Eissenstat & Caldwell, 1988). Differential effects on water resources can be exerted on different layers of the soil. Stands of the introduced grass *Agropyron cristatum* have been found to use less water from deep soil layers than native shortgrass steppe species (Trlica & Biondini, 1990).

Many tussock grasses display centrifugal growth, resulting in a continual expansion of the circumference of the tussock. Experiments with *Bouteloua gracilis* revealed that tillers retain this directional growth when isolated under experimental conditions (Hook & Lauenroth, 1994). This directional growth resulted in higher water consumption in front of and under advancing tillers than in the trailing direction. Consequently, expanding fronts of *Bouteloua* will achieve preemption of soil water because leading root systems can achieve root length densities and rates of water use as high as the central root system (Hook & Lauenroth, 1994). The

220   W. K. Lauenroth and M. O. Aguilera

competitive effects of grasses related to water extraction are associated with
both the physiological and morphological attributes of their root systems.
Soil water extraction is species specific and is dependent on the size, number
and spatial location of roots.

Competitive responses of grasses

Responses to aboveground resources: light

The architecture of grasses, particularly tussock grasses, raises questions
about the role of self-shading in affecting photosynthesis and water use.
Caldwell *et al.* (1983) evaluated the effect of self-shading in two tussock
grasses *Agropyron spicatum* and *Agropyron desertorum* and concluded that
despite their differences in architecture and amount of foliage shaded, the
ratio of net photosynthesis to transpiration did not differ under field conditions. Calculations comparing widely spaced tussock grasses with a rhizomatous grass suggested that to intercept the same amount of light as
the tussock grass, the rhizomatous grass would have to occupy an area six
times greater than that occupied by the tussock grass. Daily incident photosynthetic photon flux density and net photosynthesis were simulated to
evaluate the responses to light heterogeneity in tussock species (Ryel,
Beyschlag & Caldwell, 1994). Self-shading, within-tussock light attenuation for tillers, resulted in reduction in potential carbon gains. However,
differences between tussock and uniform tiller distributions were greater
for *Agropyron desertorum* than for *Pseudoroegneria spicata*. The more
open arrangement of *Pseudoroegneria* resulted in more efficient use of
light. On the other hand, *Agropyron* tussocks were able to minimize light
resources for establishing potential competitors, better sequestering
belowground resources under the tussock. Furthermore, competitive
effects of species with tussock growth form on the light environment
differed resulting in trade-offs associated with the exploitation of belowground resources.

   Responses of grasses to the balance of R/FR radiation involve increases
of tillering rates in response to high R/FR (Deregibus, Sanchez & Casal,
1983; Deregibus *et al.*, 1985; Casal, Deregibus & Sanchez, 1985), increase
of shoot extension to low R/FR (Casal, Sanchez & Deregibus, 1987), and
reduction of shoot zenith angles (with respect to the vertical) to low R/FR
(Casal, Sanchez & Gibson, 1990). Grass species react differently to light
quality, although few studies attempted to sort competitive responses of
species or genets. Skálová & Krahulec (1992) found that clones of *Festuca*

rubra displayed different sensitivity to light quality, suggesting that clones of sparse canopies were more sensitive to shading.

Grasses respond to both the quantity and the quality of light in their environment. Responses to both quantity of light and quality are species specific and related to physiological and morphological characteristics.

Responses to belowground resources: soil nutrients

The classic work of Weaver (1919) characterized root systems of many common North American grasses and provided the foundation for much subsequent work. Drew, Saker & Ashley (1975) found greater lateral root extension due to nutrient enrichment in localized soil volumes explored by *Hordeum vulgare*. Grime (1994) reported that in an experimental evaluation of 36 herbaceous species, the 16 grasses were found to have a lesser root proliferation response than the dicots. The significance of these differences in root response will depend on the diffusive properties of the particular nutrients (Caldwell, 1994). For effective acquisition of immobile phosphate ions, prolific branching is required, whereas for mobile ions such as nitrate less branching should be necessary (Hutchings & de Kroon, 1994).

Grasses respond to spatial heterogeneity in nutrient distribution by differential proliferation of root systems and elevation of uptake capacity, although additional factors can influence these such as presence of neighbour roots, soil equilibrium chemistry for key nutrients and individual ability to provide energy supply for uptake and assimilation (Caldwell, 1994). Responses associated with root physiological activity included more rapid shifts in phosphate acquisition for *Agropyron desertorum* than *Pseudoroegneria spicata* demonstrated by uptake of phosphate isotope (Caldwell *et al.*, 1985; Caldwell *et al.*, 1987; Jackson, Manwaring & Caldwell, 1990). In addition, differential root proliferation in enriched patches of nutrients further increased the competitive responses of both grasses to nutrient competition (Jackson & Caldwell, 1989). Differences in the abilities of these two grasses to invade root-free soil gaps was related to differences in their competitive effectiveness (Eissenstat & Caldwell, 1989). By contrast, Larigauderie & Richards (1994) evaluated the root proliferation response of seven perennial grasses including *Agropyron desertorum* and *Pseudoroegneria spicata* and found similar root length relative growth rates into adjacent nutrient- enriched microsites and suggested that differences in the capacity for root proliferation may not be related to productivity or competitive ability.

Further advances in the evaluation of competitive abilities of species due

to root responses are both needed and impending because of the development of manipulations of nutrient concentrations in sharply defined
sectors of the root volume without the confounding effects of barriers to
root proliferation (Campbell & Grime, 1989). Competitive ability is determined by differential root proliferation ability and differential uptake
capacity and interactions occur in relation to the type of mineral nutrient.
However it is not uncommon to find different competitive abilities among
species with similar root proliferation capacities. Consequently, either
uptake or aboveground resources can overwhelm similarities in root proliferation.

Responses to belowground resources: soil water

Soil water uptake by plants requires two conditions to be met. A demand
for water must be created by the physiological response of aboveground
tissues and the plant must have active roots in wet soil. Periodic droughts
which are common in all grasslands reduce the ability of grasses to respond
to increases in water availability. Following a dry period, increases in soil-
water availability do not result in immediate increases in plant physiological status. Sala, Lauenroth & Reid, (1982) evaluated the aboveground
physiological response of *Bouteloua gracilis* to a 56-day drought followed
by a 32-mm rainfall event. Leaf water potential recovered within three days
but leaf conductance required 8–10 days to recover to pre-drought levels.
Is the rapidity of the response related to the size of the wetting event? After
a small (5 mm) experimental rainfall event, Sala & Lauenroth (1982) found
rapid (12 h) increases in leaf water potential and conductance for drought-
stressed *Bouteloua gracilis* that lasted up to two days. Mid-day leaf water
potential previous to the experimental watering was -6.2 MPa. This result
suggested that small rainfall events that are very frequent in semi-arid
regions may be important to species such as *Bouteloua gracilis* but it also
raised questions about root responses in the recovery of water status. In a
similar experiment, Lauenroth *et al.* (1987) found that existing roots
accounted for the rapid response to rewetting but that within 40 h of watering new roots were initiated. Simulation analysis suggested that the carbon
cost of producing new roots in response to a 5-mm rainfall event exceeded
the potential gain.

   Analysis of the stomatal behavior of two potential competitors,
Bouteloua gracilis and *Agropyron smithii*, suggested that the observed
differences were related to competitive responses to water supply. Sala *et al.*
(1982) proposed that physiological responses of *Bouteloua gracilis* were

more closely coupled with environmental variability of water supply typical of the shortgrass steppe, whereas *Agropyron smithii* was adapted to respond to a less variable water supply typical of the northern mixed prairie. Differential responses of roots across soil layers have been found in grasslands. Fitter (1986) studied differential seasonal root activity between shallow and deep soil in two guilds of grassland species. Grasses belonging to the spring guild were more active at shallower levels in the soil than those of the summer guild. Differences were associated with the seasonality of soil water availability of the site (Fitter, 1986). Horizontal spatial variability due to microsite or habitat distribution can increase the probability of species coexistence. Plant responses in *Stipa capensis* to experimental rainfall fluctuations were habitat dependent (Kadmon, 1993). Besides coexistence, examples of competitive displacement have been reported. Highly competitive responses to water stress account for mechanisms of exclusion of native grasses by introduced African grasses at some sites in neotropical savannas of Venezuela (Baruch & Fernández, 1993).

Scaling up from plant–plant interactions to community or ecosystem process

What significance do plant–plant interactions have for community and ecosystem dynamics? Answers to this question require scaling up from individuals to patches that include many individuals and from patches to landscapes that include many patches. A number of studies have suggested that the significance of plant–plant interactions decreases as one moves from the neighbourhood scale to patches and landscapes (Singh, Bourgeron & Lauenroth, 1996) The reason is that at large scales other processes besides those explicitly involving plant–plant interactions become important and in many cases overwhelm the small-scale processes. At the scale of plant–plant interactions, the dynamics are controlled by birth, growth, and mortality which are dependent upon species-specific characteristics. Disturbance is an example of perhaps the most ubiquitous and important process, external to the system made up of plant–plant interactions, that influences the dynamics of communities and landscapes and decreases the significance of small-scale processes (White & Pickett, 1985). Does this mean that plant–plant scale interactions are irrelevant to community and ecosystem dynamics?

Watt (1947) introduced the concept of communities as mosaics of patches and the idea that patches are related. The significance of this work for the issue of scaling-up plant–plant interactions is that it provides a conceptual framework for linking the processes and dynamics of neighbourhoods to the processes and dynamics of communities and ecosystems. Watt provided evidence from seven plant communities to support his ideas. In the past 20 years, these ideas have been used as the foundation for simulation models that simulate the dynamics of patches by representing competition among plants for resources on a small plot (neighbourhood). The major development of these models has occurred in forested systems (Shugart, 1984; Pastor & Post,1988; Urban, 1990; Pacala *et al.*, 1996) but recently several models have been used in grasslands (Coffin & Lauenroth, 1990; Wu & Levin, 1994). An important characteristic of all of these models is that they explicitly contain representations of plant–plant interactions as important determinants of the dynamics of communities and ecosystems. In addition, they often contain representations of the disturbance regime as well as stochastic weather and distributions of soil characteristics (Coffin & Lauenroth, 1993). To date, these models have made important contributions to both theoretical developments in community and ecosystem ecology as well as providing an important tool to help us evaluate potential human-induced effects on communities and ecosystems such as climatic and land-use change.

Our conclusion about the answer to the question of whether plant–plant interactions matter for dynamics at larger scales is yes they do. While other processes such as disturbances, weather variability, and landscape-scale variability in soil development and parent materials have large effects on community and ecosystem dynamics, understanding plant–plant interactions is fundamental to being able to predict how grasses will respond to these other processes.

Acknowledgements

Support for this work was provided by grants from the National Science Foundation (BSR-90–11659, DEB-9416815) and the Colorado Experiment Station (1–50661). MOA also acknowledges the support of the Instituto Nacional de Tecnología Agropecuaria of Argentina (PEI 80–032).

References

Aguiar, M. R. & Sala, O. E. (1994). Competition, facilitation, seed distribution and the origin of patches in a Patagonian steppe. *Oikos*, **70**, 26–34.
Aguiar, M. R., Soriano, A. & Sala, O. E. (1992). Competition and facilitation in the recruitment of seedlings in Patagonian steppe. *Functional Ecology*, **6**, 66–70.

Aguilera, M. O. & Lauenroth, W. K. (1993*a*). Seedling establishment in adult neighbourhoods: intraspecific constraints in the regeneration of the bunchgrass *Bouteloua gracilis. Journal of Ecology*, **81**, 253–61.

Aguilera, M.O. & Lauenroth, W. K. (1993*b*). Neighborhood interactions in a natural population of the perennial bunchgrass *Bouteloua gracilis. Oecologia*, **94**, 595– 602.

Aguilera, M. O. & Lauenroth, W. K. (1995). Influence of gap disturbances and type of microsites on seedling establishment in *Bouteloua gracilis. Journal of Ecology*, **83**, 87–97.

Albertson, F. W. & Tomanek, G. W. (1965). Vegetation changes during a 30-year period on grassland communities near Hays, Kansas. *Ecology*, **46**, 714–20.

Allen, R. B. & Lee, W. G. (1989). Seedling establishment microsites of exotic conifers in *Chionochloa rigida* tussock grassland, Otago, New Zealand. *New Zealand Journal of Botany*, **27**, 491–8.

Aphalo, P. J. & Ballaré, C. L. (1995). On the importance of information-acquiring systems in plant–plant interactions. *Functional Ecology*, **9**, 5–14.

Baldocchi, D. & Collineau, S. (1994). The physical nature of solar radiation in heterogeneous canopies: spatial and temporal attributes. In *Exploitation of Environmental Heterogeneity by Plants* ed. M. M. Caldwell and R. W. Pearcy, pp. 21–71. San Diego: Academic Press.

Baruch, Z. & Fernández, D. S. (1993). Water relations of native and introduced C_4 grasses in a neotropical savanna. *Oecologia*, **96**, 179–85.

Belsky, A. J. (1986). Population and community processes in a mosaic grassland in the Serengeti, Tanzania. *Journal of Ecology*, **74**, 841–56.

Bergelson, J. (1990). Life after death: site pre-emption by the remains of *Poa annua. Ecology*, **71**, 2157–65.

Brougham, R. W. (1958). Interception of light by the foliage of pure and mixed stands of pasture plants. *Australian Journal of Agricultural Research*, **9**, 39–52.

Burke, I. C., Lauenroth, W. K. & Coffin, D. P. (1995). Soil organic matter recovery in semiarid grasslands: implications for the Conservation Reserve Program. *Ecological Applications* **5**, 793–801.

Caldwell, M. M. (1994). Exploiting nutrients in fertile microsites. In *Exploitation of Environmental Heterogeneity by Plants* ed. M. M. Caldwell & R. W. Pearcy. pp. 21–71. San Diego: Academic Press.

Caldwell, M. M., Dean, T. J., Nowak, R. S., Dzurec, R. S. & Richards, J. H. (1983). Bunchgrass architecture, light interception, and water-use efficiency: assessment by fiber optic point quadrats and gas exchange. *Oecologia*, **59**, 178– 84.

Caldwell, M. M., Eissenstat, D. M., Richards, J. H. & Allen, M. F. (1985). Competition for phosphorus: differential uptake from dual-isotope-labeled soil interspaces between shrub and grass. *Science*, **229**, 384–86.

Caldwell, M. M., Richards, J. H., Manwaring, J. H. & Eissenstat, D. M. (1987). Rapid shifts in phosphate acquisition show direct competition between neighbouring plants. *Nature*, **327**, 615–16.

Campbell, B. D. & Grime, J. P. (1989). A new method of exposing developing root systems to controlled patchiness in nutrient mineral supply. *Annals of Botany*, **63**, 395–400.

Casal, J. J., Deregibus, V. A. & Sanchez, R. A. (1985). Variations in tiller dynamics and morphology in *Lolium multiflorum* Lam. vegetative and reproductive plants as affected by differences in Red/Far-Red irradiation. *Annals of Botany*, **56**, 553–9.

226   W. K. Lauenroth and M. O. Aguilera

Casal, J. J., Sanchez, R. A. & Deregibus, V. A. (1986). The effect of plant density
on tillering; the involvement of R/FR ratio and the proportion of radiation
intercepted per plant. *Environmental and Experimental Botany*, 26, 365–71.
Casal, J. J., Sanchez, R. A. & Deregibus, V. A. (1987). The effect of light quality
on shoot extension growth in three species of grasses. *Annals of Botany*, 59,
1–7.
Casal, J. J., Sanchez, R. A. & Gibson, D. (1990). The significance of changes in
the red/far-red ratio, associated with either neighbour plants or twilight, for
tillering in *Lolium multiflorum* Lam. *New Phytologist*, 116, 565–72.
Cheplick, G. P. (1993). Sibling competition is a consequence of restricted
dispersal in an annual cleistogamous grass. *Ecology*, 74, 2161–4.
Chippindale, H. G. (1948). Resistance to inanition in grass seedlings. *Nature*, 161,
65.
Coffin, D. P. & Lauenroth, W. K. (1988). The effects of disturbance size and
frequency on a shortgrass plant community. *Ecology*, 69, 1609–17.
Coffin, D. P. & Lauenroth, W. K. (1989). Spatial and temporal variation in the
seed bank of a semiarid grassland. *American Journal of Botany*, 76, 53–8.
Coffin, D. P. & Lauenroth, W. K. (1990). A gap dynamics simulation model of
succession in a semiarid grassland. *Ecological Modelling*, 49, 229–66.
Coffin, D. P. & Lauenroth, W. K. (1992). Spatial variability in seed production of
the perennial bunchgrass *Bouteloua gracilis* (Gramineae). *American Journal
of Botany*, 79, 347–53.
Cook, S. J. & Ratcliff, D. (1984). A study of the effects of root and shoot
competition on the growth of green panic (*Panicum maximum* var.
trichoglume) seedlings in an existing grassland using root exclusion tubes.
Journal of Applied Ecology, 21, 971–82.
Deregibus, V. A., Sanchez, R. A. & Casal, J. J. (1983). Effects of light quality on
tiller production in *Lolium* spp. *Plant Physiology*, 72, 900–2.
Deregibus, V. A., Sanchez, R. A., Casal, J. J. & Trlica, M. J. (1985). Tillering
responses to enrichment of red light beneath the canopy in a humid natural
grassland. *Journal of Applied Ecology*, 22, 199–206.
Drew, M. C. , Saker, L. R. & Ashley, T. W. (1975). Nutrient supply and the
growth of the seminal root system in barley. I. The effect of nitrate
concentration on the growth of axes and laterals. *Journal of Experimental
Botany*, 24, 1189–1202.
Erikkson, O. (1989). Seedling dynamics and life histories in clonal plants. *Oikos*,
55, 231–8.
Eissenstat, D. M. & Caldwell, M. M. (1988). Competitive ability is linked to rates
of water extraction. A field study of two aridland tussock grasses. *Oecologia*,
75, 1–7.
Eissenstat, D. M. & Caldwell, M. M. (1989). Invasive root growth into disturbed
soil of two tussock grasses that differ in competitive effectiveness. *Functional
Ecology*, 3, 345–53.
Facelli, J. M. (1994). Multiple indirect effects of plant litter affect the
establishment of woody seedlings in old fields. *Ecology*, 75, 1727–35.
Facelli, J. M. &. Pickett, S. T. A. (1991). Plant litter: dynamics and effects in plant
community structure and dynamics. *Botanical Review*, 57, 1–32.
Fair, J. (1996). *Demography of Bouteloua gracilis in shortgrass steppe and mixed
grass prairie*. Unpublished Master's Thesis, Fort Collins.
Fenner, M. (1978). A comparison of the abilities of colonizers and closed-turf
species to establish from seed in artificial swards. *Journal of Ecology*, 66,
953–63.

Fitter, A. H. (1986). Spatial and temporal patterns of root activity in a species-rich alluvial grassland. *Oecologia*, **69**, 594–9.

Fowler, N. L. (1988). What is a safe site?: Neighbor, litter, germination date, and patch effects. *Ecology*, **69**, 947–61.

Franco, A. C. & Nobel, P. S. (1988). Interactions between seedlings of *Agave deserti* and the nurse plant *Hilaria rigida*. *Ecology*, **69**, 1731–40.

Gartner, B. L., Chapin, F. S. & Shaver, G. R. (1983). Demographic patterns of seedling establishment and growth of native graminoids in an Alaskan tundra disturbance. *Journal of Applied Ecology*, **20**, 965–80.

Gibson, D. J. (1988). The maintenance of plant and soil heterogeneity in dune grassland. *Journal of Ecology*, **76**, 497–508.

Goldberg, D. E. (1990). Components of resource competition in plant communities. In *Perspectives on Plant Competition*, ed. J. B. Grace & D. Tilman, pp. 27–49. San Diego: Academic Press.

Goldberg, D. E. & Landa, K. (1991). Competitive effect and response: hierarchies and correlated traits in the early stages of competition. *Journal of Ecology*, **79**, 1013–30.

Grime, J. P. (1994). The role of plasticity in exploiting environmental heterogeneity. In *Exploitation of Environmental Heterogeneity by Plants*. ed. M. M. Caldwell & R. W. Pearcy, pp. 1–19. San Diego: Academic Press.

Gurevitch, J. (1986). Competition and the local distribution of the grass *Stipa neomexicana*. *Ecology*, **67**, 46–57.

Harris, G. A. (1967). Some competitive relationships between *Agropyron spicatum* and *Bromus tectorum*. *Ecological Monographs*, **37**, 89–111.

Hook, P. B. & Lauenroth, W. K. (1994). Root system response of a perennial bunchgrass to neighbourhood-scale soil water heterogeneity. *Functional Ecology*, **8**, 738–45.

Hook, P. B., Lauenroth, W. K. & Burke, I. C. (1991). Heterogeneity of soil and plant N and C associated with individual plants and openings in North American shortgrass steppe. *Plant and Soil*, **138**, 247–56.

Hook, P. B., Lauenroth, W. K & Burke, I. C. (1994). Spatial patterns of roots in a semiarid grassland: abundance of canopy openings and regeneration gaps. *Journal of Ecology*, **82**, 485–94.

Howe, C. D. & Snaydon, R. W. (1986). Factors affecting the performance of seedlings and ramets of invading grasses in established ryegrass swards. *Journal of Applied Ecology*, **23**, 139–46.

Hutchings, M. J. & de Kroon, H. (1994). Foraging in plants: the role of morphological plasticity in resource acquisition. *Advances in Ecological Research*, **25**, 159–238.

Jackson, R. B. & Caldwell, M. M. (1989). The timing and degree of root proliferation in fertile-soil microsites for three cold-desert perennials. *Oecologia*, **81**, 149–53.

Jackson, R. B. & Caldwell, M. M. (1993). Geostatistical patterns of soil heterogeneity around individual perennial plants. *Journal of Ecology*, **81**, 683–90.

Jackson, R. B., Manwaring, J. H. & Caldwell, M. M. (1990). Rapid physiological adjustment of roots to localized soil enrichment. *Nature*, **344**, 58–60.

Jordan, P. W. & Nobel, P. S. (1979). Infrequent establishment of seedlings of *Agave deserti* (Agavaceae) in the northwestern Sonoran Desert. *American Journal of Botany*, **66**, 1079–84.

Kadmon, R. & Shmida, A. (1990). Patterns and causes of spatial variation in the reproductive success of a desert annual. *Oecologia*, **83**, 139–44.

Kadmon, R. (1993). Population dynamic consequences of habitat heterogeneity: an experimental study. *Ecology*, **74**, 816–25.

Kays, S. & Harper, J. L. (1974). The regulation of plant and tiller density in a grass sward. *Journal of Ecology*, **62**, 97–105.

Kelly, R. H. & Burke, I. C. (1997). Heterogeneity of soil organic matter following death of individual plants in shortgrass steppe. *Ecology*,78, 1256–61.

Larigauderie, A. & Richards, J. H. (1994). Root proliferation characteristics of seven perennial arid-land grasses in nutrient-enriched microsites. *Oecologia*, **99**, 102–11.

Lauenroth, W. K. & Milchunas, D. G. (1992). The shortgrass steppe. In R. T. Coupland (ed.) *Natural grasslands*. Vol 8a. *Ecosystems of the world*. Amsterdam: Elsevier Scientific Press.

Lauenroth, W. K., Sala, O. E., Coffin, D. P. & Kirchner, T. B. (1994). The importance of soil water in the recruitment of *Bouteloua gracilis* in the shortgrass steppe. *Ecological Applications*, **4**, 741–9.

Lauenroth, W. K., Sala, O. E., Milchunas, D. G. & Lathrop, R. W. (1987). Root dynamics of *Bouteloua gracilis* during short-term recovery from drought. *Functional Ecology*, **1**, 117–24.

Liddle, M. J. Butt, C. S. J. & Hutchings, M. J. (1982). Population dynamics and neighbourhood effects in establishing swards of *Festuca rubra*. *Oikos*, **38**, 52–9.

Lord, J. M. (1993). Does clonal fragmentation contribute to recruitment in *Festuca novae-zelandiae*? *New Zealand Journal of Botany*, **31**, 133–8.

Mahmoud, A. & Grime, J. P. (1974). A comparison of negative relative growth rates in shaded seedlings. *New Phytologist*, **73**, 1215–19.

Matlack, G. R. & Harper, J. L. (1986). Spatial distribution and the performance of individual plants in a natural population of *Silene dioica*. *Oecologia*, **70**, 121–7.

McConnaughay, K. D. M. & Bazzaz, F. A. (1990). Interactions among colonizing annuals: is there an effect of gap size? *Ecology*, **71**, 1941–51.

McNeilly, T. & Roose, M. L. (1984). The distribution of perennial ryegrass genotypes in swards. *New Phytologist*, **98**, 503–13.

Mead, R. (1966). A relationship between individual plant spacing and yeild. *Annals of Botany*, **30**, 301–9.

Mithen, R., Harper, J. L., Weiner, J. (1984). Growth and mortality of individual plants as a function of 'available' area. *Oecologia*, **62**, 57–60.

Montaña, C. (1992). The colonization of bare areas in two-phase mosaics of an arid ecosystem. *Journal of Ecology*, **80**, 315–27.

Nobel, P. S. (1989). Temperature, water availability, and nutrient levels at various soil depths. Consequences for shallow-rooted desert succulents, including nurse plants effects. *American Journal of Botany*, **76**, 1486–92.

Pacala, S. W., Canham, C. D., Saponara, J., Silander, J. A., Kobe, R. K. & Ribbens, E. (1996). Forest models defined by field measurements: estimation, error analysis and dynamics. *Ecological Monographs*, **66**, 1–43.

Pastor, J. & Post, W. M. (1988). Response of northern forest to CO_2 induced climate change. *Nature*, **334**, 55–8.

Potvin, M. A. (1993). Establishment of native grass seedlings along a topographic/moisture gradient in the Nebraska Sandhills. *American Midland Naturalist*, **130**, 248–61.

Ross, M. A. & Harper, J. L. (1972). Occupation of biological space during seedling establishment. *Journal of Ecology*, **60**, 77–88.

Ryel, R. J., Beyschlag W. & Caldwell, M. M. (1994). Light field heterogeneity among tussock grasses: theoretical considerations of light harvesting and seedling establishment in tussocks and uniform tiller distributions. *Oecologia,* **98**, 241–6.

Sala, O. E. & Lauenroth, W. K. (1982). Small rainfall events: an ecological role in semi-arid regions. *Oecologia,* **53**, 301–4.

Sala, O. E., Lauenroth, W. K. & Reid, C. P. P. (1982). Water relations: a new dimension for niche separation between *Bouteloua gracilis* and *Agropyron smithii* in North American semi-arid grasslands. *Journal of Applied Ecology,* **19**, 647–57.

Shugart, H. H. (1984). *A Theory of Forest Dynamics.* New York: Springer-Verlag.

Silvertown, J. W. (1981). Micro-spatial heterogeneity and seedling demography in species-rich grassland. *New Phytologist,* **88**, 117–28.

Silvertown, J. W. & Lovett Doust, J. (1993). *Introduction to Plant Population Biology.* Oxford: Blackwell Scientific Publications.

Singh, J. S., Bourgeron, P. & Lauenroth, W. K. (1996). Plant species richness and species- area relations in a shortgrass steppe. *Journal of Vegetation Science,*7, 645–5.

Skálová, H. & Krahulec, F. (1992). The response of three *Festuca rubra* clones to changes in light quality and plant density. *Functional Ecology,* **6**, 282–90.

Soriano, A., Sala, O. E. & Perelman, S. B. (1994). Patch structure and dynamics in a Patagonian arid steppe. *Vegetatio,* **111**, 127–35.

Thompson, L. & Harper, J. L. (1988). The effects of grasses on the quality of transmitted radiation and its influence on the growth of white clover *Trifolium repens. Oecologia,* **75**, 343–7.

Tilman, D. (1988). *Plant Strategies and the Dynamics and Structure of Plant Communities.* Princeton: Princeton University Press.

Tilman, D. (1994). Competition and biodiversity in spatially structured habitats. *Ecology,* **75**, 2–16.

Tilman, D. & Wedin, D. (1991). Plant traits and resource reduction for five grasses growing on a nitrogen gradient. *Ecology,* **72**, 685–700.

Thórhallsdótir, T. E. (1990). The dynamics of a grassland community: a simultaneous investigation of spatial and temporal heterogeneity at various scales. *Journal of Ecology,* **78**, 884–908.

Tremmel, D. C. & Bazzaz, F. A. (1993). How neighbor canopy architecture affects target plant performance. *Ecology,* **74**, 2114–24.

Trlica, M. J. & Biondini, M. E. (1990). Soil water dynamics, transpiration, and water losses in a crested wheatgrass and native shortgrass ecosystem. *Plant and Soil,* **126**, 187–201.

Urban, D. L. (1990). *A versatile model to simulate forest pattern: a users guide to ZELIG version 1.0.* Charlottesville: Environmental Science Department.

Vinton, M. A. & Burke, I. C. (1995). Interactions between individual plant species and soil nutrient status in shortgrass-steppe. *Ecology,* **76**, 1116–33.

Walter, H. (1979). *Vegetation of the Earth.* New York: Springer-Verlag.

Watt, A. S. (1947). Pattern and process in the plant community. *Journal of Ecology,* **35**, 1–22.

Weaver, J. E. (1919). *The Ecological Relations of Roots.* Carnegie Institute of Washington Publication 286.

White, P. S. & Pickett, S. T. A. (1985). Natural disturbance and patch dynamics: an introduction. In *The Ecology of Natural Disturbance and Patch Dynamics,* ed. S. T. A. Pickett & P. S. White, pp 3–13. Orlando: Academic Press.

Wilson, S. D. (1993). Competition and resource availability in heath and grassland in the Snowy Mountains of Australia. *Journal of Ecology*, **81**, 445–51.

Wu, J. & Levin, S. A. (1994). A spatial patch dynamic modeling approach to pattern and process in an annual grassland. *Ecological Monographs*, **64**, 447–64.

9
Competition between grasses and woody plants

SCOTT D. WILSON

Introduction

Grasses are good models for studying the role of competition in forming ecological patterns because they contribute to obvious patterns, such as the boundary between prairie and forest. Further, grasses are suitable for short-term experiments because they achieve full size and reproductive maturity quickly. In this review I examine competition in the context of the replacement of grasses by trees along gradients of increasing evapotranspiration or productivity. Grasses are also replaced by shrubs as grassland merges into desert, and I have drawn examples from that system where appropriate.

The boundary between grass and woody vegetation is especially interesting because its causes are unknown. The boundary is associated with very small differences in climate, making it likely to be highly responsive to global change (Emanuel, Shugart & Stevenson, 1985). Its abrupt nature may reflect feedbacks between plants and resources (Wilson & Agnew, 1992). Temporal shifts between grass and woody vegetation are also often sudden, prompting Westoby, Walker & Noy-Meir (1989) to suggest that grassland vegetation dynamics are best described using state and transition models. In spite of many examples of competition between grasses and trees, however, it is not at all clear how competition contributes to the boundaries between them.

Here I review what features distinguish grass competition from competition involving other kinds of plants. Because the grass plant has a unique morphology, I examine the possible contribution of niche partitioning to the distribution of grasses. General models relating species distributions to competition are evaluated using mostly experimental evidence. The role of feedbacks is assessed. Finally, I try to identify promising avenues for future research.

Grasses as competitors

The growth form of perennial grasses is characterized by extensive tillering, high root:shoot ratios, buds near the soil surface, and a lack of vertical stems. Buds near the soil surface may be protected from aboveground disturbances such as fire or grazing, and following such disturbances, grasses might enjoy a competitive advantage over woody plants with elevated, susceptible meristems (White, 1983; Milchunas, Sala & Lauenroth, 1988). In contrast, both fire and grazing have been cited as factors allowing woody plants to establish in grassland by suppressing grass and creating establishment sites for shrubs (McClaran & Bartolome, 1989; Brown & Archer, 1989; Bush & Van Auken, 1995).

The impacts of fire and grazing are variable and may be secondary to the influence of water availability (Leigh et al., 1987; Teague & Smit, 1992; Sala, Lauenroth & Golluscio, 1996). The high root:shoot ratios of grasses might confer an advantage over trees in competition for water and nutrients (Tilman, 1988); root:shoot ratios are typically 6:1 in grasslands, vs 1:4 in forests (Caldwell & Richards, 1986). These ratios are for mass, however, and because uptake rates are proportional to root length, and because grass roots are always fine whereas roots of woody plants are often coarse (Caldwell & Richards, 1986), the ratio of root length to total mass may be even higher for grasses than is suggested by mass ratios. Lastly, the fibrous nature of grass roots means that they can pre-empt infiltrating soil water before it reaches the deeper tap roots characteristic of woody plants (Walter, 1971). In spite of the high root:shoot ratio of grasses, grasslands may have lower rates of root turnover than forests (30–60% in grasslands vs 50–250% for fine roots in forests; Dahlman & Kucera, 1965; Anderson & Coleman, 1985; Aber et al., 1985; George & Marschner, 1996), which could affect the relative abilities of grasses and trees to garner patchy soil resources (Grime, 1994).

As a result of high root:shoot ratios and relatively low soil resource availabilities, competition in alpine and prairie grasslands is primarily belowground (Wilson, 1993a,b; Gerry & Wilson, 1995). Light competition increases with soil fertility (Lauenroth & Coffin, 1992; Wilson & Tilman, 1993).

Another important difference between grasses and woody plants may be in their relative nutrient requirements (Berendse, 1994a). For example, woody tissue contains little N (Aerts, 1995) and requires no C for respiration (Schulze, 1982) but allows a woody plant to pre-empt herbaceous species at the start of each growing season. On the other hand, water-use efficiencies are high in grasses with the C_4 photosynthetic pathway, which may assist them in competing with C_3 woody plants in dry periods.

The differences between grasses and woody plants are so great that, to some extent, they might avoid competition by using resources differently.

Niche differentiation

Grasses and trees may partition soil depth if fibrous roots allow grasses to efficiently exploit shallow soil wetted by rainfall, and tap roots allow trees access to deeper soil water (Walter, 1971). Walker (1981) developed a Lotka–Volterra type competition model that describes coexistence of grasses and trees under these conditions. This hypothesis is supported by differences in tissue ^{15}N concentrations among tree, shrub, and grass species in Alaska, which suggests that these life forms differ in their N sources, possibly in terms of uptake depth (Schulze, Chapin & Gebauer, 1994). Further, rocky outcrops in Australian and South African grasslands are often dominated by trees, presumably because there is no shallow water available to grasses on such outcrops and the trees have access to deeper water in cracks (Costin, 1954; Teague & Smit, 1992). On finer Patagonian soils, grass removal increased the water content of soils <60 cm deep whereas shrub removal did not, suggesting that the shrubs obtained water from deeper layers (Sala *et al.*, 1989). Alternatively, shrub mass might have been too small to affect soil moisture in that experiment. Semi-arid Colorado grasslands characterized by frequent light rains rarely have water penetration to deeper layers (Sala, Lauenroth & Parton, 1992), suggesting that both grasses and woody plants should allocate roots near the soil surface (Caldwell & Richards, 1986). The depth distributions of grass and tree roots usually overlap (Knoop & Walker, 1985; Wilson & Kleb, 1996). In summary, the plasticity of allocation probably causes the roots of grasses and woody plants to intermingle in shallow soil where water is most available.

Another potential axis for differentiation involves the variability of resources (Stark, 1994). Because of their large size relative to grasses, woody plants may increase spatial heterogeneity as they colonize grasslands (Schlesinger *et al.*, 1990; Rejmanek & Rosen, 1992). The variability of biomass is higher in forests than in adjacent prairies (Briggs, Seastedt & Gibson, 1989) as is the variability of soil moisture (Wilson & Kleb, 1996). Patchiness in soil resources under woody vegetation could be enhanced by stemflow (Parker, 1983) or uptake by roots distributed on larger scales. Large trees could benefit from an increased scale of patchiness by averaging over the resource-rich and poor locations by virtue of their relatively large size. A smaller grass, by comparison, would have to deal with a relatively uniform, frequently poor, environment.

Plants might also affect and react to the temporal variability of resources. Trees might reduce the temporal variability of moisture by intercepting small rainfalls with their canopies and causing water to evaporate before it reaches the soil. Thus, invading trees might decrease both the amount and temporal variability of water available to grasses. Grasses could benefit from relatively high levels of temporal variability. They have lower rates of root turnover than trees (Dahlman & Kucera, 1965; Aber et al., 1985), and their persistent roots should allow them to exploit brief periods of resource availability (Grime 1994). A slow growing grass derived more benefit from pulsed resources than did a faster growing grass (Campbell & Grime, 1989), and herbs are reported to be better able to exploit resource pulses than are woody plants (Schulze & Chapin, 1987). On the other hand, root proliferation rates in the shrub Artemisia tridentata in response to nutrient addition were intermediate to those of two grasses, Agropyron desertorum and A. spicatum (Jackson & Caldwell, 1989). It is interesting that grasses with relatively low rates of root turnover are associated with climates characterized by great temporal variability in water availability. Trees, in contrast, have low rates of shoot turnover and occur in forests with great temporal variability in light (Pearcy et al., 1994). This field deserves further attention, but the experiments required to address constant and fluctuating resources in time and space are very large (Naeem & Colwell, 1991).

Sala et al. (1996) have incorporated both temporal variability in water and distinct rooting depths into a model of grass–shrub interactions. They argue that grasses should dominate sites where water tends to be available at the soil surface. This could be the case where precipitation falls during the growing season, or on soils with low infiltration rates. Shrubs should dominate sites with deeper soil water, which could be caused by winter precipitation or by coarse soils. Their model predicted the balance of shrubs and grasses at a variety of sites in the USA.

A further difference between grasses and woody plants is their distinct mycorrhizae. Grasses are typically associated with vesicular-arbuscular mycorrhizae whereas most trees are colonized by ectomycorrhizae (Newman & Reddell, 1987; Johnson et al., 1991). The distinction is not universal, however, and grasses and woody plants can even be linked by mycorrhizae (Newman et al., 1994). An absence of ectomycorrhizae from grasslands could hinder the establishment of tree seedlings that are able to disperse and germinate there. Tree seedlings grew very poorly in

Tasmanian grassland soils unless they were supplied with inoculum from forest soils (Ellis & Pennington, 1992). In contrast, trees established on forest soil and transplanted into alpine grasslands grew very well (Wilson, 1993a). The possibility that trees are excluded from grasslands by a lack of suitable ectomycorrhizae deserves further attention. Similarly, the colonization of grasses by vesicular-arbuscular mycorrhizae may be negatively affected by the soil community under trees – the proportion of roots of the prairie grass *Schizachyrium scoparium* infected by mycorrhizae decreased with increasing tree cover (Benjamin, Anderson & Liberta, 1989).

In spite of several possible mechanisms for reducing competition by niche specialization, there is ample evidence that grasses and trees interact.

Grass–tree interactions

There is a large literature on grass–tree interactions, much of it related to forestry and range management (Westoby *et al.*, 1989; Milchunas & Lauenroth, 1993; Harvey, Mohammed & Noland, 1993). There are both observational and experimental studies, and many experiments are now carried out in the field. I give field experiments more attention than greenhouse or garden studies as they are more likely to be relevant to natural communities (Hairston, 1989). Similarly, I give more attention to removal than substitutive experiments, as the outcomes of substitutive experiments are sensitive to initial densities and conditions which can differ from field conditions (Herben & Krahulec, 1990).

Woody species suppress grasses in many cases. The experimental removal of woody plants increases grass or herb productivity both in forests (Ellison & Houston, 1958; Tamm, 1991; Riegel, Miller & Krueger, 1992) and grasslands (Stuart-Hill & Tainton, 1989; Harrington & Johns, 1990; Callaway, Nadkarni & Mahall, 1991; Aguiar, Soriano & Sala, 1992; Wilson, 1993a). Controlling shrub competition allows herbs to establish in both British heath (Miles, 1974) and in American chaparral (Swank & Oechel, 1991). Grass abundance was decreased by moving soil blocks from an annual California grassland to adjacent oak stands (Marañón & Bartolome, 1993). Oak leaves added to a New Jersey old field suppressed herbs and indirectly enhanced tree seedlings (Facelli, 1994). Removing tree litter from a subalpine forest produced a 15-fold increase in the density of monocot seedlings (Wilson & Zammit, 1992).

In contrast, woody vegetation can also enhance grass establishment (Wilson, 1993a) and production (Barth & Klemmedson 1978; Weltzin &

Coughenour, 1990; Callaway, 1995). Matorral shrubs protect herbs from grazing (Jaksic & Fuentes, 1980). Belsky (1994) concluded that higher soil fertility under Kenyan savanna trees was primarily responsible for the relatively high grass productivity there. In general, woody plants seem to benefit grass in warm places like Africa or Arizona (Barth & Klemmedson, 1978), but suppress it in cooler places such as central North America (Bailey & Wroe, 1974; Ko & Reich, 1993; Wilson & Kleb, 1996). At intermediate temperatures, the effect of woody plants may be neutral or variable (Jackson et al., 1990; Callaway, et al., 1991). The direction of impacts may differ between aboveground and belowground: Patagonian shrubs benefited grass seedlings by shading them from intense sunlight, but the benefit was negated by intense root competition, so that the net effect of shrubs on grasses was negative (Aguiar, et al., 1992).

Bertness & Callaway (1994) suggest that neighbour effects vary from positive to negative as productivity increases. This is supported by experiments in Wisconsin old fields, where the effects of trees on grasses shifted from positive to negative as resource limitations shifted from belowground to aboveground (Ko & Reich, 1993). Similarly, trees and herbs sown together on poor soil in England showed positive correlations in performance, whereas they were negatively correlated on rich soils (Buckley, 1984).

Grass can limit tree establishment: grass removal doubled oak seedling survivorship in California (Adams et al., 1992). Both fire and grazing can promote tree or shrub establishment by removing grass (McClaran & Bartolome, 1989; Harrington, 1991; Milchunas & Lauenroth, 1993; McPherson, 1993; Belsky & Blumenthal, 1997). Grasses can also have non-competitive negative effects on tree establishment by harbouring grazers (Gill & Marks, 1991; Myster & Pickett, 1993; Wilson, 1994) or carrying fires (Harrington, 1991; D'Antonio & Vitousek, 1992). On the other hand, subalpine grass tussocks and tree seedlings are positively associated in Australia (Noble, 1980) and New Zealand (Allen & Lee, 1989). Old-field grasses can promote tree seedling establishment by increasing soil moisture (Gill & Marks, 1991) and providing shelter from grazers (Harrison & Werner, 1984; De Steven, 1991).

Tree or shrub growth is frequently suppressed by grass (Clements, Weaver & Hanson, 1929; Parker & Salzman, 1985; Brown & Archer, 1989; Stuart-Hill & Tainton, 1989; Elliot & White, 1989; Bush & Van Auken, 1990, 1995; Wilson, 1993a,b; Egerton & Wilson, 1993; Boeken & Canham, 1995). While there are examples of trees facilitating grass establishment and growth, and grass facilitating tree establishment, I found no instances of grass facilitating the growth of established trees. This is probably due to differences in morphology: trees can benefit grass and grass can benefit small trees by offering shelter from sun, but grasses cannot have this positive effect on established trees, and can only have negative effects on large trees by belowground competition.

In summary, there are many examples of both facilitative and competitive effects between grass and woody plants. Organizing these results requires models.

General models of grass-tree competition

An advantage of any model over no model is that it provides a logical structure with which we can organize our assumptions, experiments and predictions. Models specific to grass–tree interactions are reviewed by Belsky (1990), and others exist (e.g. McMurtrie & Wolf, 1983; Riechert & Hammerstein, 1983; Walker *et al.*, 1989; Menaut *et al.*, 1990; Sala *et al.*, 1996; grass and tree models are compared by Coffin & Urban, 1993). Here I evaluate the possible contribution of general models of plant competition to understanding grass–tree relationships.

One general set of models suggests that competition may be most important in communities with high levels of standing crop, either due to a lack of disturbance or to high fertility (Fig. 9.1*A*, *B*; Grime, 1973; Connell, 1978; Huston, 1979). Forests may represent such communities, and treeless areas may be less fertile, or more disturbed, and experience less competition (Wardle, 1985; Keddy & MacLellan, 1990). Trees might be absent from grasslands because they are intolerant of drought, fire or grazing (e.g. Axelrod, 1985).

Drought might affect trees more than grasses because of their low root:shoot ratios and, at mid-latitudes, their lack of the C_4 photosynthetic pathway. On the other hand, both tree and grass populations were severely reduced by the droughts of the 1930s on the North American Great Plains. The mortality of trees was as high as 80% (Albertson & Weaver, 1945), but native perennial grasses also declined sharply: the basal covers of communities dominated by *Schizachyrium scoparium* and *Bouteloua gracilis* fell from 50% to 10% and from 90% to 35% respectively (Albertson & Tomanek, 1965). Thus, trees are not necessarily more susceptible to drought than grasses. In fact, the invasion of prairie by the clonal tree *Populus tremuloides* was positively correlated with drought (Bailey & Wroe, 1974), probably because of a proliferation of suckers when drought killed meristems and removed apical dominance. On the other hand, given the difference in stem density between trees and grasses, removing a large but

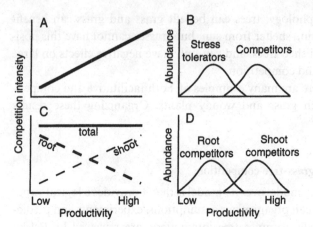

Fig. 9.1. Contrasting views of interactions between competition and productivity. One view is that competition increases with productivity (*A*), so that plants in less productive habitats (such as grasses) tend to be tolerant of stress and are replaced by more competitive species (such as trees) as productivity increases (*B*). Another view is that competition works at all levels of productivity because root competition declines and shoot competition increases as productivity increases (*C*). As a result, plants in less productive habitats are superior root competitors and plants in more productive habitats are superior shoot competitors (*D*).

equal proportion of each from a site will leave far more grass than trees (Inouye, Allison & Johnson, 1994).

Fire may reduce woody plants, especially if it kills small trees (White, 1983) or shrubs (Leigh *et al.*, 1987). Repeated burning, however, does not eliminate clonal woody species such as *P. tremuloides* or *Symphoricarpos occidentalis* (Svedarsky, Buckley & Feiro, 1986; Bailey, Irving & Fitzgerald, 1990; Anderson & Bailey, 1980), and fire may also promote non-clonal woody plants (McClaran & Bartolome, 1989).

Grazing might favour grasses with protected meristems over taller woody plants (Milchunas *et al.*, 1988, Bailey *et al.*, 1990; Belsky, 1990). Fossorial mammals may also exclude trees from grasslands (Cantor & Whitham, 1989; Inouye *et al.*, 1994). On the other hand, the direct effect of grazing on shrubs is only one aspect of these complex interactions, and the net effect is often an increase of woody plants associated with grazing (Brown & Archer, 1989; Milchunas & Lauenroth, 1993; Belsky & Blumenthal, 1997). Thus, while it is possible to argue that woody plants should be more sensitive to drought, fire, and grazing, there are few comparative data to support the idea (e.g. Reader *et al.*, 1983).

A second general set of models assumes that competition is no less intense in grasslands than forests (Fig. 9.1*C*, *D*). Trees invading forest clearings experienced the most intense competition from the grasses dominating the poorest soils (Hill, Canham & Wood, 1995), and the exclusion of trees from subalpine grasslands was associated with competition from grass, but not with any abiotic factor (Fensham & Kirkpatrick, 1992). Direct comparisons of competitive effects in forest and prairie suggest that root competition is most intense in prairie (Wilson, 1993*b*), and that the total effect of competition is comparable between woody and grassy habitats (Wilson, 1993*a*; D. A. Peltzer & S. Wilson *unpublished data*). This might be because of a shift from root to shoot competition as productivity increases (Newman, 1983; Tilman, 1988; Lauenroth & Coffin, 1992; Wilson & Tilman, 1993). There may also be a shift in the species which are competitive dominants at different levels of resource availability (Walter, 1985; Tilman, 1988). Grasses with high root:shoot ratios, for example, might reduce soil resources to levels below which species with lower root allocation can obtain sufficient resources for growth (Tilman & Wedin, 1991). At the same time, high root:shoot ratios should make grasses poor competitors for light relative to trees. Morphology should impose trade-offs between competitive abilities for light and soil resources. Removal experiments, however, reveal only small variations in competitive responses between environments differing in nutrient supply (Wilson, 1993*a*; Wilson & Tilman, 1995; D. A. Peltzer & S. Wilson *unpublished data*). Plants may differ little in response because of plasticity: they tend to respond to resource depletion in similar ways, regardless of their size or life form (Gerry & Wilson, 1995).

Grasses and trees might diverge more in competitive effect than in competitive response (Goldberg, 1990). The effect of trees on light is, of course, likely to be much greater than the effect of grasses. What is not clear is whether the effect of grasses on soil resources is greater than the effect of trees. On a per-gram basis, a grass with a high root:shoot ratio should have greater competitive effects than a tree with a low root:shoot ratio. Competitive effects of different life forms (e.g. grasses and trees) cannot be compared using experiments in different natural environments, because the effect of life forms is confounded by environment (e.g. Wilson & Tilman, 1995; Hill *et al.*, 1995). Competitive effects need to be measured in similar environments. Changes in competitive effects between environments could be tested for by establishing pure stands of each growth form in controlled and contrasting environments. Such experiments might need to run several years for their different impacts to be detectable. A one-year experiment in the south-eastern USA found no difference between a grass and a tree in their effects on soil water (Mitchell *et al.*, 1993). In contrast, a three-year experiment found significant differences among five prairie grasses in their effects on soil available N (Tilman & Wedin, 1991).

Comparisons of competitive effects also need to account for possible differences in per-gram effects of neighbours (Goldberg, 1987). *Pinus ponderosa* seedlings, for example, were suppressed much more by the grass *Dactylis glomerata* than by *Bouteloua gracilis* (Elliott & White, 1989), but this might simply reflect differences between the grasses in their mass at the start of the experiment. Differences among neighbour species in per-gram effects would suggest that differences in competitive effects are related to allocation or physiology. Specifically, we might expect grass to have larger per-gram effects than trees on soil resources, whereas trees should have larger per-gram effects than grasses on light. The per-gram effects of New South Wales trees on grasses was about twice that of shrubs (Harrington & Johns, 1990), presumably because of the greater stature of trees.

I have discussed these general models in relation to systems where taller woody plants replace grasses as productivity increases. The models do not account for the replacement of grasses by taller shrubs as productivity decreases and grassland merges into desert.

Another case in which shrubs are restricted to nutrient poor sites, and are apparently out-competed by grasses under more nutrient rich circumstances, is in the *Calluna*-dominated heaths of the Netherlands. Heathlands mined for peat are initially colonized by the ericaceous shrubs *Calluna* or *Erica*. Over several decades, shoot mass, soil organic matter and soil N all increase and the shrubs are overtopped by the grass *Molinia* (Berendse, 1994a). Succession follows a sequence from shrubs to grasses, with concomitant increases in community mass, nutrient availability and light attenuation. Aerts & van der Peijl (1993) and Berendse (1994b) propose that, on soil with low nutrient availability, a species with a relatively low nutrient requirement will out-compete a species with a higher nutrient requirement, by producing more mass per unit of limiting nutrient, a concept similar to Tilman's R^* models (Tilman, 1988; Huisman, 1994). On soil with high nutrient availability, the species with low nutrient requirements may be replaced by a species with higher nutrient requirements, if the species with the high nutrient requirement has a sufficiently high growth rate (at high nutrient availability) to attain more mass and resources than the more efficient species (Berendse, 1994b).

This model describes the heathland system. The shrubs have a low growth rate and, by virtue of their evergreen leaves and woody stems, have low nutrient loss rates and high nutrient use efficiencies. They may even contribute to maintaining low nutrient availability by depositing litter containing very little nitrogen. The grass, in contrast, has a higher growth rate, and loses most of its tissue N each year (Berendse, 1994a). An eight-year experiment provided some support for the model in that the addition of either fertilizer N or litter to shrub plots resulted in a shift to dominance by the grass (Berendse, Schmitz & de Visser, 1994). The shift occurred regardless of whether the litter was from shrub or grass plots, whereas the model suggests that the shift should occur only for grass litter. The experiment might have behaved as predicted by the model if the major component of the litter, root litter, had been transferred.

Is the heathland model broadly applicable to interactions between grasses and woody plants? Perhaps not, since it describes community responses to soil disturbance (peat mining) over the course of a few decades, and is a model of primary succession. Further, it differs from most grass–shrub systems in that the grass overtops its woody neighbours over time. The rates of root turnover are higher for the grasses than for shrubs (Aerts, Bakker & de Caluwe, 1992), which is also contrary to the usual pattern. In spite of the particular nature of the system, it provides important insights into competition between contrasting life forms occupying soils of different fertility. The model explicitly incorporates positive feedback.

Tree invasions and feedbacks

Feedback occurs when the effect of one component on a second is also carried back to the first component. Negative feedbacks can maintain a biological system in its current state, whereas positive feedbacks can change a system. Negative feedbacks could account for the segregation of plant life forms and positive feedbacks could allow invasions (DeAngelis, Post & Travis, 1986; Berendse, Bobbink & Rouwenhorst, 1989; Wilson & Agnew, 1992).

Grasslands may exclude trees because of negative feedbacks on soil resources (Fig. 9.2A). The high root:shoot ratio of grasses promotes high insolation and evaporation rates. This could slow mineralization by depriving soil microbes of water. Low availabilities of soil water and nutrients should favour plants with high root allocation. Thus, the negative effects of grasses on soil resources should exclude trees (with lower root:shoot ratios) and stabilize the system. A similar argument can be made for fire (Kellman, 1984): grasslands are drier and more likely to carry a fire than are forests, which helps to exclude trees and keeps grassy areas dry (Fig. 9.2A).

Invading trees may generate positive feedback by their effects on soil

A

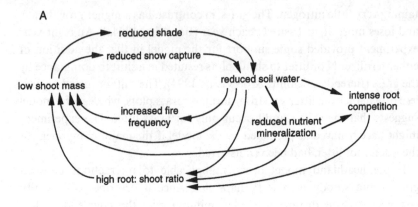

B

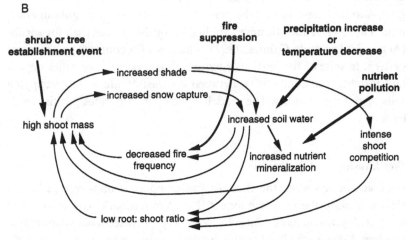

Fig. 9.2. (A) Negative feedbacks involving resources, fire and competition that may contribute to the exclusion of trees from grasslands, in cases where grasses and trees are segregated along gradients of increasing productivity. (B) Positive feedbacks that may allow trees to persist in grasslands. External factors that could initiate positive feedbacks are shown in bold.

resources (Fig. 9.2B). Trees cast shade, which cools the soil, reduces evaporation, and increases soil moisture. Trees also slow wind and capture more snow than adjacent treeless areas (Payette & Filion, 1985; Wilson & Agnew, 1992). More snow under trees contributes directly to increasing soil moisture, and may contribute to soil moisture indirectly by increasing the late winter albedo of deciduous forests, thus delaying snow melt and making water more available early in the growing season (Hare & Ritchie, 1972 make the opposite argument for arctic treelines). All these effects may be enhanced by the fact that trees typically establish in grasslands at high densities (Bailey and Wroe, 1974; Magee & Antos, 1992) or in association with shrubs (Bird, 1927; Petranka & McPherson, 1979; Callaway & D'Antonio, 1991; Callaway, 1992). On a global scale, productivity increases with nutrient mineralization (Schulze & Chapin, 1987) and forests tend to develop soils with less organic matter than do grasslands (e.g. Brady, 1990). On a local scale, soils under shrubs or trees tend to have more moisture and available N than soils under adjacent grassland (Charley & West, 1975; Petranka & McPherson, 1979; Zak *et al.*, 1990; McPherson *et al.*, 1991; Belsky *et al.*, 1989; Jackson *et al.*, 1990; Wesser & Armbruster, 1991; Wilson, 1993*b*; Vieira, Uhl & Nepstad, 1994; Köchy & Wilson, 1997). Invading trees enhance soil organic matter decomposition in northern prairies (Dormaar & Lutwick, 1966; Severson and Arneman, 1973; Schoenau & Bettany, 1987; Anderson, 1988; Fuller & Anderson, 1993) and thus increase the availability of nutrients. Mineralization could also be enhanced by litter quality, if trees produce litter that is more easily decomposed (Hobbie, 1992). Similarly, previously established shrubs may encourage tree establishment by reducing the intensity of competition from grasses (Werner & Harbeck, 1982).

Tree invasions of grasslands associated with positive feedback could be initiated by external factors (Fig. 9.2*B*). External factors might increase shoot mass directly, as in the case of reduced fire frequency (White, 1983), or by allowing woody plants to establish in high numbers following an unusual germination opportunity (Morisset & Payette, 1983). Indirect factors that increase soil resource availability, such as increased precipitation, decreased temperature, or atmospheric input of anthropogenic nutrients (Lovett 1992), could also initiate positive feedback. Positive feedbacks may be mediated by animals. Nutrient availability might be high under trees because mammals shelter under them and deposit faeces. Leigh *et al.* (1987) found that rabbit pellets were 3–9 times more abundant in Australian subalpine forest than in adjacent grassland, but Belsky (1994) found no effect of animal exclosures on soil fertility under African trees. Clumps of woody species in grasslands have been attributed to seed deposition by sheltering animals (Archer *et al.*, 1988; Teague & Smit, 1992; Vieira *et al.*, 1994).

The evidence linking woody plant invasion to positive feedbacks is equivocal, however. Co-occurring woody species might simply reflect shared habitat preferences. Further, not all the comparisons between grassy and woody vegetation are consistent with the proposed mechanism. Soils under grasslands in Turkey and alpine Australia have higher mineral N availabilities than nearby soils dominated by woody plants and supporting more biomass (Gökçeoglu, 1988; Wilson, 1993a). Leaf litter decomposition rates are higher in northern prairies than in adjacent forests, and are reduced by shading (Köchy & Wilson, 1997). Lastly, transplants of both grasses and trees grown for 1–2 summers typically have higher growth rates in grasslands than in forests (Wilson, 1993a,b; D. A. Peltzer & S. D. Wilson *unpublished data*). The last two results suggest that the net effect of trees in temperate areas is to restrict both decomposition and plant growth rates.

A critical test of the positive feedback hypothesis might involve the removal of woody neighbours from around a young target tree or shrub in a grassland. Such a target would grow less well with its woody neighbours removed if positive feedback occurs (Hunter & Aarssen, 1988; Callaway, 1995). This type of experiment, but not involving grassy neighbours, has shown that N-fixing shrubs facilitate tree establishment during primary succession in Alaska (Chapin *et al.*, 1994). Similarly, removal of the shrub *Prosopis* from an Arizona grassland produced a decrease in soil nutrient availability (Klemmedson & Tiedemann, 1986). I found no direct tests of facilitation among trees with grass neighbours (Aarssen & Epp, 1990; Goldberg, 1990; Bertness & Callaway, 1994).

As noted above, spatial and temporal heterogeneity could also be involved in positive feedback. Due to their relatively large size and morphology, trees should generate large-scale patchiness in soil moisture and mineralization via stemflow (Parker, 1983), canopy-drip, and the movement of soil resources to stems (Schlesinger *et al.*, 1990; Vieira *et al.*, 1994). Trees might also benefit from increases in spatial variability because of their ability to sample and average over larger areas (Bloom, Chapin & Mooney, 1985; Grime, 1994). Similarly, trees might decrease the temporal variability of rain by excluding light showers from the soil. Plants characteristic of relatively nutrient-poor soils, like grasses, might benefit from high temporal variability (Grime, 1994).

Lastly, mycorrhizae could also play a role in feedbacks. Grasses are colonized by vesicular-arbuscular mycorrhizae (VAM) and trees by ectomycorrhizae, which might impose an additional constraint on invading trees (Newman *et al.*, 1992). In northern prairies, tree invasions are preceded by the establishment of the shrub *Symphoricarpos occidentalis* (Bird, 1927), which probably has VAM (Gilbert, 1995). *S. occidentalis* might establish in grasslands more easily than trees, and, once established, invoke positive feedback effects that would facilitate tree establishment.

Future directions

Grass competition differs from competition between other kinds of plants in that most of it occurs underground. We need more measurements and tests of belowground mechanisms, using techniques such as rhizotrons and isotopes (Pearcy *et al.*, 1989) to measure spatial and temporal variability as well as means. The world belowground differs from that above in that ecological processes are dominated by several kingdoms, not just plants, and we should consider competition between kingdoms. For example, bacterial biomass contains far more N than does grass (Jackson, Schimel & Firestone, 1989; Bristow & Jarvis, 1991), suggesting that the relative abilities of grasses and other plants to compete against bacteria may be more important than their abilities to compete against each other. Similarly, the distinction between the mycorrhizae associated with grasses and trees needs to be examined to see what role, if any, it plays (Newman *et al.*, 1992; Ellis & Pennington, 1992). Positive feedbacks have received considerable attention and circumstantial support, but need direct tests. Changes in the rank order of competitive effects also deserve tests. Such experiments need to run for several years (e.g. Berendse *et al.*, 1994). In general, future advances in grass competition are likely to be greatest in studies which rely more on manipulating and measuring belowground processes.

Acknowledgements

I thank D. Goldberg, K. Kosola and D. Zak for recent discussions in this area, L. Ambrose and M. Friebel for typing, K. Muir for library help, G. Cheplick, A. Gerry, M. Köchy, W. Lauenroth and D. Peltzer for improving earlier drafts, and the Natural Sciences and Engineering Research Council of Canada for support.

References

Aarssen, L. W. & Epp, G. A. (1990). Neighbor manipulations in natural vegetation: a review. *Journal of Vegetation Science*, **1**, 13–30.

Aber, J. D., Melillo, J. M., Nadelhoffer, K. J., McClaugherty, C. A. & Pastor, J. (1985). Fine root turnover in forest ecosystems in relation to quantity and form of nitrogen availability: a comparison of two methods. *Oecologia*, **66**, 317–21.

Adams, T. E., Sands, P. B., Weitkamp, W. H. & McDougald, N. K. (1992). Oak seedling establishment on California rangelands. *Journal of Range Management*, **45**, 93–8.

Aerts, R. (1995). The advantages of being evergreen. *Trends in Ecology and Evolution*, **10**, 402–7.

Aerts, R., Bakker, C. & de Caluwe, H. (1992). Root turnover as determinant of the cycling of C, N, and P in a dry heathland ecosystem. *Biogeochemistry*, **15**, 175–90.

Aerts, R. & van der Peijl, M. J. (1993). A simple model to explain the dominance of low-productive perennials in nutrient-poor environments. *Oikos*, **66**, 144–7.

Aguiar, M. R., Soriano, A. & Sala, O. E. (1992). Competition and facilitation in the recruitment of seedlings in Patagonian steppe. *Functional Ecology*, **6**, 66–70.

Albertson, F. W. & Tomanek, G. W. (1965). Vegetation changes during a 30-year period in grassland communitites near Hays, Kansas. *Ecology*, **46**, 714–20.

Albertson, F. W. & Weaver, J. E. (1945). Injury and death or recovery of trees in prairie climate. *Ecological Monographs*, **15**, 393–433.

Allen, R. B. & Lee, W. G. (1989). Seedling establishment microsites of exotic conifers in *Chionocloa rigida* tussock grassland, Otago, New Zealand. *New Zealand Journal of Botany*, **27**, 491–8.

Anderson, D. W. (1988). The effect of parent material and soil development on nutrient cycling in temperate ecosystems. *Biogeochemistry*, **5**, 71–97.

Anderson, D. W. & Coleman, D. C. (1985). The dynamics of organic matter in grassland soils. *Journal of Soil and Water Conservation*, **40**, 211–16.

Anderson, H. G. & Bailey, A. W. (1980). Effects of annual burning on grassland in the aspen parkland of east-central Alberta. *Canadian Journal of Botany*, **58**, 985–96.

Archer, S., Scifres, A., Bassham, C. R. & Maggio, R. (1988). Autogenic succession in a subtropical savanna: conversion of a grassland to thorn woodland. *Ecological Monographs*, **58**, 111–27.

Axelrod, D. I. (1985). Rise of the grassland biome, central North America. *The Botanical Review*, **51**, 163–201.

Bailey, A. W., Irving, B. D. & Fitzgerald, R. D. (1990). Regeneration of woody species following burning and grazing in aspen parkland. *Journal of Range Management*, **43**, 212–15.

Bailey, A. W. & Wroe, R. A. (1974). Aspen invasion in a portion of the Alberta Parklands. *Journal of Range Management*, **27**, 263–66.

Barth, R. C. & Klemmedson, J. O. (1978). Shrub-induced spatial patterns of dry matter, nitrogen, and organic carbon. *Soil Science Society of America Journal*, **42**, 804–9.

Belsky, A. J. (1990). Tree/grass ratios in East African savanna: a comparison of existing models. *Journal of Biogeography*, **17**, 483–9.

Belsky, A. J. (1994). Influences of trees on savanna productivity: tests of shade, nutrients, and tree-grass competition. *Ecology*, **75**, 922–32.

Belsky, A. J., Amundson, R. G., Duxbury, J. M., Riha, S. J., Ali, A. R. & Mwonga, S. M. (1989). The effects of trees on their physical, chemical, and biological environments in a semi-arid savanna in Kenya. *Journal of Applied Ecology*, **26**, 1005–24.

Belsky, A. J. & Blumenthal, D. M. (1997). Effects of livestock grazing on upland forests, stand dynamics, and soils of the interior west. *Conservation Biology*, **11**, 315–27.

Benjamin, P. K., Anderson, R. C. & Liberta, A. E. (1989). Vesicular-arbuscular mycorrhizal ecology of little bluestem across a prairie–forest gradient. *Canadian Journal of Botany*, **67**, 2678–85.

Berendse, F. (1994a). Competition between plant populations at low and high nutrient supplies. *Oikos*, **71**, 253–60.

Berendse, F. (1994*b*). Litter decomposability – a neglected component of plant fitness. *Journal of Ecology*, **82**, 187–90.

Berendse, F., Bobbink, R. & Rouwenhorst, G. (1989). A comparative study on nutrient cycling in wet heathland ecosystems. II. Litter decomposition and nutrient mineralization. *Oecologia*, **78**, 338–48.

Berendse, F., Schmitz, M. & de Visser, W. (1994). Experimental manipulation of succession in heathland ecosystems. *Oecologia*, **100**, 38–44.

Bertness, M. D. & Callaway, R. (1994). Positive interactions in communities. *Trends in Ecology and Evolution*, **9**, 191–3.

Bird, R. D. (1927). A preliminary ecological survey of the district surrounding the entomological station at Treesbank, Manitoba. *Ecology*, **8**, 207–20.

Bloom, A. J., Chapin III, F. S. & Mooney, H. A. (1985). Resource limitation in plants – an economic analogy. *Annual Review of Ecology and Systematics*, **16**, 363–92.

Boeken, B. & Canham, C. D. (1995). Biotic and abiotic control of the dynamics of gray dogwood (*Cornus racemosa* Lam.) shrub thickets. *Journal of Ecology*, **83**, 569–80.

Brady, N. C. (1990). *The Nature and Properties of Soils*. New York: MacMillan.

Briggs, J. M., Seastedt, R. T. & Gibson, D. J. (1989). Comparative analysis of temporal and spatial variability in above-ground production in a Kansas forest and prairie. *Holarctic Ecology*, **12**, 130–6.

Bristow, A. W. & Jarvis, S. C. (1991). Effects of grazing and nitrogen fertilizer on the soil microbial biomass under permanent pasture. *Journal of the Science of Food and Agriculture*, **54**, 9–21.

Brown, J. R. & Archer, S. (1989). Woody plant invasion of grasslands: establishment of honey mesquite (*Prosopis glandulosa* var. *glandulosa*) on sites differing in herbaceous biomass and grazing history. *Oecologia*, **80**, 19–26.

Buckley, G. P. (1984). The uses of herbaceous companion species in the establishment of woody species from seed. *Journal of Environmental Management*, **18**, 309–22.

Bush, J. K. & Van Auken, O. W. (1990). Growth and survival of *Prosopis glandulosa* seedlings associated with shade and herbaceous competition. *Botanical Gazette*, **151**, 234–9.

Bush, J. K. & Van Auken, O. W. (1995). Woody plant growth related to planting time and clipping of a C_4 grass. *Ecology*, **76**, 1603–9.

Caldwell, M. M. & Richards, J. H. (1986). Competing root systems: morphology and models of absorption. In *On the Economy of Plant Form and Function*, ed. T. J. Givnish, pp. 251–73. Cambridge: Cambridge University Press.

Callaway, R. M. (1992). Effects of shrubs on recruitment of *Quercus douglasii* and *Quercus lobata* in California. *Ecology*, **73**, 2118–28.

Callaway, R. M. (1995). Positive interactions among plants. *Botanical Review*, **61**, 306–49.

Callaway, R. M. & D'Antonio, C. M. (1991). Shrub facilitation of coastal live oak establishment in central California. *Madroño*, **38**, 158–69.

Callaway, R. M., Nadkarni, N. M. & Mahall, B. E. (1991). Facilitation and interference of *Quercus douglasii* on understory productivity in central California. *Ecology*, **72**, 1484–99.

Campbell, B. D. & Grime, J. P. (1989). A comparative study of plant responsiveness to the duration of episodes of mineral nutrient enrichment. *New Phytologist*, **112**, 261–7.

Cantor, L. F. & Whitham, T. G. (1989). Importance of belowground herbivory:

pocket gophers may limit aspen to rock outcrop refugia. *Ecology*, **70**, 962–70.

Chapin III, F. S., Walker, L. R., Fastie, C. L. & Sharman, L. C. (1994). Mechanisms of primary succession following deglaciation at Glacier Bay, Alaska. *Ecological Monographs*, **64**, 149–75.

Charley, J. L. & West, N. E. (1975). Plant-induced soil chemical patterns in some shrub-dominated semi-desert ecosystems of Utah. *Journal of Ecology*, **63**, 945–63.

Clements, F. E., Weaver, J. E. & Hanson, H. C. (1929). *Plant Competition*. Washington: Carnagie Institute.

Coffin, D. P. & Urban, D. L. (1993). Implications of natural history traits to system-level dynamics: comparisons of a grassland and a forest. *Ecological Modelling*, **67**, 147–78.

Connell, J. H. (1978). Diversity in tropical rain forests and coral reefs. *Science*, **199**, 1302–10.

Costin, A. B. (1954). *A Study of the Ecosystems of the Monaro Region of New South Wales*. Sydney: Government Printer.

D'Antonio, C. M. & Vitousek, P. M. (1992). Biological invasions by exotic grasses, the grass/fire cycle, and global change. *Annual Review of Ecology and Systematics*, **23**, 63–87.

Dahlman, R. C. & Kucera, C. L. (1965). Root productivity and turnover in native prairie. *Ecology*, **46**, 84–9.

DeAngelis, D. L., Post, W. M. & Travis, C. C. (1986). *Positive Feedback in Natural Systems*. Berlin: Springer-Verlag.

De Steven, D. (1991). Experiments on mechanisms of tree establishment in old-field succession: seedling emergence. *Ecology*, **72**, 1066–75.

Dormaar, J. T. & Lutwick, L. E. (1966). A biosequence of soils of the Rough Fescue prairie–poplar transition in southwestern Alberta. *Canadian Journal of Earth Sciences*, **3**, 457–71.

Egerton, J. & Wilson, S. D. (1993). Overwinter competition in alpine plant communities. *Arctic and Alpine Research*, **25**, 124–9.

Elliott, K. J. & White, A. S. (1989). Competitive effects of various grasses and forbs on Ponderosa pine seedlings. *Forestry Science*, **32**, 356–66.

Ellis, R. C. & Pennington, P. I. (1992). Factors affecting the growth of *Eucalyptus delegatensis* seedlings in inhibitory forest and grassland soils. *Plant and Soil*, **145**, 93–105.

Ellison, L. & Houston, W. R. (1958). Production of herbaceous vegetation in openings and under canopies of western aspen. *Ecology*, **39**, 338–45.

Emanuel, W. R., Shugart, H. H. & Stevenson, M. (1985). Climatic change and the broad-scale distribution of terrestrial ecosystem complexes. *Climatic Change*, **7**, 29–43.

Facelli, J. M. (1994). Multiple indirect effects of plant litter affect the establishment of woody seedlings in old fields. *Ecology*, **75**, 1727–35.

Fensham, R. J. & Kirkpatrick, J. B. (1992). The eucalypt forest-grassland/grassy woodland boundary in central Tasmania. *Australian Journal of Botany*, **40**, 123–38.

Fuller, L. G. & Anderson, D. W. (1993). Changes in soil properties following forest invasion of black soils of the Aspen Parkland. *Canadian Journal of Soil Science*, **73**, 613–27.

George, E. & Marschner, H. (1996). Nutrient and water uptake by roots of forest trees. *Zeitschrift fuer Pflanzenernaehrungund Bodenkunde*, **159**, 11–21.

Gerry, A. K. & Wilson, S. D. (1995). The influence of initial size on the competitive responses of six plant species. *Ecology*, **76**, 272–9.

Gilbert, O. L. (1995). *Symphoricarpos albus* (L.) S. F. Blake (*S. rivularis* Suksd., *S. racemosus* Michaux). *Journal of Ecology*, **83**, 159–66.

Gill, D. S. & Marks, P. L. (1991). Tree and shrub seedling colonization of old fields in central New York. *Ecological Monographs*, **61**, 183–206.

Gökçeoglu, M. (1988). Nitrogen mineralization in volcanic soil under grassland, shrub and forest vegetation in the Aegean region of Turkey. *Oecologia*, **77**, 242–8.

Goldberg, D. E. (1987). Neighbourhood competition in an old-field plant community. *Ecology*, **68**, 1211–23.

Goldberg, D. E. (1990). Components of resource competition in plant communities. In *Perspectives on Plant Competition*, ed. J. B. Grace & D. Tilman, pp. 27–49. San Diego: Academic Press.

Grime, J. P. (1973). Competitive exclusion in herbaceous vegetation. *Nature*, **242**, 344–7.

Grime, J. P. (1994). The role of plasticity in exploiting environmental heterogeneity. In *Exploitation of Environmental Heterogeneity By Plants*, ed. M. M. Caldwell & R. W. Pearcy, pp. 1–19. San Diego: Academic Press.

Hairston Sr., N. G. (1989). *Ecological Experiments: Purpose, Design and Execution*. Cambridge: Cambridge University Press.

Hare, F. K. & Ritchie, J. C. (1972). The boreal bioclimates. *Geographical Review*, **62**, 333–65.

Harrison, J. S. & Werner, P. A. (1984). Colonization by oak seedlings into a heterogeneous successional habitat. *Canadian Journal of Botany*, **62**, 559–63.

Harrington, G. N. (1991). Effects of soil moisture on shrub seedling survival in a semi-arid grassland. *Ecology*, **72**, 1138–49.

Harrington, G. N. & Johns, G. G. (1990). Herbaceous biomass in a Eucalyptus savanna woodland after removing trees and/or shrubs. *Journal of Applied Ecology*, **27**, 775–87.

Harvey, E. M., Mohammed, G. H. & Noland, T. L. (1993). *A Bibliography on Competition, Tree Seedling Characteristics, and Related Topics*. Sault Ste. Marie: Ontario Forest Research Institute.

Herben, T. & Krahulec, F. (1990). Competitive hierarchies, reversals of rank order and the de Wit approach: are they compatible? *Oikos*, **58**, 254–6.

Hill, J. D., Canham, D. & Wood, D. (1995). Patterns and causes of resistance to tree invasion in rights-of-way. *Ecological Applications*, **5**, 459–70.

Hobbie, S. E. (1992). Effects of plant species on nutrient cycling. *Trends in Ecology and Evolution*, **7**, 336–9.

Huisman, J. (1994). The models of Berendse and Tilman: two different perspectives on plant competition? *Functional Ecology*, **8**, 282–8.

Hunter, A. F. & Aarssen, L. W. (1988). Plants helping plants. *Bioscience*, **38**, 34–40.

Huston, M. (1979). A general hypothesis of species diversity. *American Naturalist*, **113**, 81–101.

Inouye, R. S., Allison, T. D. & Johnson, N. C. (1994). Old field succession on a Minnesota sand plain: effects of deer and other factors on invasion by trees. *Bulletin of the Torrey Botanical Club*, **121**, 266–76.

Jackson, L. E., Schimel, J. P. & Firestone, M. K. (1989). Short-term partitioning of ammonium and nitrate between plants and microbes in an annual grassland. *Soil Biology and Biogeochemistry*, **21**, 409–415.

Jackson, L. E., Strauss, R. B., Firestone, M. K. & Bartolome, J. W. (1990). Influence of tree canopies on grassland productivity and nitrogen dynamics in deciduous oak savanna. *Agricultural Ecosystems and Environments*, **32**, 89–105.

Jackson, R. B. & Caldwell, M. M. (1989). The timing and degree of root proliferation in fertile-soil microsites of three cold-desert perennials. *Oecologia*, **81**, 149–53.

Jaksic, F. M. & Fuentes, E. R. (1980). Why are native herbs in the Chilean matorral more abundant beneath bushes: microclimate or grazing? *Journal of Ecology*, **68**, 665–9.

Johnson, N. C., Zak, D. R., Tilman, D. & Pfleger, F. L. (1991). The dynamics of vesicular-arbuscular mycorrhizae during old field succession. *Oecologia*, **86**, 349–58.

Keddy, P. A. & MacLellan, P. (1990). Centrifugal organization in forests. *Oikos*, **59**, 75–84.

Kellman, M. (1984). Synergistic relationships between fire and low soil fertility in neotropical savannas: a hypothesis. *Biotropica*, **16**, 158–60.

Klemmedson, J. O. & Tiedemann, A. R. (1986). Long-term effects of mesquite removal on soil characteristics. II. Nutrient availability. *Soil Science Society of America Journal*, **50**, 476–80.

Knoop, W. T. & Walker, B. H. (1985). Interactions of woody and herbaceous vegetation in a southern African savanna. *Journal of Ecology*, **73**, 235–53.

Ko, L. J. & Reich, P. B. (1993). Oak tree effects on soil and herbaceous vegetation in savannas and pastures in Wisconsin. *American Midland Naturalist*, **130**, 31–42.

Köchy, M. & Wilson, S. D. (1997). Litter decomposition and nitrogen dynamics in aspen forest and mixed-grass prairie. *Ecology*, **78**, 732–9.

Lauenroth, W. K. & Coffin, D. P. (1992). Belowground processes and the recovery of semiarid grasslands from disturbance. In *Ecosystem Rehabilitation*, vol. 2, *Ecosystem Analysis and Synthesis*, ed. M. K. Wali, pp. 131–50. The Hague: SBP Academic Publishing.

Leigh, J. H., Wimbush, D. J., Wood, D. H., Holgate, A. V., Stanger, M. G. & Forrester, R. I. (1987). Effects of rabbit grazing and fire on a subalpine environment. I. Herbaceous and shrubby vegetation. *Australian Journal of Botany*, **35**, 433–64.

Lovett, G. M. (1992). Atmospheric deposition and canopy interactions of nitrogen. In *Atmospheric Deposition and Forest Nutrient Cycling*, ed. D. W. Johnson & S. E. Lindberg, pp. 152–66. New York: Springer-Verlag.

Magee, T. K. & Antos, J. A. (1992). Tree invasion into a mountain-top meadow in the Oregon Coast Range, USA. *Journal of Vegetation Science*, **3**, 485–94.

Marañón, T. & Bartolome, J. W. (1993). Reciprocal transplants of herbaceous communities between *Quercus agrifolia* woodland and adjacent grassland. *Journal of Ecology*, **81**, 673–82.

McClaran, M. P. & Bartolome, J. W. (1989). Fire-related recruitment in stagnant *Quercus douglasii* populations. *Canadian Journal of Forest Research*, **19**, 580–5.

McMurtrie, R. & Wolf, L. (1983). A model of competition between trees and grass for radiation, water and nutrients. *Annals of Botany*, **52**, 449–58.

McPherson, G. R. (1993). Effects of herbivory and herb interference on oak establishment in a semi-arid temperate savanna. *Journal of Vegetation Science*, **4**, 687–92.

McPherson, G. R., Rasmussen, G. A., Wester, D. B. & Masters, R. A. (1991).

Vegetation and soil zonation associated with *Juniperus pinchotii* Sudw. trees. *Great Basin Naturalist*, **51**, 316–24.

Menaut, J. C., Gignoux, J., Prado, C. & Clobert, J. (1990). Tree community dynamics in a humid savanna of the Cote-d'Ivoire – modelling the effects of fire and competition with grass and neighbours. *Journal of Biogeography*, **17**, 471–81.

Milchunas, D. G. & Lauenroth, W. K. (1993). Quantitative effects of grazing on vegetation and soils over a global range of environments. *Ecological Monographs*, **63**, 327–66.

Milchunas, D. G., Sala, O. E. & Lauenroth, W. K. (1988). A generalized model of the effects of grazing by large herbivores on grassland community structure. *American Naturalist*, **132**, 87–106.

Miles, J. (1974). Effects of experimental interference with stand structure on establishment of seedlings in Callenetum. *Journal of Ecology*, **62**, 675–87.

Mitchell, R. J., Zutter, B. R., Green, T. H., Perry, M. A., Gjerstad, D. H. & Glover, G. R. (1993). Spatial and temporal variation in competitive effects on soil moisture and pine response. *Ecological Applications*, **3**, 167–74.

Morisset, P. & Payette, S. (1983). *Tree-line Ecology*. Quebec: Université Laval.

Myster, R. W. & Pickett, S. T. A. (1993). Effects of litter, distance, density and vegetation patch type on postdispersal tree seed predation in old fields. *Oikos*, **66**, 381–8.

Naeem, S. & Colwell, R. K. (1991). Ecological consequences of heterogeneity of consumable resources. In *Ecological Heterogeneity*, ed. J. Kolasa & S. T. A. Pickett, pp. 224–55. New York: Springer-Verlag.

Newman, E. I. (1983). Interactions between plants. In *Physiological Plant Ecology. III. Responses to the Chemical and Biological Environment*, ed. O. L. Lange, P. S. Nobel, C. B. Osmond & H. Ziegler, Berlin: Springer-Verlag.

Newman, E. I., Devoy, C. L. N., Easen, N. J. & Fowles, K. J. (1994). Plant species that can be linked by VA mycorrhizal fungi. *New Phytologist*, **126**, 691–3.

Newman, E. I., Eason, W. R., Eissenstat, D. M. & Ramos, M. I. R. (1992). Interactions between plants: the role of mycorrhizae. *Mycorrhiza*, **1**, 47–53.

Newman, E. I. & Reddell, P. (1987). The distribution of mycorrhizas among families of vascular plants. *New Phytologist*, **106**, 745–51.

Noble, I. R. (1980). Interactions between tussock grass (*Poa* spp.) and *Eucalyptus pauciflora* seedlings near treeline in south-eastern Australia. *Oecologia*, **45**, 350–3.

Parker, G. G. (1983). Throughfall and stemflow in the forest nutrient cycle. *Advances in Ecological Research*, **13**, 58–135.

Parker, M. A. & Salzman, A. G. (1985). Herbivore exclosure and competitor removal: effects on juvenile survivorship and growth in the shrub *Gutierrezia microcephala*. *Journal of Ecology*, **73**, 903–13.

Payette, S. & Filion, L. (1985). White spruce expansion at the tree line and recent climatic change. *Canadian Journal of Forest Research*, **15**, 241–51.

Pearcy, R. W., Chazdon, R. L., Gross, L. J. & Mott, K. A. (1994). Photosynthetic utilization of sunflecks: a temporally patchy resource on a time scale of seconds to minutes. In *Exploitation of Environmental Heterogeneity by Plants*, ed. M. M. Caldwell & R. W. Pearcy, pp. 175–208. San Diego: Academic Press.

Pearcy, R. W., Ehleringer, J., Mooney, H. A. & Rundel, P. A. (1989). *Plant Physiological Ecology: Field Methods and Instrumentation*. New York: Chapman and Hall.

Petranka, J. W. & McPherson, J. K. (1979). The role of *Rhus copallina* in the dynamics of the forest–prairie ecotone in north-central Oklahoma. *Ecology*, **60**, 956–65.

Reader, R. J., Mallik, A. U., Hobbs, R. J. & Gimingham, C. H. (1983). Shoot regeneration after fire or freezing temperatures and its relation to plant lifeform for some heathland species. *Vegetatio*, **55**, 181–9.

Rejmanek, M. & Rosen, E. (1992). Cycles of heterogeneity during succession: a premature generalization? *Ecology*, **73**, 2329–31.

Riechert, S. E. & Hammerstein, P. (1983). Game theory in the ecological context. *Annual Review of Ecology and Systematics*, **14**, 377–409.

Riegel, G. M., Miller, R. F. & Krueger, W. C. (1992). Competition for resources between understory vegetation and overstory *Pinus ponderosa* in northeastern Oregon. *Ecological Applications*, **2**, 71–85.

Sala, O. E., Golluscio, R. A., Lauenroth, W. K. & Soriano, A. (1989). Resource partitioning between shrubs and grasses in a Patagonian steppe. *Oecologia*, **81**, 501–5.

Sala, O. E., Lauenroth, W. K. & Golluscio, R. A. (1996). Plant functional types in temperate semi-arid regions. In *Plant Functional Types*, ed. T. M. Smith & H. H. Shugart. Cambridge: Cambridge University Press.

Sala, O. E., Lauenroth, W. K. & Parton, W. J. (1992). Long-term soil water dynamics in the shortgrass steppe. *Ecology*, **73**, 1175–81.

Schlesinger, W. H., Reynolds, J. F., Cunningham, G. L., Huenneke, L. F., Jarrell, W. M., Virginia, R. A. & Whitford, W. G. (1990). Biological feedbacks in global desertification. *Science*, **247**, 1043–8.

Schoenau, J. J. & Bettany, J. R. (1987). Organic matter leaching as a component of carbon, nitrogen, phosphorus and sulfur cycles in a forest, grassland and gleyed soil. *Soil Science Society of America Journal*, **51**, 646–51.

Schulze, E.-D. (1982). Plant life forms and their carbon, water and nutrient relations. In *Physiological Plant Ecology. II. Water Relations and Carbon Assimilation*, ed. O. L. Lange, P. S. Nobel, C. B. Osmond & H. Ziegler, pp. 615–76. Berlin: Springer-Verlag.

Schulze, E.-D., Chapin III, F. S. & Gebauer, G. (1994). Nitrogen nutrition and isotope differences among life forms at the northern treeline of Alaska. *Oecologia*, **100**, 406–12.

Schulze, E.-D. & Chapin III, F. S. (1987). Plant specialization to environments of different resource availability. In *Potentials and Limitations of Ecosystem Analysis*, ed. E.-D. Schulze & H. Zwölfer, pp. 120–48. Berlin: Springer-Verlag.

Severson, R. C. & Arneman, H. F. (1973). Soil characteristics of the forest-prairie ecotone in Northwestern Minnesota. *Soil Science Society of America Proceedings*, **37**, 593–9.

Stark, J. M. (1994). Causes of soil nutrient heterogeneity at different scales. In *Exploitation of Environmental Heterogeneity by Plants*, ed. M. M. Caldwell & R. W. Pearcy, pp. 255–84. San Diego: Academic Press.

Stuart-Hill, G. C. & Tainton, N. M. (1989). The competitive interaction between *Acacia karroo* and the herbaceous layer and how this is influenced by defoliation. *Journal of Applied Ecology*, **26**, 285–98.

Svedarsky, W. D., Buckley, P. E. & Feiro, T. A. (1986). The effect of 13 years of annual burning on an aspen–prairie ecotone in northwestern Minnesota. In *Proceedings of the Ninth North American Prairie Conference*, ed. G. K. Clambey & R. H. Pemble, pp. 118–22. Fargo, N.D: Tri-College University Center for Environmental Studies.

Swank, S. E. & Oechel, W. C. (1991). Interactions among the effects of herbivory, competition, and resource limitation on chaparral herbs. *Ecology*, **72**, 104–15.

Tamm, C. O. (1991). *Nitrogen in Terrestrial Ecosystems: Questions of Productivity, Vegetational Changes, and Ecosystem Stability.* Berlin: Springer-Verlag.

Teague, W. R. & Smit, G. N. (1992). Relations between woody and herbaceous components and the effects of bush-clearing in southern African savannas. *Tydskrif van die Weidingsvereniging van Suidelike Afrika*, **9**, 60–71.

Tilman, D. (1988). *Plant Strategies and the Dynamics and Structure of Plant Communities.* Princeton: Princeton University Press.

Tilman, D. & Wedin, D. (1991). Dynamics of nitrogen competition between successional grasses. *Ecology*, **72**, 1038–49.

Vieira, I. C. G., Uhl, C. & Nepstad, D. (1994). The role of the shrub *Cordia multispicata* Cham. as a 'succession facilitator' in an abandoned pasture, Paragominas, Amazônia. *Vegetatio*, **115**, 91–9.

Walker, B. H. (1981). Is succession a viable concept in African savanna ecosystems? In *Forest Succession: Concepts and Application*, ed. D. C. West, H. H. Shugart & D. B. Botkin, pp. 431–47. New York: Springer-Verlag.

Walker, J., Sharpe, P. J. H., Penridge, L. K. & Wu, H. (1989). Ecological Field Theory: the concept and field tests. *Vegetatio*, **83**, 81–95.

Walter, H. (1971). *Ecology of Tropical and Sub-tropical Vegetation.* Edinburgh: Oliver & Boyd.

Walter, H. (1985). *Vegetation of the Earth.* New York: Springer-Verlag.

Wardle, P. (1985). New Zealand timberlines. 3. A synthesis. *New Zealand Journal of Botany*, **23**, 263–71.

Weltzin, J. F. & Coughenour, M. B. (1990). Savanna tree influence on understorey vegetation and soil nutrients in northwestern Kenya. *Journal of Vegetation Science*, **1**, 325–34.

Werner, P. A. & Harbeck, A. L. (1982). The pattern of tree seedling establishment relative to staghorn sumac cover in Michigan old fields. *American Midland Naturalist*, **108**, 124–32.

Wesser, S. D. & Armbruster, W. S. (1991). Species distribution controls across a forest–steppe transition: a causal model and experimental test. *Ecological Monographs*, **61**, 323–42.

Westoby, M., Walker, B. & Noy-Meir, I. (1989). Opportunistic management for rangelands not at equilibrium. *Journal of Range Management*, **42**, 266–73.

White, A. S. (1983). The effect of thriteen years of annual prescribed burning on a *Quercus ellipsoidalis* community in Minnesota. *Ecology*, **64**, 1081–85.

Wilson, J. B. & Agnew, A. D. Q. (1992). Positive feedback switches in plant communities. *Advances in Ecological Research*, **23**, 263–336.

Wilson, S. D. (1993*a*). Competition and resource availability in heath and grassland in the Snowy Mountains of Australia. *Journal of Ecology*, **81**, 445–51.

Wilson, S. D. (1993*b*). Belowground competition in forest and prairie. *Oikos*, **68**, 146–50.

Wilson, S. D. (1994). The contribution of grazing to plant diversity in alpine grassland and heath. *Australian Journal of Ecology*, **19**, 137–40.

Wilson, S. D. & Kleb, H. R. (1996). The influence of prairie and forest vegetation on soil moisture and available nitrogen. *American Midland Naturalist*, **136**, 222–31.

Wilson, S. D. & Tilman, D. (1993). Plant competition in relation to disturbance, fertility and resource availability. *Ecology*, **74**, 599–611.

Wilson, S. D. & Tilman, D. (1995). Competitive responses of eight old-field plant species in four environments. *Ecology*, **76**, 1169–80.

S. D. Wilson

Wilson, S. D. & Zammit, C. A. (1992). Tree litter and the lower limits of
subalpine herbs and grasses in the Brindabella Range, A.C.T. *Australian
Journal of Ecology*, **17**, 321–7.
Zak, D. R., Grigal, D. F., Gleeson, S. & Tilman, D. (1990). Carbon and nitrogen
cycling during old-field succession: constraints on plant and microbial
biomass. *Biogeochemistry*, **11**, 111–29.

10
Fungal endophyte infection and the population dynamics of grasses

KEITH CLAY

Introduction

Grasses have many interesting features of their population biology that are explored in detail in this volume. One of the most unique features of grasses is their symbiotic association with systemic fungal endophytes (Clavicipitaceae, Ascomycota) that grow within or on aboveground plant parts. These fungi can have significant effects on all aspects of the grass life cycle from germination and establishment to growth and reproduction. The population biology of grasses therefore varies with their endophyte infection status.

Plant population biology research has focused on describing patterns of recruitment, survival and reproduction, and on investigating how resource levels, density and genetic variability affect these processes. Increasing attention is being given to microbial symbioses in plant population dynamics (Burdon, 1987; Allen, 1991; Jarosz & Davelos, 1995). Mycorrhizal fungi infecting grasses enhance nutrient uptake and plant vigour, and alter interactions with pathogens and competitors (Hartnett et al., 1993; Stanley, Koide & Shumway, 1993; Newsham, Fitter & Watkinson, 1995; Newsham & Watkinson, this volume). Other mutualistic microbes in the rhizosphere and phyllosphere of grasses also occur (Turkington et al., 1988; Bever, 1994). For example, a nitrogen-fixing bacterial endophyte in the stems of sugar cane contributes most or all of the plant's nitrogen requirements (Dong et al., 1994). Fungal pathogens have been better studied compared with mutualists, especially in agricultural systems where grasses predominate (Burdon, 1987; Dinoor & Eshed, 1984; Clay, 1995). In contrast with mutualists, pathogens reduce the survival, growth and reproduction of host grasses (Mack & Pyke, 1984; Clay, Cheplick & Marks, 1989; Van der Putten, Van Dijk & Peters, 1993; Fowler & Clay, 1995; Govinthasamy &

Cavers, 1995). Wild grasses differ from domesticated grasses in that they are more likely to exhibit genetic variation within populations, to have uneven age distributions, and to occur in more heterogeneous biotic and abiotic environments. Pathogens potentially represent an important selective force on grass populations.

Clavicipitaceous fungal endophytes of grasses are host-specific parasites that range in their effects from mutualist to pathogen (Clay, 1990a; Schardl et al., 1994; Schardl & Clay, 1997). The growth rates of plant populations are a function of their age-specific patterns of survival, growth and reproduction. The objectives of this chapter are to consider how endophyte infection affects host fitness during these stages of the life cycle, and to consider how the life history and phylogenetic history of the host affects the type of fungal interaction. Research results from a limited number of grasses suggest that the potential impact of endophytes should be considered in all future ecological studies of grasses.

Endophytes of grasses

A wide range of grasses are infected by fungi in the family Clavicipitaceae (Ascomycota) which form systemic infections of aboveground host tissues. Several reviews provide detailed information on the systematics, morphology, and population biology of these associations (Bacon & Siegel, 1988; Clay, 1990b; Leuchtmann, 1992; Hill, 1994). The major features of endophytes relevant to their effects on the population biology of grasses are outlined here.

Growth in the plant

While most species of *Balansia*, plus species of *Epichloë* and *Acremonium*, grow endophytically in intercellular spaces of vegetative and reproductive tillers, others grow epiphytically on meristems, young leaves and inflorescences (Clay & Frentz, 1993). Epiphytes include species of *Atkinsonella* and *Myriogenospora*, and remaining species of *Balansia* (Table 10.1). Dense mats of hyphae occur in the meristematic regions of infected tillers and colonize daughter tillers as they develop. Neither group produces haustoria. These two fungal growth forms are very similar from a functional perspective, suggesting that an evolutionary transition between epiphytic and endophytic growth is possible. Unless otherwise indicated, the term 'endophyte' is used hereafter to refer to a monophyletic group of fungi that differs in specific growth form.

Table 10.1. *Summary of taxonomic diversity of endophytes, their growth form, and distribution in the grass family*

Genus	No. of species	Growth form	Conidia	Distribution	Host subfamilies	Seed transmission	Type of association
Atkinsonella	2	epi[a]	macro[b], micro	NW[c]	A[h], PO	+	II
Balansia	15+	endo, epi	macro	NW, OW	A, B, C, PA, O	–	I
Balansiopsis[d]	2	endo, epi	none	NW, OW	B	–	I
Echinodothis[e]	1	epi	micro	NW	A	–	–[i]
Epichloë[f]	5+	endo	micro	NW, OW	PO	+, –	I,II
Acremonium[g]	8+	endo	micro	NW, OW	PO	+	III
Myriogenospora	2	epi	macro	NW	B, PA	–	I

Notes:

[a] epi, epiphytic; endo, endophytic.

[b] macro, macroconidia; micro, microconidia.

[c] NW, New World; OW, Old World.

[d] Distinction from *Balansia* is questionable.

[e] Molecular data do not place genus with other Balansieae (Glenn *et al.*, 1996).

[f] Does not include questionable species described from tropical grasses, as discussed in White (1994).

[g] Inclusion of this imperfect stage is to emphasize its derivation from sexual *Epichloë* species.

[h] A, Arundinoideae; B, Bambusoideae; C, Chloridoideae; PA, Panicoideae; PO, Pooideae; O, Oryzoideae.

[i] Fungus is non-systemic and has no direct effects on host reproduction.

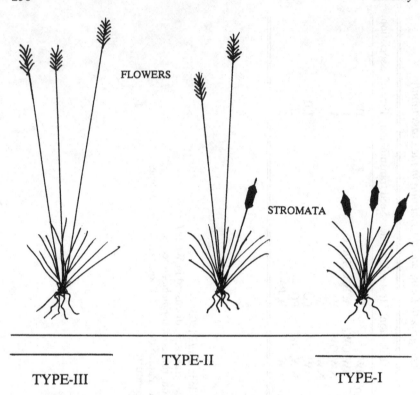

TYPE-II

TYPE-III   TYPE-I

Fig. 10.1. Flowering and stromata production in types I, II, and III grass/endophyte associations.

Reproduction

Grass/endophyte associations fall into one of three categories based on their reproduction (Fig. 10.1; White, 1988). In 'type-I' associations the host is sterilized by infection while the fungus sporulates and spreads contagiously. Infected plants do not produce flowers but fungal fruiting bodies (stromata) form on leaves or on developing inflorescences. Fungal spores can contagiously infect new plants, although the mechanisms of spread are poorly understood (White & Bultman, 1987; Leuchtmann & Clay, 1989). At the opposite extreme are type-III associations, where the host plant flowers normally and the fungus spreads vertically through seeds. These include endophyte-infected tall fescue (*Festuca arundinacea*) and perennial ryegrass (*Lolium perenne*). In general, type-I associations are pathogenic while type-III associations are mutualistic. Type-II associations are intermediate in that a host plant can produce both aborted inflorescences with fungal stromata and healthy inflorescences with infected seeds (Sampson, 1933; Leuchtmann & Clay, 1993). Molecular data confirm the close relationship between pathogenic and mutualistic endophytes, and suggest that type-III infections have been derived from type-I or type-II infections (Schardl *et al.* 1994; Schardl & Clay, 1997).

Endophyte taxonomy

The sexual stages of endophytes are grouped in the tribe Balansieae into one of six genera and about 30 species (see Table 10.1). The type of conidia and morphology of stromata have been the primary traits used for classification (Diehl, 1950). The sexual stage of the life cycle has been lost in endophytes forming type-III associations. These fungi are classified in the asexual genus *Acremonium* and are thought to have been derived from sexual species of *Epichloë* (Schardl *et al.*, 1991). The systematics of grass endophytes is currently in a state of flux and subject to ongoing research (Leuchtmann, Schardl & Siegel, 1994; White, 1993*a*; Glenn *et al.*, 1996). Much of the diversity of grass endophytes occurs in the tropics where relatively little research has been conducted.

There is a strong association between host and endophyte taxonomy (Clay, 1990*b*). Species of *Acremonium* and *Epichloë* infect only cool season grasses (subfamily Pooideae) with the C_3 photosynthetic mechanism, while species of *Balansia*, *Balansiopsis*, and *Myriogenospora* infect warm-season grasses (subfamily Panicoideae and Chloridoideae) with the C_4 photosynthetic mechanism (Table 10.1). Thus, mutualistic type-III infections are limited to cool-season grasses. The two species of *Atkinsonella* are each restricted to a single grass genus, as is *Echinodothis* (Leuchtmann & Clay, 1989; White, 1993*b*). There has also been a radiation of *Balansia* on some tropical and subtropical Cyperaceae (Diehl, 1950; Clay, 1986).

Distribution of endophytes

The significance of endophyte infection depends on both the frequency of host species and the frequency of infection within species. Clavicipitaceous endophytes infect grasses in all subfamilies and most tribes, in a wide range of ecological habitats, and on all continents except Antarctica (Diehl, 1950; Clay, 1990*b*; 1996; White, 1994). New hosts are constantly being reported from around the world (e.g. Wilson *et al.*, 1991; Clay, 1994). Nevertheless, it is difficult to estimate what fraction of grasses are infected, where they are infected, and to what frequency. Censuses from specific habitats suggest endophyte infections can be extremely common. For example, Clay & Leuchtmann (1989) found that 16 of 26 grasses from southern Indiana forests were infected, some at 100% frequency. A recent compilation showed that a large number of common grass genera and species have high levels of infection (Clay, 1997). It has been estimated that 15 million hectares of the USA is covered by tall fescue, of which 80% is infected (Coley *et al.*, 1995). In general, type-III infections of pooid grasses are more common than type-I or type-II infections. This is not unexpected given that type-III infections are more mutualistic than type-I or type-II infections (see below).

Population biology of infected grasses

Reproduction

The most obvious and dramatic effect of endophyte infection is on host reproduction. In type-I associations, where hosts are completely sterilized, seed production in the population is reduced as a function of the frequency of infection. In type-II associations hosts produce both aborted and seed-bearing inflorescences in a ratio that can vary dramatically depending on species combination, plant and fungal genotypes, and environment. For example, most inflorescences produced by infected *Festuca rubra* are normal and the endophyte is primarily seed transmitted (Sampson, 1933). In *Elymus virginicus* and *E. villosus* most infected plants have some aborted inflorescences, and some plants have only aborted inflorescences, but other infected plants are completely asymptomatic (Leuchtmann & Clay, 1993). Thus, the entire gamut of grass–endophyte interactions can be found within a single host population. Seed production by the host population is reduced as a function of both the frequency of infection and the frequency of aborted inflorescences on infected plants.

Empirical data suggest that fungal genotype is the primary determinant of stroma production and inflorescence abortion in type-II associations. When seedlings of *Brachyelytrum erectum* (which rarely produces stromata when naturally infected by its native strains of *Epichloë typhina*) were inoculated with *E. typhina* isolates from *Elymus virginicus* (which frequently produces stromata when infected) flowering plants produced aborted inflorescences with stromata (Leuchtmann & Clay, 1993). The two grasses often occur in the same site, yet cross infections apparently are rare. Schardl *et al.* (1994) described an unusual infection of perennial ryegrass by *E. typhina* that caused complete abortion of inflorescences whereas the more common endophyte of ryegrass (*A. lolii*) is completely asymptomatic and seedborne. Further, Bucheli & Leuchtmann (1996) found that stroma-producing and asymptomatic endophytes isolated from a single population of *Brachypodium sylvaticum* were genetically distinct. Genetic variation in inflorescence abortion and stromata production provides the raw material for coevolutionary interactions between grass and fungus, and illustrates how types I, II, and III associations might arise.

Plant environment and age also affects host phenotype. Sun, Clark & Funk (1990) examined red fescue over a range of soils and found that a higher proportion of inflorescences were aborted with decreasing levels of nitrogen. Endophytes may be more pathogenic in stressful environments (see Cheplick, Clay & Marks, 1989). In a field study with red fescue (J. Bier & K. Clay, unpubl.) there was virtually no inflorescence abortion the first flowering season (<1%) but in the second year there was a dramatic increase in the frequency of aborted inflorescences (50%) and in the third year the frequency dropped to <10%. It is unlikely that this shift was due to soil nutrient changes but may reflect overall plant vigour; more inflorescences, both healthy and aborted, were produced in the second year than in the first or third years.

The detrimental effect of sterility in type-I and type-II associations is mitigated in many host grasses by several compensatory mechanisms. Infected plants of rhizomatous species can spread aggressively despite the loss of seed production (Diehl, 1950; Bradshaw, 1959; Harberd, 1961). Infection by *Balansia cyperi* stimulates tuber production by purple nutsedge (*Cyperus rotundus*), through which both the plant and fungus spreads (Stovall & Clay, 1988). Viviparous grasses can also spread by vegetative plantlets as in the case of *Poa bulbosa* infected by *E. typhina* (Sampson & Western, 1954). The sedge *C. virens* produces only aborted inflorescences when infected by *B. cyperi* but a large proportion of infected plants produce viviparous plantlets from the aborted inflorescences, suggesting that the fungus has induced novel structures for its dissemination (Clay, 1986). Viviparous plantlets developing on the aborted inflorescences of two *Andropogon* species infected by *Myriogenospora atramentosa* were also described in the same study. A similar case of a rust inducing 'pseudoflowers' for spore dispersal on its cruciferous host has been described (Roy, 1993). The case of *Atkinsonella hypoxylon* infecting eastern North American *Danthonia* grasses deserves special mention. Unlike other type-II associations where plants produce healthy and aborted inflorescences, infected *Danthonia* produce only partially aborted inflorescences. *Danthonia* species are characterized by a floral dimorphism where potentially outcrossing flowers are produced at the apex of an inflorescence while obligately self-fertilized cleistogamous flowers are produced in lower leaf sheaths (Cheplick & Clay, 1989). In infected plants the terminal flowers are aborted but the cleistogamous flowers in the leaf sheaths are not and the fungus is transmitted through the seeds (Clay, 1994). Thus, infection enforces selfing, and could give rise to lineages of highly inbred plants infected by the same fungal genotype (Clay & Kover, 1996). Moreover, given that the cleistogamous seeds have reduced dispersal potential compared with the terminal seeds, uninfected plants are likely to have a greater probability of founding new populations. The congener *A. texensis* is known only to infect *Stipa leucotricha* (Leuchtmann & Clay, 1989). Interestingly, this is the only *Stipa* species of many in the genus to produce cleistogamous flowers in the leaf axils, which suggests that they have been induced by infection.

In type-III associations the endophyte is vertically transmitted and infection has no direct effect on host reproduction. Infection could have an indirect effect on plant reproduction if the vigour of host plants was altered by infection. A vertically transmitted symbiont that reduces host fitness should not persist in the population (Ewald, 1987). Instead, most studies suggest that inflorescence and seed production in type-III associations are enhanced by infection (Clay, 1990a; Rice et al., 1990; Bier, 1995). For example, Rice et al. (1990) conducted common garden experiments with infected and uninfected tall fescue clones. The increase in performance of infected (vs uninfected) plants ranged from 20% up to 79% for various measures of fecundity (Table 10.2). In contrast, Siegel et al. (1984) found no difference in seed production between infected and uninfected 'Kenhy' tall fescue. The production of inflorescences, albeit aborted, is enhanced in some type-I and type-II associations as well. For example, in an extant population of *Danthonia spicata*, plants infected by *A. hypoxylon* produced an average of 18 inflorescences compared with five by uninfected plants (Clay, 1984). Similarly, plants of *Holcus lanatus* infected by *Epichloë clarkii* in an extant population produced an average of 21 inflorescences vs 12 for uninfected plants (Clay & Brown, 1997). On the other hand, *Panicum agrostoides* infected by *Balansia henningsiana*, *Cyperus virens* infected by *B. cyperi*, and *Sporobolus poiretti* infected by *B. epichloë* produced similar numbers of inflorescences to uninfected plants (Clay et al., 1989; Clay, 1990a). Moreover, in the greenhouse, *C. rotundus* infected by *B. cyperi* produced significantly fewer inflorescences, but more tubers, than uninfected plants (Stovall & Clay, 1988).

To examine the correlation of grass life history and the type of associa-

Table 10.2. *Relative fecundity of infected and uninfected tall fescue grown in a common garden environment*

Difference is the percentage increase or decrease in trait of infected plants relative to uninfected plants. An asterisk indicates a significant difference ($P<0.05$).

		Per plant measures		
Experiment	Endophyte	Inflorescences	Seed number	Seed weight (g)
A – 1987	+	30	4,474	6.9
	−	25	2,795	3.8
	difference	+20%*	+60%*	+79%*
B – 1987	+	194	60,398	80.3
	−	143	43,665	60.9
	difference	+36%*	+38%*	+32%*
B – 1988	+	330	44,450	52.5
	−	250	35,312	40.1
	difference	+32%*	+26%*	+31%*

Note:
Data modified from Rice *et al.* (1990).

tion, Clay (1988) compiled a list of 178 endophyte hosts from the literature and then classified hosts as either annual or perennial, and either caespitose or rhizomatous. A total of 38 species formed type-III associations with asymptomatic and seed-transmitted endophytes. Over 80% of the hosts are perennial caespitose grasses, 16% are annuals, while only 3% are perennial rhizomatous grasses (Table 10.3). Further, all host species occur in the subfamily Pooideae. A total of 140 hosts were infected by endophytes that produce stromata and sterilize at least some inflorescences. In this group 65% are perennial caespitose grasses, 26% are rhizomatous perennials, and 9% are annuals. Most hosts are in the subfamily Panicoideae and form type-I associations. However, from the literature it was not always possible to distinguish between type-I and type-II associations in some pooid hosts of *Epichloë.*

A significantly higher percentage of type-I and type-II associations were with rhizomatous grasses while a higher percentage of type-III associations involved caespitose and annual grasses (Table 10.3). The cost of infection is likely to be lower in rhizomatous grasses than in caespitose grasses because rhizomatous grasses sterilized by infection can still spread vigorously. The cost of infection is likely to be especially high in annuals and proportionally more annuals formed type-III associations in this survey.

264   K. Clay

Table 10.3. *Distribution of annual, perennial, rhizomatous, and caespitose grasses forming type-I and type-II vs. type-III associations*

	Type of association		
Type of host	Type I and II	Type III	Total
Perennial, caespitose*	91 (65%)	31 (82%)	122
Perennial, rhizomatous	37 (26%)	1 (3%)	38
Annual	12 (9%)	6 (16%)	18
Total	140	38	178

Notes:
*$G=12.58$, $P<0.001$ (G test of null hypothesis that type of association is independent of type of host (perennial hosts only)). See Clay (1988) for details.
Modified from Clay (1988).

However, it should be recognized that these patterns may be confounded by host taxonomy. For example, rhizomatous perennials might be more common in the Panicoideae than in the Pooideae. Because type-II and type-III associations do not occur in the Panicoideae, this could cause a spurious correlation between infection type and grass life history.

An analysis of these correlations within subfamilies would address this problem.

Seed germination
Infected seeds are produced by host plants in type-II and type-III associations. Hosts are limited entirely to the subfamily Pooideae, or the genus *Danthonia* (subfamily Arundinoideae). A few studies, primarily with *Festuca*, indicate that germination can be altered by infection (see below). The dynamics of grass populations may be affected insofar as germination rates affect population growth rates. The endophyte has the highest viability in fresh seeds. In both tall fescue and perennial ryegrass, storage of seed at room temperature and high humidity results in rapid loss of endophyte infection (Rolston et al., 1986; Welty, Azevedo & Cooper, 1987).

Germination of seeds from *Festuca obtusa* and *Poa sylvestris*, two common woodland grasses in the eastern United States, was investigated by Bier (1995). Populations of both species typically comprised only infected plants so uninfected plants were generated by fungicide treatments

Table 10.4. *Percentage germination of endophyte infected and uninfected tall fescue seed at different temperatures and water potentials*

Data are means from ten fescue genotypes. Difference is the percentage increase (or decrease) in germination of infected seeds relative to uninfected seeds; an asterisk indicates a significant difference ($P<0.05$). Data not shown for 1.0 MPa where no seeds germinated.

Water potential (MPa)	Endophyte	Temperature (°C)		
		16	24	32
0.00	+	64.2	65.1	9.6
	−	57.2	61.1	11.7
	difference	+12.2*	+6.5*	−21.9
0.50	+	29.2	40.0	0
	−	25.4	32.7	0
	difference	+15.0*	+22.3*	0
0.75	+	13.8	24.8	0
	−	10.7	19.6	0
	difference	+29.0*	+26.5*	0

Note:
Data modified from Pinkerton *et al.* (1990).

and seeds were then obtained from field transplants. Resulting seeds were planted into a native site for two consecutive years. Although infected seeds had higher germination (defined as emergence above the soil) both years in both species, the differences were not statistically significant (Bier, 1995). It is possible that there may have been some mortality of seedlings before they emerged from the soil.

Several germination studies have been conducted with tall fescue and perennial ryegrass in artificial environments. Pinkerton, Rice & Undersander (1990) germinated tall fescue seeds in incubators at either 16, 24 or 32 °C where PEG-8000 solutions were used to produce osmotic pressures of 0, 0.5, 0.75, and 1.0 MPa (Table 10.4). Seeds were obtained from infected and uninfected clones of 10 fescue genotypes. No germination occurred at 1.0 MPa (data not shown) and at 32 °C germination occurred only at 0 MPa. For six out of seven remaining combinations, infected seeds had significantly higher germination rates than uninfected seeds. Infected seeds from five genotypes had significantly greater germination than uninfected seeds, there was no difference in four genotypes, and infected seeds had lower germination in one genotype.

Germination rates of tall fescue and perennial ryegrass did not differ significantly between infected and uninfected seeds in a greenhouse study (Clay, 1987). However, final germination was about 10% higher for infected seeds and they had higher germination than uninfected seeds at each of five sample dates in both species. Rice *et al.* (1990) also reported a higher germination rate of tall fescue seed when endophyte infected. In contrast, Keogh & Lawrence (1987) and Bacon (1993) found that germination of perennial ryegrass and tall fescue, respectively, was unaffected by endophyte infection. Similarly, there was no overall difference in germination of cleistogamous seeds from infected vs uninfected *Danthonia epilis* (Clay, 1994).

In the aforementioned studies there was generally either enhanced germination of infected seeds or no difference, in contrast to many seed-borne pathogens (Neergaard, 1977; Govinthasamy & Cavers, 1995). The vigour of seed germination may reflect the vigour and physiological state of the infected or uninfected maternal plant (Roach & Wulff, 1987). The loss of endophyte viability in dormant seeds and the relative germination of infected vs uninfected seeds will interact to determine the frequency of infection in progeny of infected plants.

Host plant vigour

Unlike the variable effects of endophyte infection on plant reproduction, infected plants typically exhibit greater tillering and/or biomass production than uninfected plants over a wide range of environmental conditions. One measure of vigour is size distributions in static population samples. Infected plants of many species tend to be bigger than uninfected plants in natural populations (Diehl, 1950; Bradshaw, 1959; Harberd, 1961; Clay *et al.*, 1989; Leuchtmann & Clay, 1993; Clay & Brown, 1997). These studies are on type-I and type-II associations where infection status is easily detected by visual examination. Two typical examples are *Danthonia spicata* infected by *Atkinsonella hypoxylon* and *Cyperus virens* infected by *Balansia cyperi* (Fig. 10.2). In the first case, mean tiller number was nearly three times higher in infected plants while in the second case it was over two times higher. While increased size is consistent with enhanced growth, it is also consistent with an increased probability of larger plants and/or older plants becoming infected through contagious spread.

In type-III associations there is no contagious spread, so that differences in sizes of infected and uninfected plants reflect differences in growth and survival. In a greenhouse experiment, tall fescue and perennial ryegrass were grown from seed and sequential harvests were conducted to follow plant growth rate over time (Clay, 1987). For tall fescue, there was no effect of infection on tillering and biomass production after six weeks of growth, but harvests at 10 and 14 weeks revealed an accelerating growth advantage for infected plants (Fig. 10.3a). At 14 weeks infected plants were nearly 60% heavier than uninfected plants. Infected plants of perennial ryegrass also exhibited faster growth than uninfected plants with statistically significant differences at every harvest (Fig. 10.3b). In another study with perennial ryegrass, infected clones produced 38% more biomass than uninfected clones after eight weeks in a growth chamber (Latch, Hunt & Musgrave, 1985). In contrast, Keogh & Lawrence (1987), Lewis & Clements (1990) and Lewis (1992) did not find any consistent effect of infection on tiller and biomass production in perennial ryegrass.

Clay (1997) recently summarized studies on tall fescue comparing the growth of infected vs uninfected plants in a range of environments. In 14 of 15 total comparisons, infected plants outyielded uninfected plants (see also Read & Camp, 1986; Arechavaleta *et al.*, 1989; De Battista *et al.*, 1990b; Hill, Belesky & Stringer, 1991; Bouton *et al.*, 1993; West *et al.*, 1993). For example, Hill *et al.* (1991) found that infected tall fescue clones produced more biomass than uninfected clones in both spring and fall harvests over two growing seasons. Schmidt (1993) also reported enhanced growth of endophyte infected meadow fescue (*Festuca pratensis*). Growth differences vary with plant environment. Arechavaleta *et al.* (1989) and Cheplick *et al.* (1989) found that relative growth of infected tall fescue was equal to or less than uninfected fescue at low soil nitrogen levels but increased with increasing nutrient level. Bouton *et al.* (1993) also found that the relative performance of infected vs uninfected fescue varied among experimental stations in Georgia that differed in soils and climate. Thus, while most studies have demonstrated a growth advantage for infected plants, it is not universal and depends in part on host plant environment.

Increased growth also occurs in other grasses infected by species of *Atkinsonella, Balansia* and *Epichloë* (Harberd, 1961; Clay, 1990b; Fowler & Clay, 1995). For example, in greenhouse experiments infected plants of *Panicum agrostoides* infected by *B. henningsiana* produced 50% more tillers and 25% more biomass after 15 weeks of growth (Clay *et al.*, 1989) while the sedge *Cyperus rotundus* infected by *B. cyperi* produced significantly more tillers and tubers than uninfected plants (Stovall & Clay, 1988). In a field study *Danthonia spicata* infected by *A. hypoxylon* produced 42% more tillers over a two year period (Clay, 1984). Inoculation experiments demon-

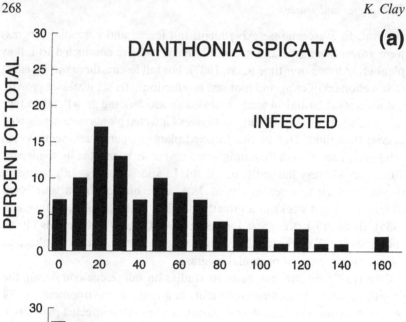

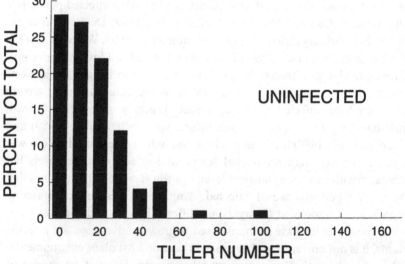

Fig. 10.2. Size distributions of (a) *Danthonia spicata* plants infected by *Atkinsonella hypoxylon* vs uninfected plants and (b) *Cyperus virens* plants infected by *Balansia cyperi* vs uninfected plants. The x-axis is in groups of 10 tillers (e.g. 0=0–9 tillers, 20=20–29 tillers). Data from Clay (1984) and Clay (1986).

strate that endophyte infection can directly enhance plant growth. When *D. spicata* seedlings were inoculated with two strains of *Atkinsonella*, a compatible strain caused a 50% increase in tillering relative to controls after 10 weeks while an incompatible strain had no effect on growth (Leuchtmann

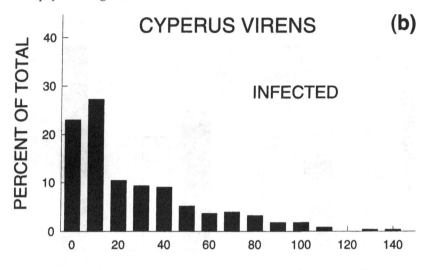

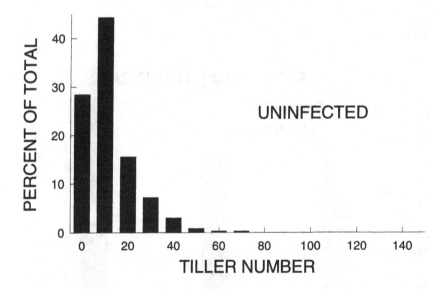

& Clay, 1988). However, there was no significant effect of infection on tiller production in *Sporobolus poiretti* (Clay, 1990*a*) and Hill (1994) reported that plants of *Paspalum notatum* infected by *Myriogenospora atramentosa* were stunted and grew less than uninfected plants. Thus, within type-I and type-II associations both beneficial and detrimental effects on plant growth can be found.

The survival of infected tall fescue and perennial ryegrass is often greater

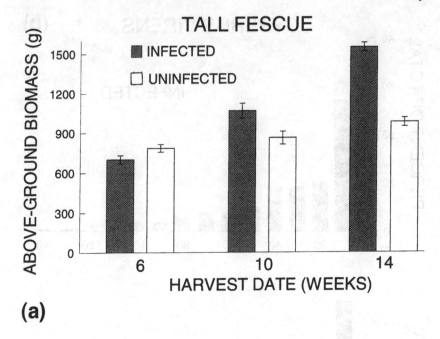

(a)

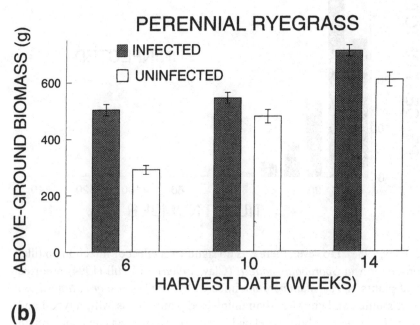

(b)

Fig. 10.3. Growth of (*a*) tall fescue and (*b*) perennial ryegrass in the greenhouse. Data from Clay (1987).

than uninfected plants in field experiments (tall fescue – Read & Camp, 1986; Clay, 1990*a*; Read & Walker, 1990; Bouton *et al.*, 1993; perennial rye-grass – Funk *et al.*, 1983; Francis & Baird, 1989; Cunningham, Foot & Reed, 1993). Survival differences often appear to be related to abiotic or biotic stress tolerance (see below). Increased survival has also been reported from other endophyte-infected grasses forming type-I and type-II associations. These include *Festuca longifolia* (Saha *et al.*, 1987), *Danthonia spicata* (Clay, 1984) and *Cyperus virens* (Clay, 1990*a*). There were no differences in survival of infected and uninfected *Stipa leucotricha* and *Sporobolus poiretti* (Clay, 1990*a*; Fowler & Clay, 1995).

Changes in the growth and vigour of host plants may affect competitive interactions within and between grass species. Competition, in turn, may result in long-term changes in the frequency of infection in grass popula-tions and communities. For example, Kelley & Clay (1987) compared the performance of infected and uninfected *D. spicata* in competition with the co-occuring grass *Anthoxanthum odoratum* under field conditions. While there was no effect of infection on intraspecific competition, there was a highly significant effect on interspecific competition, with infected *Danthonia* producing twice as many inflorescences and 56% more vegeta-tive tillers than uninfected plants when grown with *A. odoratum*.

Other published competition experiments involve fescue and ryegrass. In a greenhouse study with tall fescue and perennial ryegrass, endophyte infec-tion generally enhanced the competitive ability of tall fescue but reduced the competitive ability of perennial ryegrass (Marks, Clay & Cheplick, 1991). In another greenhouse experiment with tall fescue, plant density, soil moisture level, and planting combination (all infected, all uninfected, or mixture) were manipulated. There was a highly significant effect of infec-tion with infected fescue outyielding uninfected fescue in each of the eight combinations of experimental variables (Clay, 1993). Hill *et al.* (1990, 1991) examined competition between infected and uninfected clones of KY-31 tall fescue in greenhouse and field experiments. In the greenhouse, there was a great deal of variation and no effect of infection the first year but by the second year there was a trend for increased productivity of infected clones. In the field, infected clones produced more biomass and seed than unin-fected clones when averaged over genotypes and densities. They also found differences among genotypes in response to infection, indicating that certain plant/fungal combinations are more competitive than others.

Several studies have been conducted on competitive interactions between clover (*Trifolium repens*) and perennial ryegrass. While Sutherland & Hoglund (1990) found that endophyte infection resulted in a significant suppression of clover in New Zealand, Prestidge *et al.* (1992) did not. Lewis (1992) also did not find any effect of endophyte infection on competition with clover in the UK.

In another series of greenhouse experiments, there was a general increase in the performance of endophyte-infected tall fescue, perennial ryegrass and red fescue in competition, but the specific effect of infection was herbivory dependent (Clay, Marks & Cheplick, 1993). In general, the competitive ability of endophyte-infected plants was increased by herbivore pressure.

The limited number of competition experiments thus far conducted demonstrate that endophyte infection can enhance competitive ability. However, the effect of infection depends on both plant genotype and environment. Only a single wild grass species has been examined experimentally (*Danthonia spicata*, Kelley & Clay, 1987), so additional research is required before drawing any general conclusions.

Host plant stress tolerance

Several mechanisms may cause a growth advantage for endophyte-infected grasses. Reallocation of energy from flowering could account for the increased growth of hosts in type-I and type-II associations (Clay, 1991*a*). Phytohormone production by endophytic fungi, which are most densely concentrated near meristematic regions, could also affect plant growth (De Battista *et al.*, 1990*a*). The best documented mechanism, however, is the greater tolerance of infected plants to abiotic and biotic stresses (Bacon, 1993; Latch, 1993).

Several studies have demonstrated that infected tall fescue is more drought tolerant than uninfected plants. This might explain the greater survival of infected fescue in southern and western portions of its range in the USA where high summer temperatures and soil water deficits are common (Bouton *et al.*, 1993; West, 1994). West *et al.* (1993) found that endophyte infection enhanced plant density and productivity in experimental plots subject to water deficits but not in frequently irrigated plots. Also, infected tall fescue plants exhibit more rapid leaf rolling, a greater osmotic adjustment, and better survival than uninfected plants under drought stress (Arechavaleta *et al.*, 1989; West, 1994). Recently, Hill, Pachon & Bacon (1996) found that drought acclimation of tall fescue was partially dependent on endophyte genotype with some genotypes enhancing drought acclimation and others having no effect. Infected fescue plants are also more heat tolerant than uninfected plants and are able to maintain higher photosynthetic rates as temperatures increase (Marks & Clay, 1996). Endophyte-infected fescue may tolerate waterlogged conditions better than uninfected fescue (Bacon, 1993).

Most research on abiotic stress tolerance has focused on tall fescue but Cunningham *et al.* (1993) indicated that endophyte-infected perennial rye-grass is more persistent and has greater yields in low rainfall sites at the margin of the species range in Australia. Infection rates in two cultivars in these sites increased from 80% and 78% respectively up to 100% in four years. Grand-Ravel *et al.* (1995) also reported that endophyte infection increased with evapotranspiration rate in France.

Considering biotic stress, interest in the resistance of endophyte-infected grasses to herbivory was stimulated by a report that toxic tall fescue pastures in the south-eastern USA were highly infected by a fungal endophyte while non-toxic pastures were endophyte free (Bacon *et al.*, 1977). Shortly thereafter, poisoning of sheep in New Zealand was linked to an endophyte in perennial ryegrass (Fletcher & Harvey, 1981) and Prestidge *et al.* (1982) reported increased resistance of endophyte-infected ryegrass to the Argentine stem weevil, a major insect pest. Many studies now show that endophyte-infected grasses are more resistant to a wide range of vertebrate and invertebrate herbivores (Clay, 1991*b*). For example, Latch (1993) and Rowan & Latch (1994) list 38 insect species that are deterred by endophyte infection of tall fescue and perennial ryegrass. Affected mammals include cattle, horses, sheep, deer, goats, rabbits, voles, rats and mice, as well as a number of birds (Clay, 1991*b*; Madej & Clay, 1991; Coley *et al.*, 1995). At least 21 grass species (plus three sedges) exhibit increased resistance to herbivory when infected (Clay, 1991*b*). It is likely that more cases will be added to the list as further studies are conducted. While exceptions exist with some grasses showing no difference in herbivore resistance (Siegel *et al.*, 1990), and with some herbivores unresponsive to endophyte presence (Latch, Christensen & Gaynor, 1985; Kirfman, Brandenburg & Garner, 1986; Lewis & Clements, 1986), the majority of studies have found enhanced resistance of endophyte-infected grasses to herbivory.

Differential predation of infected and uninfected seeds could alter the infection frequency of seedlings. In type-II and type-III associations the endophyte is seed transmitted and toxic to seed-feeding mammals, birds and insects (Neal & Schmidt, 1985; Cheplick & Clay, 1988; Madej & Clay, 1991; Knoch, Faeth & Arnott, 1993). Choice experiments indicate that seed predators can discriminate between infected and uninfected seeds and avoid the former (Madej & Clay, 1991; Knoch *et al.*, 1993).

Herbivore toxicity is caused by a variety of alkaloids, some of which can be produced by the endophyte alone whereas others are produced only in infected plants. Related *Claviceps* (ergot) species are also known for the production of toxic alkaloids. Ergot alkaloids are one of the four major groups of endophyte alkaloids, along with lolines, lolitrems and peramine (Porter, 1994). A single plant can contain a variety of alkaloids in different concentrations. For example, one collection of endophyte-infected sleepygrass (*Stipa robusta*) from New Mexico contained five ergot alkaloids and one loline alkaloid (Petroski, Powell & Clay, 1992). Up to 1% of the dry weight of tall fescue seeds, and 0.4% of leaves, is endophyte alkaloids (Bush *et al.*, 1982; Jones, *et al.*, 1983). In general, alkaloids are most concentrated in the crown and inflorescences where fungal hyphae are most concentrated.

The ecology of host grasses is affected in two ways by endophyte toxins. Due to avoidance by herbivores, infected plants incur less damage than uninfected conspecifics. A range of vertebrate and invertebrate herbivores can discriminate between infected and uninfected plants (Bacon & Siegel, 1988; Madej & Clay, 1991; Latch, 1993). Even large herbivores like cattle can distinguish between individual plants in mixed stands (Van Santen, 1992). Differential herbivory can alter competitive interactions among plants. In a greenhouse experiment with tall fescue and orchardgrass (*Dactylis glomerata*), the presence of an herbivorous insect shifted the competitive balance between the species (Clay *et al.*, 1993). In the absence of herbivory, orchardgrass was a stronger competitor than infected tall fescue but fescue outcompeted orchardgrass when herbivores were present.

A second consequence of endophyte toxins is reduced herbivore population sizes. Grazing on infected grasses causes a reduction in survival, growth rate, and fecundity of domestic livestock (Bacon & Siegel, 1988; Clay, 1991*b*; Latch, 1993). Recent research has shown that small mammal populations are lower in infected tall fescue populations than in uninfected populations (Giuliano, Elliott & Sole, 1994; Coley *et al.*, 1995). Insect populations are also smaller in tall fescue and perennial ryegrass where infection frequencies are high (Prestidge *et al.*, 1982; Funk *et al.*, 1983; Kirfman *et al.*, 1986). Lower herbivore populations should translate into less damage to infected grasses. Indeed, observations on wild grasses suggest that herbivore pressure may drive the dominance of endophyte-infected species within grassland communities (Clay, 1997; Miles *et al.*, 1996).

Endophyte infection may also provide protection against certain fungal and viral pathogens (Clay *et al.*, 1989; Ford & Kirkpatrick, 1989; Gwinn & Gavin, 1992; Burpee & Bouton, 1993; Mahmood *et al.*, 1993), although others report no effects (Welty, Barker & Azevedo, 1993). Several studies have also found that endophyte infection reduces nematode populations in tall fescue plants and surrounding soil (Pederson, Rodriquez-Kabana & Shelby, 1988; West *et al.*, 1988; Kimmons, Gwinn & Bernard, 1990) although other studies have not found any effect of endophyte infection in perennial ryegrass (Cook, Lewis & Mizen, 1991).

In total, considerable evidence exists that endophyte-infected plants exhibit greater tolerance to a range of biotic and abiotic stresses. However, the degree of protection can vary with host and endophyte, and with the identity of the pest species. Toxicity of infected plants to herbivores is especially well documented in domestic livestock where it is a major economic problem. For infected grasses, stress tolerance may provide a strong selective advantage in mixtures with uninfected grasses.

Infection dynamics

The dynamics of infection within host grass populations provides a direct measure of the benefit or detriment of endophyte infection. Unlike type-I and type-II associations, purely seed transmitted endophytes, such as found in tall fescue and perennial ryegrass, increase or decrease in host populations as a function of relative host fitness. A selective advantage for infected plants should eventually result in 100% infection frequencies while a reduction in host fitness should purge the endophyte from the population. This assumes, however, that the advantage of infection is not frequency dependent. A computer simulation model of type-III infection dynamics predicted that, with a two-fold fitness advantage for infected plants, endophyte infection should go from 50% to 100% frequency in less than 10 generations (Clay, 1993). Similarly, a 50% fitness reduction would eliminate endophyte infection in seven or eight generations. However, the selective advantage of infection is likely to vary over time with drought stress, herbivore pressure, interspecific competition, etc.

Grasses forming type-I and type-II associations will not exhibit the same dynamics because the endophyte can spread contagiously among plants. Infection frequencies could increase even if endophyte infection reduces host fitness. Several reports have been published documenting increases in infection frequencies over time in populations or marked cohorts of grasses infected by *Atkinsonella*, *Balansia* or *Epichloë* (Diehl, 1950; Large, 1952; Clay, 1984; 1990*a*). In all of these cases, increasing infection frequency resulted from differential survival of infected vs uninfected plants and/or contagious spread.

Of greater interest are type-III associations. Several grass species have

Table 10.5. *Changes in endophyte infection frequency of tall fescue populations in the USA over time since establishment*

Site	Initial (%)	Final (%)	Difference (%)	Time (y)	Reference
Alabama	90	94	+4	4	Shelby & Dalrymple,
	58	92	+34	4	1993
	27	83	+56	4	
Georgia	<1	<1	0	3	Bouton et al., 1993
	<1	11	11	3	
Tennessee	75	74	−1	2	Thompson, Fribourg &
	75	83	+8	2	Reddick, 1989
	60	66	+6	2	
	45	72	+27	2	
	30	63	+33	2	
	15	59	+44	2	
	0	12	+12	2	
	0	5	+5	2	
Texas	60	73	+13	5	Read & Walker, 1990
	43	83	+40	5	
	7	28	+21	5	
	5	23	+18	5	
	0	0	0	5	
	0	0	0	5	

been examined. Increases in infection frequency over time were reported in hard fescue (*F. longifolia*) (Saha *et al.*, 1987). In perennial ryegrass, Lewis & Clements (1986) found that most endophyte infection occurred in pastures in the UK that were older than 15 years. However, this pattern could also reflect grass breeding which resulted in the inadvertent elimination of infection in many European cultivars over the past 30 years (G. Lewis, pers. comm.). In Tasmania, infection frequency in 30 perennial ryegrass pastures was positively correlated with pasture age (Cunningham *et al.*, 1993). Changes in infection frequency often occurred very rapidly during seedling establishment (Francis & Baird, 1989; Cunningham *et al.*, 1993) or during periods of environmental stress (Funk *et al.*, 1983).

The results of several published studies that followed infection frequencies in tall fescue populations over time are summarized in Table 10.5. There was a general tendency for infection frequencies to increase; only one plot showed a 1% decrease in infection. The high levels of infection found in natural populations of other type-III associations imply that displacement of uninfected plants has occurred in the past, perhaps repeatedly. Increases tended to be greater in plots with lower initial infection frequencies. For example, in Alabama, USA, an initial infection frequency of 90% climbed to 94% after four years while an initial infection frequency of 27% climbed to 83% (Shelby & Dalrymple, 1993). Finally, while some plots remained at 0% infection over several years, others increased substantially from 0% over time. In these latter cases, the initial seed stocks presumed to be 0% infected must have had a very low level of infection. This presents a problem for pasture management if a low level of infection in seeds or residual plants can soon give rise to highly infected pastures.

Summary and conclusions

The population dynamics of many grasses are intimately coupled with infection by systemic fungal endophytes. While many endophyte species act as typical plant pathogens, reducing host survival and stunting growth, other endophyte species are mutualists that enhance host plant survival, growth and reproduction. Other endophytes have both pathogenic and mutualistic effects, reducing host fecundity but enhancing vegetative vigour and stress tolerance. Type-III associations, which are mutualistic, are also the hardest to detect because no visible symptoms of infection are produced. Any cool-season pooid grass is a strong candidate for endophyte infection and should be examined microscopically prior to field or greenhouse studies.

Unlike other mutualisms with microbes (e.g. mycorrhizas), mixtures of endophyte infected and uninfected grasses can occur in a single population. Therefore, any mutualistic effects of endophyte infection will not be uniformly distributed over the population and may exacerbate fitness differences among individuals. Further, endophytes (at least those forming type-II and type-III associations) have a stronger potential for coevolution with their grass hosts. Where endophytes are seed transmitted, a single fungal genotype may become associated with a lineage of host plants. Even in type-I associations vegetative reproduction of hosts may maintain specific plant/fungal combinations over time. This contrasts with many plant pathogens where a single plant may be infected by dozens of genotypes that do not persist for more than a single growing season.

The importance of grasses in agriculture has provided a strong motivation for research on endophyte associations. The importance of grasses in ecological communities provides an equally strong motivation for understanding how endophyte infection affects the ecology of individuals, populations and entire grassland communities. While most research has been conducted on just two widespread species, the effects of endophyte infection on tall fescue and perennial ryegrass are not fundamentally different from those found in other grasses. These and other endophyte associations provide a model for understanding how plant–microbe interactions evolve and how they affect present day ecological processes.

Allen, M. F. (1991). *The Ecology of Mycorrhizae.* Cambridge: Cambridge University Press.
Arechavaleta, M., Bacon, C. W., Hoveland, C. S. & Radcliffe, D. E. (1989). Effect of tall fescue endophyte on plant response to environmental stress. *Agronomy Journal*, 81, 83–90.
Bacon, C. W. (1993). Abiotic stress tolerances (moisture, nutrients) and photosynthesis in endophyte-infected fescue. *Agriculture, Ecosystems and Environment*, 44, 123–41.
Bacon, C. W., Porter, J. K., Robbins, J. D. & Luttrell, E. S. (1977). *Epichloë typhina* from toxic tall fescue grasses. *Applied and Environmental Microbiology*, 34, 576–81.
Bacon, C. W. & Siegel, M. R. (1988). Endophyte parasitism of tall fescue. *Journal of Production Agriculture*, 1, 45–55.
Bever, J. D. (1994). Feedback between plants and their soil communities in an old field community. *Ecology*, 75, 1965–77.
Bier, J. (1995). Relationship between transmission mode and mutualism in the grass-endophyte system. Ph.D. Dissertation, Indiana University, Bloomington.
Bouton, J. H., Gates, R. N., Belesky, D. P. & Owsley, M. (1993). Yield and persistence of tall fescue in the southeastern coastal plain after removal of its endophyte. *Agronomy Journal*, 85, 52–5.
Bradshaw, A. D. (1959). Population differentiation in *Agrostis tenuis* Sibth II. The incidence and significance of infection by *Epichloë typhina*. *New Phytologist*, 58, 310–5.
Bucheli, E. & Leuchtmann, A. (1996). Evidence for genetic differentiation between choke-inducing and asymptomatic strains of the *Epichloë* grass endophyte from *Brachypodium sylvaticum*. *Evolution*, 50, 1879–87.
Burdon, J. J. (1987). *Diseases and Plant Population Biology.* Cambridge: Cambridge University Press.
Burpee, L. L. & Bouton, J. H. (1993). Effect of eradication of the endophyte *Acremonium coenophialum* on epidemics of Rhizoctonia blight in tall fescue. *Plant Disease*, 77, 157–9.
Bush, L. P., Cornelius, P. L., Buckner, R. C., Varney, D. R., Chapman, R. A., Burrus, P. B., II, Kennedy, C. W., Jones, T. A. & Saunders, M. J. (1982). Association of N-acetyl loline and N-formyl loline with *Epichloë typhina* in tall fescue. *Crop Science*, 22, 941–3.
Cheplick, G. P. & Clay, K. (1988). Acquired chemical defenses of grasses: the role of fungal endophytes. *Oikos*, 52, 309–18.
Cheplick, G. P. & Clay, K. (1989). Convergent evolution of cleistogamy and seed heteromorphism in two perennial grasses. *Evolutionary Trends in Plants*, 3, 127–36.
Cheplick, G. P., Clay, K. & Marks, S. (1989). Interactions between fungal endophyte infection and nutrient limitation in the grasses *Lolium perenne* and *Festuca arundinacea*. *New Phytologist*, 111, 89–97.
Clay, K. (1984). The effect of the fungus *Atkinsonella hypoxylon* (Clavicipitaceae)

on the reproductive system and demography of the grass *Danthonia spicata*. *New Phytologist*, **98**, 165–75.

Clay, K. (1986). Induced vivipary in the sedge *Cyperus virens* and the transmission of the fungus *Balansia cyperi* (Clavicipitaceae). *Canadian Journal of Botany*, **64**, 2984–8.

Clay, K. (1987). Effects of fungal endophytes on the seed and seedling biology of *Lolium perenne* and *Festuca arundinacea*. *Oecologia*, **73**, 358–62.

Clay, K. (1988). Clavicipitaceous fungal endophytes of grasses: coevolution and the change from parasitism to mutualism. In *Co-evolution of Fungi with Plants and Animals*, ed. D. L. Hawksworth & K. Pirozynski, pp. 79–105. London: Academic Press.

Clay, K. (1990a). Comparative demography of three graminoids infected by systemic, clavicipitaceous fungi. *Ecology*, **71**, 558–70.

Clay, K. (1990b). Fungal endophytes of grasses. *Annual Review of Ecology and Systematics*, **21**, 275–97.

Clay, K. (1991a). Parasitic castration of plants by fungi. *Trends in Ecology and Evolution*, **6**, 141–72.

Clay, K. (1991b). Endophytes as antagonists of plant pests. In *Microbial Ecology of Leaves*, ed. J. H. Andrews & S. S. Hirano, pp. 331–57. New York: Springer-Verlag.

Clay, K. (1993). The ecology and evolution of endophytes. *Agriculture, Ecosystems and Environment*, **44**, 39–64.

Clay, K. (1994). Hereditary symbiosis in the grass genus *Danthonia*. *New Phytologist*, **126**, 223–31.

Clay, K. (1995). Correlates of pathogen species richness in the grass family. *Canadian Journal of Botany*, **73**, s 42–9.

Clay, K. (1997). Fungal endophytes, herbivores and the structure of grassland communities. In *Multitrophic Interactions in Terrestrial Systems*, ed. A. C. Gange & V. K. Brown, pp. 151–69. Oxford: Blackwell Science.

Clay, K. & Brown, V. K. (1997). Infection of *Holcus lanatus* and *H. mollis* by *Epichloë* in experimental grasslands. *Oikos*, **79**, 363–70.

Clay, K., Cheplick, G. P. & Marks, S. M. (1989). Impact of the fungus *Balansia henningsiana* on the grass *Panicum agrostoides*: frequency of infection, plant growth and reproduction, and resistance to pests. *Oecologia*, **80**, 374–80.

Clay, K. & Frentz, I. C. (1993). *Balansia pilulaeformis*, an epiphytic species. *Mycologia*, **85**, 527–34.

Clay, K. & Kover, P. (1996). Evolution and stasis in plant/pathogen associations. *Ecology*, **77**, 997–1003.

Clay, K. & Leuchtmann, A. (1989). Infection of woodland grasses by fungal endophytes. *Mycologia*, **81**, 805–11.

Clay, K., Marks, S. & Cheplick, G. P. (1993). Effects of insect herbivory and fungal endophyte infection on competitive interactions among grasses. *Ecology*, **74**, 1767–77.

Coley, A. B., Fribourg, H. A., Pelton, M. R. & Gwinn, K. D. (1995). Effects of tall fescue infestation on relative abundance of small mammals. *Journal of Environmental Quality*, **24**, 472–5.

Cook, R., Lewis, G. C. & Mizen, K. A. (1991). Effects of plant parasitic nematodes on infection of perennial ryegrass, *Lolium perenne*, by the endophytic fungus *Acremonium lolii*. *Crop Protection*, **10**, 403–7.

Cunningham, P. J., Foot, J. Z. & Reed, K. F. M. (1993). Perennial ryegrass (*Lolium perenne*) endophyte (*Acremonium lolii*) relationships: the Australian experience. *Agriculture, Ecosystems and Environment*, **44**, 157–68.

DeBattista, J. P., Bacon, C. W., Severson, R., Plattner, R. D. & Bouton, J. H. (1990a). Indole acetic acid production by the fungal endophyte of tall fescue. *Agronomy Journal*, **82**, 878–80.

DeBattista, J. P., Bouton, J. H., Bacon, C. W. & Siegel, M. R. (1990b). Rhizome and herbage production of endophyte-removed tall fescue clones and populations. *Agronomy Journal*, **82**, 651–4.

Diehl, W. W. (1950). Balansia *and the Balansiae in America*. Washington, D.C.: U.S.D.A.

Dinoor, A. & Eshed, N. (1984). The role and importance of pathogens in natural plant communities. *Annual Review of Phytopathology*, **22**, 443–66.

Dong, Z., Canny, M. J., McCully, M. E., Roboredo, M. R., Cabadilla, C. F., Ortega, E. & Rodes, R. (1994). A nitrogen-fixing endophyte of sugarcane stems. *Plant Physiology*, **105**, 1139–47.

Ewald, P. (1987). Transmission modes and evolution of the parasitism–mutualism continuum. *Annals of the New York Academy of Sciences*, **503**, 295–306.

Fletcher, L. R. & Harvey, I. C. (1981). An association of a *Lolium* endophyte with ryegrass staggers. *New Zealand Veterinary Journal*, **29**, 185–6.

Ford, V. L. & Kirkpatrick, T. L. (1989). Effects of *Acremonium coenophialum* in tall fescue on host disease and insect resistance and allelopathy to *Pinus taeda* seedlings. *Proceedings of the Arkansas Fescue Toxicosis Conference*, **140**, 29–34.

Fowler, N. L. & Clay, K. (1995). Environmental heterogeneity, fungal parasitism and the demography of the grass *Stipa leuchotricha*. *Oecologia*, **103**, 55–62.

Francis, S. M. & Baird, D. B. (1989). Increase in the proportion of endophyte-infected perennial ryegrass plants in over-drilled pastures. *New Zealand Journal of Agricultural Research*, **32**, 437–40.

Funk, C. R., Halisky, P. M., Johnson, M. C., Siegel, M. R., Stewart, A. V., Ahmad, S., Hurley, R. H. & Harvey, I. C. (1983). An endophytic fungus and resistance to sod webworms: association in *Lolium perenne*. *Bio/Technology*, **1**, 189–91.

Giuliano, W. M., Elliott, C. L. & Sole, J. D. (1994). Significance of tall fescue in the diet of the eastern cottontail. *Prairie Naturalist*, **26**, 53–60.

Glenn, A. E., Bacon, C. W., Price, R. & Hanlin, R. T. (1996). Molecular phylogeny of *Acremonium* and its taxonomic implications. *Mycologia*, **88**, 369–83.

Govinthasamy, T. & Cavers, P. B. (1995). The effects of smut (*Ustilago destruens*) on seed production, dormancy, and viability in fall panicum (*Panicum dichotomiflorum*). *Canadian Journal of Botany*, **73**, 1628–34.

Grand-Ravel, C., Astier, C., Nafaa, W. & Guillaumin, J. (1995). Relationships between climatic data and endophyte infection in France. *The 2nd Conference on Harmful and Beneficial Microorganisms in Grassland, Pastures and Turf*, **2**, 122. (abstract)

Gwinn, K. D. & Gavin, A. M. (1992). Relationship between endophyte infection level of tall fescue seed lots and *Rhizoctonia zeae* seedling disease. *Plant Disease*, **76**, 911–4.

Harberd, D. J. (1961). Note on choke disease of *Festuca rubra*. *Scottish Plant Breeding Station Report*, 1961, 47–51.

Hartnett, D. C., Hetrick, A. D., Wilson, G. W. T. & Gibson, D. J. (1993). Mycorrhizal influence on intra-and interspecific neighbour interactions among co-occurring prairie grasses. *Journal of Ecology*, **81**, 787–96.

Hill, N. S. (1994). Ecological relationships of Balansiae-infected graminoids. In *Biotechnology of Endophytic Fungi of Grasses*, ed. C. W. Bacon & J. F. White, pp. 59–71. Boca Raton: CRC Press.

Hill, N. S., Belesky, D. P. & Stringer, W. C. (1991). Competitiveness of tall fescue and influenced by *Acremonium coenophialum*. *Crop Science*, **31**, 185–90.

Hill, N. S., Pachon, J. G. & Bacon, C. W. (1996). *Acremonium coenophialum*-mediated short- and long-term drought acclimation in tall fescue. *Crop Science*, **36**, 665–72.

Hill, N. S., Stringer, W. C., Rottinghaus, G. E., Belesky, D. P., Parrot, W. A. & Pope, D. D. (1990). Growth, morphological, and chemical component responses of tall fescue to *Acremonium coenophialum*. *Crop Science*, **30**, 156–61.

Jarosz, A. M. & Davelos, A. L. (1995). Effects of disease in wild plant populations and evolution of pathogen aggressiveness. *New Phytologist*, **129**, 371–87.

Jones, T. A., Buckner, R. C., Burrus II, P. B. & Bush, L. P. (1983). Accumulation of pyrrolizidine alkaloids in benomyl-treated tall fescue parents and their untreated progenies. *Crop Science*, **23**, 1135–40.

Kelley, S. E. & Clay, K. (1987). Interspecific competitive interactions and the maintenance of genotypic variation within the populations of two perennial grasses. *Evolution*, **41**, 92–103.

Keogh, R. G. & Lawrence, T. M. (1987). Influence of *Acremonium lolii* presence on emergence and growth on ryegrass seedlings. *New Zealand Journal of Agricultural Research*, **30**, 507–10.

Kimmons, C. A., Gwinn, K. D. & Bernard, E. C. (1990). Nematode reproduction on endophyte-infected and endophyte-free tall fescue. *Plant Disease*, **74**, 757–61.

Kirfman, G. W., Brandenburg, R. L. & Garner, G. B. (1986). Relationship between insect abundance and endophyte infestation level in tall fescue in Missouri. *Journal of the Kansas Entomological Society*, **59**, 552–4.

Knoch, T. R., Faeth, S. H. & Arnott, D. L. (1993). Endophytic fungi alter foraging and dispersal by desert seed-harvesting ants. *Oecologia*, **95**, 470–5.

Large, E. C. (1952). Surveys for choke (*Epichloë typhina*) in cocksfoot seed crops, 1951. *Plant Pathology*, **1**, 23–8.

Latch, G. C. M. (1993). Physiological interactions of endophytic fungi and their hosts. Biotic stress tolerance imparted to grasses by endophytes. *Agriculture, Ecosystems and Environment*, **44**, 143–56.

Latch, G. C. M., Christensen, M. J. & Gaynor, D. L. (1985). Aphid detection of endophytic infection in tall fescue. *New Zealand Journal of Agricultural Research*, **28**, 129–32.

Latch, G. C. M., Hunt, W. F. & Musgrave, D. R. (1985). Endophytic fungi affect growth of perennial ryegrass. *New Zealand Journal of Agricultural Research*, **28**, 165–8.

Leuchtmann, A. (1992). Systematics, distribution and host specificity of grass endophytes. *Natural Toxins*, **1**, 150–62.

Leuchtmann, A. & Clay, K. (1988). Experimental infection of host grasses and sedges with *Atkinsonella hypoxylon* and *Balansia cyperi* (Balansiae, Clavicipitaceae). *Mycologia*, **81**, 692–701.

Leuchtmann, A. & Clay, K. (1989). Morphological, cultural and mating studies on *Atkinsonella*, including *A. texensis*. *Mycologia*, **81**, 692–701.

Leuchtmann, A. & Clay, K. (1993). Nonreciprocal compatibility interactions between *Epichloë typhina* and four host grasses. *Mycologia*, **85**, 157–63.

Leuchtmann, A., Schardl, C. L. & Siegel, M. R. (1994). Sexual compatibility and taxonomy of a new species of *Epichloë* symbiotic with fine fescue grasses. *Mycologia*, **86**, 802–12.

Lewis, G. C. (1992). Effect of ryegrass endophyte in mixed swards of perennial ryegrass and white clover under two levels of irrigation and pesticide treatment. *Grass and Forage Science*, **47**, 302–5.

Lewis, G. C. & Clements, R. O. (1986). A survey of ryegrass endophyte (*Acremonium loliae*) in the U.K. and its apparent ineffectuality on a seedling pest. *Journal of Agricultural Science*, **107**, 633–8.

Lewis, G. C. & Clements, R. O. (1990). Effect of *Acremonium lolii* on herbage yield of *Lolium perenne* at three sites in the United Kingdom. In *Proceedings of the International Symposium on* Acremonium/*Grass Interactions*, ed. S. A. Quisenberry & R. E. Joost, pp. 160–2. Baton Rouge: Louisiana Agricultural Experiment Station.

Mack, R. N. & Pyke, D. A. (1984). The demography of *Bromus tectorum*: the role of microclimate, grazing and disease. *Journal of Ecology*, **72**, 731–48.

Madej, C. W. & Clay, K. (1991). Avian seed preference and weight loss experiments: the effect of fungal endophyte-infected tall fescue seeds. *Oecologia*, **88**, 296–302.

Mahmood, T., Gergerich, R. C., Milus, E. A., West, C. P. & D'Arcy, C. J. (1993). Barley yellow dwarf viruses in wheat, endophyte-infected and endophyte-free tall fescue, and other hosts in Arkansas. *Plant Disease*, **77**, 225–8.

Marks, S. & Clay, K. (1996). Physiological responses of *Festuca arundinacea* to fungal endophyte infection. *New Phytologist*, **133**, 727–33.

Marks, S., Clay, K. & Cheplick, G. P. (1991). Effects of fungal endophytes on interspecific and intraspecific competition in the grasses *Festuca arundinacea* and *Lolium perenne*. *Journal of Applied Ecology*, **28**, 194–204.

Miles, C. O., Lane, G. A., di Menna, M. E., Garthwaite, I., Piper, E. L., Ball, O. J. P., Latch, G. C. M., Allen, J. M., Hunt, M. B., Bush, L. P., Min, F. K., Fletcher, I. & Harris, P. S. (1996). High levels of ergonovine and lysergic acid amide in toxic *Achnatherum inebrians* accompany infection by an *Acremonium*-like endophytic fungus. *Journal of Agricultural and Food Chemistry*, **44**, 1285–90.

Neal, W. D. & Schmidt, S. P. (1985). Effects of feeding Kentucky 31 tall fescue seed infected with *Acremonium coenophialum* to laboratory rats. *Journal of Animal Science*, **61**, 603–11.

Neergaard, P. (1977). *Seed Pathology*. New York: John Wiley and Sons.

Newsham, K. K., Fitter, A. H. & Watkinson, A. R. (1995). Arbuscular mycorrhiza protect an annual grass from root pathogenic fungi in the field. *Journal of Ecology*, **83**, 991–1000.

Pedersen, J. F., Rodriquez-Kabana, R. & Shelby, R. A. (1988). Ryegrass cultivars and endophyte in tall fescue affect nematodes in grass and succeeding soybean. *Agronomy Journal*, **80**, 811–4.

Petroski, R. J., Powell, R. G. & Clay, K. (1992). Alkaloids of *Stipa robusta* (Sleepygrass) infected with an *Acremonium* endophyte. *Natural Toxins*, **1**, 84–8.

Pinkerton, B. W., Rice, J. S. & Undersander, D. J. (1990). Germination in *Festuca arundinacea* as affected by the fungal endophyte, *Acremonium coenophialum*. In *Proceedings of the International Symposium on* Acremonium/*Grass Interactions*, ed. S. S. Quisenberry & R. E. Joost, pp. 176–80. Baton Rouge: Louisiana Agricultural Experiment Station.

Porter, J. K. (1994). Chemical constituents of grass endophytes. In *Biotechnology of Endophytic Fungi of Grasses*, ed. C. W. Bacon & J. F. White, pp. 103–23. Boca Raton: CRC Press.

Prestidge, R. A., Lauren, D. R., van der Zijpp, S. G. & di Menna, M. E. (1982). An association of *Lolium* endophyte with ryegrass resistance to Argentine stem weevil. *Proceedings of the New Zealand Weed Pest Control Conference,* **35,** 199–202.

Prestidge, R. A., Thom, E. R., Marshall, S. L., Taylor, M. J., Willoughby, B. & Wildermoth, D. D. (1992). Influence of *Acremonium lolii* infection in perennial ryegrass on germination, emergence, survival and growth of white clover. *New Zealand Journal of Agricultural Research,* **35,** 225–34.

Read, J. C. & Camp, B. J. (1986). The effect of fungal endophyte *Acremonium coenophialum* in tall fescue on animal performance, toxicity, and stand maintenance. *Agronomy Journal,* **78,** 848–50.

Read, J. C. & Walker, D. W. (1990). The effect of the fungal endophyte *Acremonium coenophialum* on dry matter production and summer survival of tall fescue. In *Proceedings of the International Symposium on* Acremonium/*Grass Interactions,* ed. S. S. Quisenberry & R. E. Joost, pp. 181–4. Baton Rouge: Louisiana Agricultural Experiment Station.

Rice, J. S., Pinkerton, B. W., Stringer, W. C. & Undersander, D. J. (1990). Seed production in tall fescue as affected by fungal endophyte. *Crop Science,* **30,** 1303–5.

Roach, D. A. & Wulff, R. D. (1987). Maternal effects in plants. *Annual Review of Ecology and Systematics,* **18,** 209–35.

Rolston, M. P., Hare, M. D., Moore, K. K. & Christensen, M. J. (1986). Viability of *Lolium* endophyte fungus in seed stored at different moisture contents and temperatures. *Journal of Experimental Agriculture,* **14,** 297–300.

Rowan, D. D. & Latch, G. C. M. (1994). Utilization of endophyte-infected perennial ryegrasses for increased insect resistance. In *Biotechnology of Endophytic Fungi of Grasses,* ed. C. W. Bacon & J. F. White, pp. 169–83. Boca Raton: CRC Press.

Roy, B. A. (1993). Floral mimicry of a plant pathogen. *Nature,* **362,** 56–8.

Saha, D. C., Johnson-Cicalese, J. M., Halisky, P. M., Van Heemstra, M. I. & Funk, C. R. (1987). Occurrence and significance of endophytic fungi in the fine fescues. *Plant Disease,* **71,** 1021–4.

Sampson, K. (1933). The systematic infection of grasses by *Epichloë typhina* (Pers.) Tul. *Transactions of the British Mycological Society,* **18,** 30–47.

Sampson, K. & Western, J. H. (1954). *Diseases of British Grasses and Herbage Legumes.* Cambridge: Cambridge University Press.

Schardl, C. L. & Clay, K. (1997). Evolution of mutualistic endophytes from plant pathogens. In *The Mycota V, Part B, Plant Relationships,* ed. G. Carroll & P. Tudzynski, pp. 221–38. Berlin: Springer-Verlag.

Schardl, C. L., Leuchtmann, A., Tsai, H.-F., Collet, M. A., Watt, D. M. & Scott, D. B. (1994). Origin of a fungal symbiont of perennial ryegrass by interspecific hybridization of a mutualist with the ryegrass choke pathogen *Epichloë typhina. Genetics,* **136,** 1307–17.

Schardl, C. L., Liu, J., White, J. F., Finkel, R. A., An, Z. & Siegel, M. R. (1991). Molecular phylogenetic relationships of nonpathogenic grass mycosymbionts and clavicipitaceous plant pathogens. *Plant Systematics and Evolution,* **178,** 27–41.

Schmidt, D. (1993). Effects of *Acremonium uncinatum* and a *Phialophora*-like endophyte on vigour, insect and disease resistance of meadow fescue. In *Proceedings of the Second International Symposium on* Acremonium/*Grass Interactions,* ed. D. E. Hume, G. C. M. Latch & H. S. Easton, pp. 185–8. Palmerston North: AgResearch.

Shelby, R. A. & Dalrymple, L. W. (1993). Long-term changes of endophyte infection in tall fescue stands. *Grass and Forage Science,* **48,** 356–61.

Siegel, M. R., Johnson, M., Varney, D. R., Nesmith, W. C., Buckner, R. C., Bush, L. P. & Burrus, P. B. (1984). A fungal endophyte in tall fescue: incidence and dissemination. *Phytopathology*, **74**, 932–7.

Siegel, M. R., Latch, G. C. M., Bush, L. P., Fannin, N. F., Rowan, D. D., Tapper, B. A., Bacon, C. W. & Johnson, M. C. (1990). Fungal endophyte-infected grasses: alkaloid accumulation and aphid response. *Journal of Chemical Ecology*, **16**, 3301–15.

Stanley, M. R., Koide, R. T. & Shumway, D. L. (1993). Mycorrhizal symbiosis increases growth, reproduction and recruitment of *Abutilon theophrasti* Medic. in the field. *Oecologia*, **94**, 30–5.

Stovall, M. E. & Clay, K. (1988). The effect of the fungus *Balansia cyperi* on the growth and reproduction of purple nutsedge, *Cyperus rotundus*. *New Phytologist*, **109**, 351–9.

Sun, S., Clarke, B. & Funk, C. R. (1990). Effect of fertilizer and fungicide applications on choke expression and endophyte transmission in chewings fescue. *Proceedings of the International Symposium on* Acremonium/*Grass Interactions*, ed. S. S. Quisenberry & R. E. Joost, pp. 62–3. Baton Rouge: Louisiana Agricultural Experiment Station.

Sutherland, B. L. & Hoglund, J. H. (1990). Effects of ryegrass containing the endophyte *Acremonium lolii* on associated white clover. In *Proceedings of the International Symposium on* Acremonium/*Grass Interactions*, ed. S. S. Quisenberry & R. E. Joost, pp. 67–71. Baton Rouge: Louisiana Agricultural Experiment Station.

Thompson, R. W., Fribourg, H. A. & Reddick, B. B. (1989). Sample intensity and timing for detecting *Acremonium coenophialum* incidence in tall fescue pastures. *Agronomy Journal*, **81**, 966–71.

Turkington, R., Holl, F. B., Chanway, C. P. & Thompson, J. D. (1988). The influence of microorganisms, particularly *Rhizobium*, on plant competition in grass–legume communities. In *Plant Population Ecology*, ed. A. J. Davy, M. J. Hutching & A. R. Watkinson, pp. 343–66. Oxford: Blackwell Scientific Publications.

Van der Putten, W. H., Van Dijk, C. & Peters, B. A. M. (1993). Plant-specific soil-borne diseases contribute to succession in foredune vegetation. *Nature*, **362**, 53–6.

Van Santen, E. (1992). Animal preference of tall fescue during reproductive growth in the spring. *Agronomy Journal*, **84**, 979–82.

Welty, R. E., Azevedo, M. D. & Cooper, T. M. (1987). Influence of moisture content, temperature, and length of storage on seed germination and survival of endophytic fungi in seeds of tall fescue and perennial ryegrass. *Phytopathology*, **77**, 893–900.

Welty, R. E., Barker, R. E. & Azevedo, M. D. (1993). Response of field-grown tall fescue infected by *Acremonium coenophialum* to *Puccinia graminis* subsp. *graminicola*. *Plant Disease*, **77**, 574–5.

West, C. P. (1994). Physiology and drought tolerance of endophyte-infected grasses. In *Biotechnology of Endophytic Fungi of Grasses*, ed. C. W. Bacon & J. F. White, pp. 87–101. Boca Raton: CRC Press.

West, C. P., Izekor, E., Oosterhuis, D. M. & Robbins, R. T. (1988). The effect of *Acremonium coenophialum* on the growth and nematode infestation of tall fescue. *Plant and Soil*, **112**, 3–6.

West, C. P., Izekor, E., Turner, K. E. & Elmi, A. A. (1993). Endophyte effects on growth and persistence of tall fescue along a water-supply gradient. *Agronomy Journal*, **85**, 264–70.

White, J. F. (1988). Endophyte–host associations in forage grasses. XI. A proposal concerning origin and evolution. *Mycologia*, **80**, 442–6.

White, J. F. (1993*a*). Endophyte–host associations in grasses. XIX. A systematic study of some sympatric species of *Epichloë* in England. *Mycologia*, **85**, 444–55.

White, J. F. (1993*b*). Structure and mating system of the graminicolous fungal epibiont *Echinodothis tuberiformis* (Clavicipitales). *American Journal of Botany*, **80**, 1465–71.

White, J. F. (1994). Taxonomic relationships among the members of the Balansieae (Clavicipitales). In *Biotechnology of Endophytic Fungi of Grasses*, ed. C. W. Bacon & J. F. White, pp. 3–20. Boca Raton: CRC Press.

White, J. F. & Bultman, T. L. (1987). Endophyte–host associations in forage grasses. VIII. Heterothallism in *Epichloë typhina*. *American Journal of Botany*, **74**, 1716–21.

Wilson, A. D., Clement, S. L., Kaiser, W. J. & Lester, D. G. (1991). First report of clavicipitaceous anamorphic endophytes in *Hordeum* species. *Plant Disease*, **75**, 215.

11
Arbuscular mycorrhizas and the population biology of grasses

K. K. NEWSHAM AND A. R. WATKINSON

Introduction

The majority of herbaceous plant roots growing in natural ecosystems are colonized by the arbuscular mycorrhizal (AM) fungi, otherwise known as the vesicular-arbuscular mycorrhizal (VAM) fungi (Trappe, 1987). These are relatively primitive fungi belonging to six genera (*Acaulospora, Entrophospora, Gigaspora, Glomus, Sclerocystis* and *Scutellospora*) in the order Glomales and class Zygomycetes (Morton & Benny, 1990). Despite their simplicity, these fungi have become the main mutualistic organisms associated with grasses, with the possible exception of clavicipitaceous leaf endophytic fungi (see Clay, this volume). AM fungi typically enter plant roots via root hairs from either chlamydospores in the root zone or more commonly from other colonized roots (Read, Koucheki & Hodgson, 1976) and form comparatively broad and typically non-septate hyphae within roots. These give rise to finely branched haustoria, termed arbuscules, which have a high surface area in contact with the plant plasmalemma and which are usually considered to be the point of nutrient exchange between the fungus and its host (Smith & Gianinazzi-Pearson, 1988). Large terminal or intercalary structures termed vesicles develop at later stages in the colonization of roots (Vietti & van Staden, 1990). These contain fatty acids, lipids and sterols (Jabaji-Hare, Deschene & Kendrick, 1984) and are usually thought to be storage organs.

Since AM fungi are heterotrophic, they rely upon their plant host for up to 20% of carbon fixed in leaves (Wang *et al.*, 1989) in what Harley (1975) referred to as 'a direct short-circuiting of photosynthesis'. In return, the fungus generally confers benefits on its plant host, most notably in the acquisition of phosphorus (P), a highly immobile element in soil which is therefore frequently limiting to plant growth. This is the most widely accepted

286

function of AM fungi in a wide variety of plant species and biomes (Read, 1991). The hyphae of AM fungi grow into soil beyond the P depletion zone created around roots and generally transport a much greater amount of P to the host than could be gathered by roots alone (Sanders & Tinker, 1973). Many other functions of AM fungi to host plants are known (Newsham, Fitter & Watkinson, 1995a), but these are usually considered to be subsidiary functions to P acquisition. Most plant species can survive under controlled conditions without AM colonization, but it is generally recognized that mycorrhizas are beneficial to plant growth, particularly under the nutrient-stressed conditions that can be encountered in the natural environment.

Mutualistic interactions, such as those between AM fungi and their host plants, are potentially important factors controlling the dynamics of plant populations (Law, 1988). Yet by comparison with our understanding of the roles of competition (Grace & Tilman, 1990), herbivory (Crawley, 1983) or pathogens (Burdon, 1987) in controlling plant fecundity, mortality, immigration and emigration, the study of mutualistic effects on plant populations is still very much in its infancy (Law, 1988). There is not even a sound theoretical framework for the treatment of mutualistic interactions. One of the reasons for this is that the Lotka–Volterra models that have formed the basis for theoretical studies of competition and predator–prey interactions lead to both populations undergoing unbounded growth, in what May (1981) has referred to as 'an orgy of mutual benefaction'. To counter this, most models of mutualisms involve the density of the host plant species being limited by factors other than the mutualistic partner, where the equilibrium density of the latter is directly proportional to the density of the former (Law, 1988). In contrast, Crawley (1986) includes mutualism in his model as a negative term, arguing that the full capacity for population growth by the plant can only be exhibited when the individual has its full complement of obligate mutualists; when these mutualists are in limited supply the actual rate of increase is reduced.

In this paper we explore the impacts of mutualistic AM fungi on the population biology of grasses. Our aims are to (1) review the occurrence of mycorrhizas in the Poaceae, (2) assess the scope of mycorrhizal benefits, and (3) explore the population consequences of the mutualistic interaction between grasses and AM fungi.

The occurrence of mycorrhizas in the Poaceae

Table 11.1 shows the distribution of arbuscular mycorrhizas in UK grasses across different life forms, habitats and soil fertilities. The main feature to

Table 11.1. *Distribution of AM colonization in British grass species across different life forms, habitats and soil fertilities**

	Percentage of grass species	
	Normally/ occasionally colonized	Rarely/ never colonized
Life-form		
Annual	100	0
Perennial	90	10
Habitats		
Agricultural/managed land	100	0
Sand dunes/beaches	96	4
Inland cliffs/sands	96	4
Scrub/grassland	92	8
Forests (omitting wet forests)	92	8
Forests (with wet forests)	84	16
Bogs/marshes	83	17
Mud flats/salt marshes	80	20
Soil fertility		
Very infertile	93	7
Infertile	86	14
Fertile	78	22
Very fertile	80	20

Note:
*Data from The Ecological Flora Database (Fitter & Peat, 1994).

emerge from these data is that grasses, certainly in the UK flora, are usually colonized by AM fungi. Colonization commonly occurs in both annual and perennial species, and the only grasses in which AM colonization never or only rarely occurs are perennials. Those species that are rarely or never colonized tend to be perennials which occur in wet forests, mud flats, salt marshes or bogs where mycorrhizal growth is generally limited by anaerobic soil conditions (Table 11.1; Peat & Fitter, 1993). By contrast, the roots of grass species that normally occur in agricultural or managed land, dunes and beaches, inland cliffs and sand, scrub and grassland and forests (omitting wet forests) are consistently colonized (Table 11.1; Peat & Fitter, 1993). The only non-mycotrophic perennial grasses in the UK flora that do not commonly occur in wet habitats are *Festuca altissima* and *F. gigantea* (Peat & Fitter, 1993). Soil fertility also appears to have an effect on the colonization of roots by AM fungi: a greater proportion of grass species that are normally or occasionally colonized exist in very infertile habitats than in very fertile habitats (Table 11.1).

Surveys in Europe (Read & Haselwandter, 1981; Read *et al.*, 1976), the Americas (Pendleton & Smith, 1983; Molina, Trappe & Strickler, 1978) and Australasia (Armstrong, Helyar & Christie, 1992) also indicate that mycorrhizal colonization is widespread in the Poaceae. Given that AM colonization can be a considerable cost in terms of fixed carbon to the plant host, it follows from its ubiquity that it probably has beneficial effects in the grasses, since natural selection should have selected out an association that was non-beneficial, to the same extent to which pathogens are excluded from natural plant populations (Burdon, 1987).

Measuring the scope of mycorrhizal benefits in grasses

There are considerable difficulties in studying and demonstrating impacts of mycorrhizal fungi on individual plants and plant populations. Numerous experiments have been carried out in which plants have been grown with and without AM fungi under controlled conditions, and whilst this approach has proved useful for providing estimates of biomass changes as a result of colonization, it has limited relevance to plant population response in the natural environment. Plant response to mycorrhizas under controlled conditions is often contrived by the application of no P and elevated levels of nitrogen (N) and potassium (K) in nutrient solutions, which forces the host to respond to any extra P that is supplied by the mycorrhiza, however insignificant this might be. A further problem with this approach is that often the AM fungal species which naturally occurs with a plant population will not be that which is used in controlled environment experiments: this is usually caused by problems with obtaining sufficient inoculum of natural isolates. Several AM fungal species are known to be efficient in improving plant P nutrition and biomass production, notable examples being *Glomus etunicatum* and *G. intraradices* (Morton, Bentivenga & Wheeler, 1993), hence these are frequently used isolates in laboratory or glasshouse studies.

Under field conditions, three approaches have been used for studying mycorrhizal benefits derived by plants. The first involves the application of fungicides to natural populations of plants in order to eliminate or reduce colonization by AM fungi. This technique is currently the only available method for directly comparing the population responses of plants growing with and without mycorrhizas in the field. However, the fungicides that are used are typically broad ranging and therefore affect other fungi in soil and roots (West, Fitter & Watkinson, 1993). A less commonly used method is to grow plants on disturbed sites such as strip mines, which are typically low in mycorrhizal propagules, and to inoculate them with AM fungi (Allen & Allen, 1988). A more refined approach is to inoculate plants with AM fungi isolated from a natural plant population prior to transplantation back into that population, but uninoculated control plants typically become colonized by indigenous AM fungi, which can complicate data interpretation (e.g. Wallace, 1987).

Direct effects: growth, survival and fecundity

Table 11.2 shows the fecundity and seedling emergence responses of 14 grass species to mycorrhizal colonization. It is apparent that although AM fungi do indeed improve the performance of most of these species, null responses to colonization are as common as positive responses, and that there are also growth reductions associated with AM fungi. These mixed responses are also reflected in a survey of biomass changes in a wide range of grass species. Positive responses to colonization were recorded in 27 species, null in 33 and negative responses in 12 species (K. K. Newsham, unpublished data). A number of possible mechanisms have been put forward to explain these direct impacts of AM fungi on host plants. Improvements in host performance are usually ascribed to the ability of mycorrhizas to accumulate phosphorus, although other aspects of host physiology are known to be affected by colonization.

Phosphorus relations

Many studies have demonstrated that the P acquisition of grasses can be improved by AM colonization and that this can, in turn, lead to improved biomass and fecundity. The most convincing evidence for this phenomenon comes from work on C_4 warm season grasses which inhabit tallgrass prairie communities of North America. These studies have typically used *Glomus etunicatum* as mycorrhizal inoculum in prairie soils of low P availability (<10 mg kg^{-1}), with zero P and excess N ($c.$ 35 mg g^{-1}) and K ($c.$ 30 mg g^{-1}) added in nutrient solutions. Most notable of the grasses to benefit from AM colonization under these conditions is *Andropogon gerardii*, in which total biomass has been shown to be increased by as much as 170-fold in mycorrhizal vs non-mycorrhizal plants (Hetrick, Kitt & Wilson, 1988). The response of *A. gerardii* to AM colonization is consistent with that of other

Table 11.2. Grass species for which fecundity or seedling emergence responses to mycorrhizal colonization have been recorded

Host species	Response to colonization		Mechanism attributed to response	Experimental location	Reference
Andropogon gerardii	+/0	Fec	nd	F	Hartnett et al. (1994)
	0	SE	nd	F	Hartnett et al. (1994)
Avena barbata	+	Fec	nd	Gh	Nelson & Allen (1993)
Avena fatua	+	Fec	P acquisition (maternal)	Gc	Koide & Lu (1992)
	+	Fec	P acquisition (maternal)	Gc	Lu & Koide (1991)
	0/–	Fec	–	Gh	Koide et al. (1988)
Avena sativa	+	Fec	P acquisition	Gh	Koide et al. (1988)
Bouteloua gracilis	0/–	Fec	nd	Gc	Hays et al. (1982)
Dicanthelium oligosanthes	–	Fec	nd	F	Hartnett et al. (1994)
Elymus canadensis	+	SE	nd	F	Hartnett et al. (1994)
Hordeum distichon	0/–	Fec	nd	F	Jensen (1983)
	0/–	Fec	nd	F	Black & Tinker (1979)
Hordeum vulgare	+/0	Fec	P, Zn & Cu acquisition	Gh	Jensen (1982)
Panicum virgatum	0	SE	nd	F	Hartnett et al. (1994)
Sorghastrum nutans	+/0	Fec	nd	F	Hartnett et al. (1994)
Triticum aestivum	+	Fec	P acquisition	F	Khan (1975)
Vulpia ciliata	+/–	Fec	Pathogen protection	F	Newsham et al. (1994)
	+/–	Fec	nd	F	Carey et al. (1992)
Zea mays	+	Fec	P acquisition	F	Khan (1972)

Note:
Abbreviations: +, significant positive effect; 0, no effect; –, significant negative effect. Fec, fecundity; SE, seedling emergence. nd, not determined; Gc, growth cabinet; Gh, glasshouse; F, field.

tallgrass prairie warm season C_4 grasses: the total biomass of mycorrhizal *Bouteloua curtipendula*, *Panicum virgatum* and *Sorghastrum nutans* is also stimulated by up to 60-, 200- and 220-fold, respectively, when compared with non-mycorrhizal plants (Hetrick *et al.*, 1988). However, in natural prairie soils of high P availability (*c.* 40 mg kg⁻¹), C_4 grasses generally remain unresponsive to AM colonization (Anderson, Hetrick & Wilson, 1994).

The responses of warm season C_4 grasses to AM fungi in tallgrass prairie communities are in stark contrast to those recorded for cool season C_3 grasses from these habitats. For example, Hetrick *et al.* (1988) found that the biomass of the C_3 grasses *Bromus inermis*, *Elymus canadensis*, *Koeleria cristata* and *Lolium perenne* did not respond to AM colonization. The positive growth response of C_4 warm season grasses to AM colonization has been attributed to an ability to respond positively to increases in tissue P concentrations, whereas C_3 species generally remain unresponsive to any increases in tissue P (Hetrick, Wilson & Schwab, 1994). An alternative explanation relates to root system architecture (Newsham *et al.*, 1995a); C_4 grasses typically have coarser roots than C_3 species and are therefore more dependent on AM fungi for nutrient uptake (Hetrick *et al.*, 1988).

Some of the most consistent effects of AM colonization on the P relations and hence biomass and fecundity of grasses have come from studies on tropical species. This can generally be ascribed to the low P fertility of heavily leached soils which may be encountered in the tropics (Janos, 1983). For example, Cooperband, Boerner & Logan (1994) found that the total biomass and shoot biomass of the C_4 grass *Homolepsis aturensis* were increased by 8.5-fold and 10.5-fold, respectively, in mycorrhizal vs non-mycorrhizal plants grown in pots in a P-deficient soil (8 mg kg⁻¹). Similarly, the total biomass of *Paspalum notatum* was shown to be increased by 7.5 fold in heavily P-deficient Brazilian soils in plants inoculated with a mixture of AM fungi compared with non-mycorrhizal plants (Mosse, Hayman & Arnold, 1973). Dramatic effects of AM colonization on the biomass, grain yield and grain weight of *Zea mays* have also been reported (Khan, 1972).

It is clear from the above studies that AM fungi have considerable potential to influence the per capita rate of increase of grass populations. Further evidence for this phenomenon has been provided by Nelson & Allen (1993), who found improved seed production in mycorrhizal vs non-mycorrhizal *Avena barbata*. Clear effects of mycorrhizal colonization on the vigour of *Avena fatua* populations by impacts on offspring generations have also been demonstrated. Koide & Lu (1992) showed that the growth of wild oats with *Glomus intraradices* did not improve the growth of parent plants, but consistently improved the vigour of their offspring. Non-mycorrhizal offspring in general had greater leaf areas and elevated nutrient contents, but seeds produced in turn by these offspring plants were lighter with increased P concentrations (Koide & Lu, 1992). Greater endosperm P reserves in seeds from mycorrhizal parent plants were also generally associated with increased P accumulation in offspring seedlings (Lu & Koide, 1991). These studies were made with an AM fungus that is effective in transporting P to the host plant, and so parental mycorrhizal effects under natural conditions may not occur quite so consistently, but this is a clear case where AM fungi could have a potentially significant influence over grass population dynamics.

Other nutrients

In addition to influencing the P relations of plants, AM fungi have also been shown to improve the N nutrition of some plant species. Examples of this in the grasses are scarce, but Wallace (1981) showed increases in leaf N concentrations of *Panicum coloratum* colonized by AM fungi. Similarly, Trent, Svejcar & Bethlenfalvay (1993) demonstrated that shoot N concentrations of *Agropyron desertorum* and *Oryzopsis hymenoides* were improved by *Glomus pallidum* and *Glomus mosseae*. In neutral and alkaline soils it is perhaps unlikely that AM fungi will influence N nutrition, as nitrate ions which dominate in these soils are highly mobile. However, in more acid soils, the less mobile ammonium ion is the dominant form of N, so it is more plausible that the hyphal transport of this ion influences plant growth. The uptake of Zn and Cu (Jensen, 1982), K (Hamel, Furlan & Smith, 1991) or Mn (Trent *et al.*, 1993) may also be improved in grasses by AM colonization, but it is unclear whether mycorrhizal facilitation of the uptake of these nutrients can significantly improve plant growth under normal field conditions.

Water relations

Many grasslands occur in semi-arid or seasonally dry biomes in which water can be a limiting resource. Since the short-term productivity of a plant relies heavily upon maintaining photosynthetic tissue at a high water status, plants need to maintain a positive turgor pressure to survive under such conditions. It has been demonstrated on a number of occasions that AM colonization can improve the water relations of grasses and hence potentially lead to increased survival of individuals. For example, Allen *et* al. (1981) and Allen (1982) found that colonization by Glomus fasciculatum doubled the transpiration rate and more than halved the leaf water resistance of Bouteloua gracilis compared with non-mycorrhizal plants, but found no effects of colonization on biomass accumulation. Mycorrhizal plants did not have significantly different leaf areas or root lengths, and Allen (1982) suggested that the improved water relations of B. gracilis were due to transport through mycorrhizal hyphae. Similarly, decreased stomatal resistance and increased leaf water potentials were observed in mycorrhizal vs non-mycorrhizal Agropyron smithii but not in A. dasystachyum in a field study by Allen & Allen (1986). More recently, Zea mays has been shown to have improved drought tolerance when colonized by AM fungi (Augé et al., 1994; Subramanian et al., 1995). The precise mechanism by which this occurs remains unclear. Direct hyphal transport of water seems unlikely, as the cross-sectional area of AM fungal hyphae connected to a root is typically too small for enough water to stop wilting to pass through hyphae alone (Fitter, 1985). This suggests that amelioration of water relations by AM fungi may only be important under the most severely droughted conditions.

Growth reductions and null effects

The inoculation of plants with AM fungi can result in decreased biomass or fecundity of the host. Growth reductions associated with mycorrhizal colonization generally occur when mycorrhizal drain of photosynthates outweighs mycorrhizal benefit to the host (Buwalda & Goh, 1982) and are usually associated with growing plants of a low photosynthetic capacity with a physiologically very active AM fungus. Their occurrence is not uncommon in studies on grasses. For example, Hays et al. (1982) grew Bouteloua gracilis with and without Glomus fasciculatum in washed sand with a low cation exchange capacity in a growth chamber for six weeks and found that colonized plants were consistently smaller than those that were uncolonized, and that the only plants to flower were those that were not colonized by AM fungi. In a study of the relationship between the intensity of colonization by naturally occurring AM fungi and the productivity of Hordeum distichon crops, Black & Tinker (1979) found in one year that the intensity of AM colonization was negatively related to grain yield, but in the following year that no relationship between colonization and fecundity existed. Similarly, in a glasshouse study with natural inocula, Sparling & Tinker (1978) found that the growth of the upland grasses Anthoxanthum odoratum, Nardus stricta, Festuca rubra, Cynosurus *cristatus* and *Agrostis tenuis* was frequently depressed by AM colonization.

A small number of studies have attempted to assess the natural mycorrhizal dependence of grasses by growing plants with and without AM fungi prior to transplanting them into the field. These studies have typically not shown clear mycorrhizal growth responses, apart from those conducted in tropical ecosystems (e.g. Khan, 1972, 1975). For example, Wallace (1987) grew mycorrhizal and non-mycorrhizal *Panicum virgatum*, *Schizachyrium scoparium*, *Dicanthelium oligosanthes* var. *scribnerianum*, *Bouteloua gracilis* and *Paspalum setaceum* var. *stramineum* for six months in a mixed grass prairie and found no effect of AM colonization on plant growth rates or biomass. Similarly, Trent *et al.* (1993) grew mycorrhizal and non-mycorrhizal *Oryzopsis hymenoides* and *Agropyron desertorum* for three months in a semi-arid environment and recorded no effects of mycorrhizal colonization on plant biomass. Jensen (1983) found reduced fecundity but no effects of AM colonization on the biomass of transplanted *Hordeum distichon*. In each of these studies, indigenous AM fungi colonized the roots of uninoculated control plants, which may have resulted in less pronounced effects of colonization on plants.

Effects on species interactions

In addition to directly affecting host nutrient and water relations, AM fungi may also have impacts on competition, herbivory and pathogenic interactions in natural plant populations. In some cases these effects may be direct, as for example in the case of AM fungi providing protection against root pathogens, but in others the benefit may be indirect through improved or altered biomass accumulation.

Competition

Successional studies suggest that mycorrhizal fungi increase the competitive ability of host species to the detriment of non-mycorrhizal species (Allen, 1991). Experimental studies in the field and glasshouse also provide support for this hypothesis. For example, Allen & Allen (1984) demonstrated a decrease in the competitive ability of the non-mycotrophic chenopod *Salsola kali* when grown in interspecific competition with *Agropyron smithii* and *Bouteloua gracilis* in the presence of AM fungi. Similarly, in a study using a de Wit replacement design to investigate the effect of AM fungi on competitive interactions between the mycorrhizal grass *Agropyron*

dasystachyum and S. kali, Benjamin & Allen (1987) found that in the presence of mycorrhiza, the former species dominated in terms of biomass, but
that without mycorrhiza, the latter species became dominant (see also
Allen & Allen, 1992). Evidence from experimental studies indicates that
AM fungi influence the outcome of competition when neighbouring
species have differential responses to mycorrhizal fungi (Allen & Allen,
1990). For example, Crush (1974) and Hall (1978) found that AM fungi
improved the competitive ability of legumes when grown with grasses.
Furthermore, Fitter (1977) found that AM colonization reduced the
growth of Lolium perenne grown in interspecific competition with Holcus
lanatus. By contrast, West (1996) found that AM colonization could
increase the aggressiveness of H. lanatus growing in competition with
Dactylis glomerata by improving the ability of the former species to
compete for limiting resources.

One of the most detailed studies of the influence of AM fungi on competition in grasses has been carried out by Hartnett et al. (1993) on
Andropogon gerardii and Elymus canadensis. Both species were grown at a
range of plant densities in prairie soil of very low P (6 mg kg^{-1}) in pots in
a glasshouse to examine the effects of natural AM colonization on intra-
and interspecific competition. The biomass of mycorrhizal vs non-
mycorrhizal A. gerardii was increased by some 14-fold when grown without
competition from neighbouring plants, but this effect was absent at the
highest level of neighbour density (Fig. 11.1). In stark contrast, the
biomass of the less mycotrophic E. canadensis was unaffected by AM fungi
(Fig. 11.1). Moreover, in mixtures of A. gerardii and E. canadensis, the presence of mycorrhizal fungi increased the competitive success of A. gerardii
but had little effect on E. canadensis (Fig. 11.2).

These results show that the competitive dominance of A. gerardii in tallgrass prairie is highly dependent upon mycorrhizal associations. In cases
where AM fungi influence the competitive abilities of plants, this is probably owing to impacts on the growth, nutrient or water status of one plant
relative to another (Allen, 1991). Indeed, altering the resource levels under
which competing plants are grown may have dramatic effects on the impact
of mycorrhizal fungi on plants (see Hall, 1978; Hartnett et al., 1993).
Complicating the interpretation of AM fungal effects on competition,
however, are the findings that AM fungi may in some cases inhibit the
growth and survival of non-mycotrophic species, while in other cases non-
mycotrophic species may inhibit mycorrhizal formation of host plants
(Allen, 1991). For example, Iqbal & Qureshi (1976) found that Brassica
campestris inhibited mycorrhizal formation in Triticum aestivum.

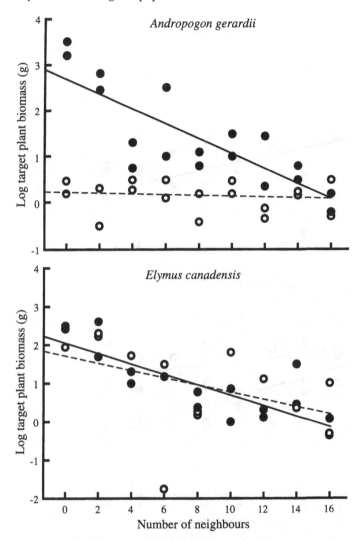

Fig. 11.1. Log target biomass of *Andropogon gerardii* and *Elymus canadensis* grown in intraspecific competition as a function of number of neighbours. Plants were grown without phosphorus fetilization and with (●) and without (○) mycorrhizal inocula. Solid and dashed lines represent best linear fits for responses of plants grown with and without inocula, respectively. Redrawn from Hartnett *et al.* (1993).

The limited amount of data that exists from multispecies competition studies indicates that AM fungi may have a considerable influence on the composition of plant communities (Allen & Allen, 1990). In an artificial mixture of forbs and grasses, Grime *et al.* (1987) found that the presence of

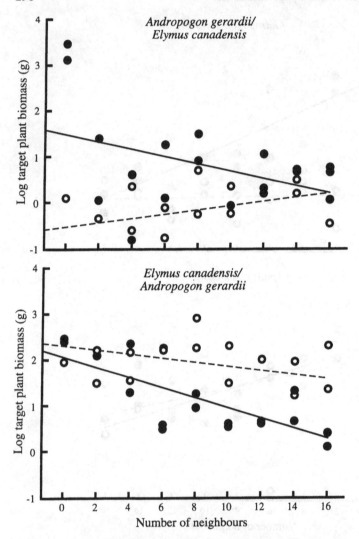

Fig. 11.2. Log target biomass of *Andropogon gerardii* grown in interspecific competition with *Elymus canadensis* and of *E. canadensis* grown with *A. gerardii*, as a function of number of neighbours. Plants were grown without phosphorus fertilization and with (●) and without (○) mycorrhizal inocula. Solid and dashed lines represent best linear fits for responses of plants grown with and without inocula, respectively. Redrawn from Hartnett *et al.* (1993).

mycorrhizal fungi reduced the dominance of grasses. This result is consistent with the finding that AM fungi improve the competitive ability of legumes when grown with grasses (Crush, 1974; Hall, 1978). Another potentially important impact of AM colonization on plants, however, is the export of assimilate from 'source' (canopy dominants) to 'sink' (understorey) components of the community through a common mycelial network. This has the potential to raise the biomass of subordinate species relative to the dominants in grassland communities and to thereby increase species diversity. These so-called 'nurse' effects have been shown to exist in mixed grassland communities by Grime *et al.* (1987) and within intraspecific (Whittingham & Read, 1982) and interspecific (Francis, Finlay & Read, 1986) mixtures of *Festuca ovina* and *Plantago lanceolata*. The occurrence of hyphal interconnections enables the transfer of nutrients, notably P, between plants, and may result in improved growth of receiver plants. Significant transfer of ^{32}P from donor to receiver plants has been demonstrated within 24 hours in *Festuca ovina*, and it has also been shown that the application of nutrients to donor plants grown in split pots can result in fourfold increases in the biomass of receivers after 30 weeks (Whittingham & Read, 1982).

Herbivory

Several studies have demonstrated that AM fungi may influence herbivory in grass populations through effects on biomass allocation. For example, mycorrhizal colonization in *Panicum coloratum* has been shown to promote prostrate shoot morphology and increased root growth, and as such, to confer grazing tolerance by maximizing plant biomass below the grazing zone (Wallace, 1981). Similarly, Miller, Jarstfer & Pillai (1987) grew *Agropyron smithii* with *Glomus mosseae* at a range of P levels (2–20 mg kg^{-1}) and found that mycorrhizal plants were shorter with more tillers at the highest P level. Miller *et al.* (1987) also concluded that mycorrhiza may be important in determining patterns of grazing in natural populations of grasses. Increases in tiller production resulting from mycorrhizal colonization have also been recorded in *Dactylis glomerata* and *Lolium perenne* by Powell (1975) and Hall, Johnstone & Dolby (1984). However, mycorrhizal benefits on grasses through effects on biomass allocation and hence effects on herbivory may be offset by increases in leaf N (see p. 293), which probably in turn lead to greater grazing pressure, as N-rich tissues are generally more palatable to herbivores. This would help to explain the positive relationship between AM colonization and grazing intensity recorded in the field by Wallace (1981) and Wallace, McNaughton & Coughenour (1982).

Whilst mycorrhizal fungi may have a direct impact on the structure and biomass allocation of grasses and may subsequently influence grazing, grazing may in turn influence mycorrhizal activity. Soil invertebrates graze mycorrhizal hyphae and can consequently inhibit mycorrhizal activity. For example, Boerner & Harris (1991) have demonstrated that grazing by collembola can reduce the ability of mycorrhizal *Panicum virgatum* to compete with the non-mycotrophic *Brassica nigra*. Grazing of mycorrhizal hyphae by animals may in part explain why some plants show a lack of response to mycorrhizal colonization in the field (McGonigle & Fitter, 1988). Grazing of aboveground plant parts by ungulates has been shown to increase (Reece & Bonham, 1978), decrease (Bethlenfalvay & Dakessian, 1984) or to not affect (Wallace, 1987) mycorrhizal colonization.

Pathogens

The roots of grasses are often densely infected by the hyphae of fungi. Some of these may be mycorrhizal fungi, but they are frequently formed by either opportunistic organisms having no or poorly defined impacts, such as the 'dark septate' fungi (Read & Haselwandter, 1981) or by pathogenic fungi (Skipp, Christensen & Caradus, 1982; Newsham, Watkinson & Fitter, 1995c). The most infamous pathogens are those causing the large-scale destruction of crop grasses, such as *Gaeumannomyces graminis*, the take-all fungus, but wilt diseases caused by *Fusarium* spp. (typically *F. oxysporum*) are also frequent in grasses in a wide variety of habitats (Garrett, 1970). In addition to pathogenic fungi in roots, pathogenic nematodes such as species of *Heterodera*, *Hoplolaimus* and *Pratylenchus* also occur, forming what are known as 'pathocomplexes', associations of pathogens that can lead to the decline of plant populations. Two recent examples from the literature illustrate the potentially vital role that AM fungi may play in reducing the impacts of pathogens in natural populations of grasses. In a recent study, Newsham, Fitter & Watkinson (1995b) grew seedlings of *Vulpia ciliata* in a factorial combination with a mycorrhizal fungus (a *Glomus* sp.) and *F. oxysporum*, both of which had been isolated from a natural population of *V. ciliata*. The seedlings were then transplanted back into the same population and were sampled after 62 and 90 days of growth in the field. The data indicated that AM fungi significantly improved the shoot biomass of *V. ciliata*, but only in the presence of the pathogen: those plants grown with *Glomus* sp. and *F. oxysporum* prior to transplantation were larger than those grown with *F. oxysporum* alone, but plants grown with *Glomus* sp. alone were no larger than those grown with neither fungus. The plants in this study did not set seed, but the number of seeds per plant has been estimated from shoot biomass values. These data show the clear benefit derived by *V. ciliata* from AM colonization in protection from *F. oxysporum*

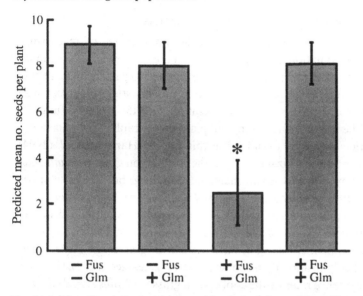

Fig. 11.3. The effects of a factorial combination of the pathogen *Fusarium oxysporum* (Fus) and an AM fungus, *Glomus* sp. (Glm) on predicted seed number of *Vulpia ciliata*. Plants were inoculated with fungi in the laboratory, transplanted into the field and sampled after 62 days. Values are means of 16 replicates ± SEM. Asterisk denotes a mean that differed at $P < 0.05$ after a Fisher's Pairwise comparisons test. Data from Newsham *et al.* (1995*b*).

(Fig. 11.3). Mycorrhizal colonization also did not affect plant P concentrations, confirming the view that natural populations of *V. ciliata* do not rely upon AM fungi for P acquisition (see p. 304).

In a more recent study, Little & Maun (1996) have attempted to define the role played by AM fungi in the amelioration of the large-scale decline of marram grass populations, in what has become known as 'the *Ammophila* problem'. Little & Maun (1996) grew *Ammophila breviligulata* in sterile sand with a factorial combination of AM fungi and root endoparasitic nematodes in the glasshouse for 20 weeks. Mycorrhizal colonization was shown to be an important factor in reducing the deleterious effects of the nematodes on plant biomass, but only when plants were buried with sand after inoculation. Little & Maun (1996) concluded that AM fungi were potentially significant in ameliorating the large-scale decline of *Ammophila* populations. In a response similar to that recorded by Newsham *et al.* (1995*b*), AM colonization was only beneficial to *A. breviligulata* when pathogens were present. However, P analyses were not carried out on plant tissues, so it cannot be ruled out that mycorrhizal benefits were not related to P acquisition in this case.

Impacts of mycorrhizas in natural grass populations

Few studies have specifically examined the direct effects of mycorrhizal fungi on the population biology of grasses in the field, reflecting the inherent difficulties of studying microorganisms in heterogeneous natural environments. The most commonly used approach for studying mycorrhizal impacts in the field is to reduce colonization with the systemic fungicide benomyl. The impact of AM fungi on plant demographic relations in tallgrass prairies have been examined in this way by Hartnett et al. (1994). Suppression of mycorrhizal activity in the cool season grass *Dichanthelium oligosanthes* by benomyl led to increased flowering, but by contrast, the flowering of the warm season grasses *Andropogon gerardii* and *Sorghastrum nutans* was significantly improved by the presence of AM fungi in burned prairie, but not in unburned sites. This corroborates conclusions from controlled environment studies that warm season C_4 grasses are more dependent on AM colonization than cool season C_3 grasses (see p. 292). However, seedling emergence rates of the cool season grasses *Elymus canadensis* and *Koeleria cristata* were reduced in benomyl-treated areas, implying that mycorrhizal fungi may be important for stimulating seedling germination and survival in these species (Fig. 11.4). There was, however, no effect of benomyl treatment on seedling emergence in the warm season grasses *A. gerardii* and *Panicum virgatum*. The results obtained by Hartnett et al. (1994) indicated that mycorrhizal fungi may possibly have different effects on the demographic responses of grasses at different life history stages and that mycorrhizas may influence demographic relations of tallgrass prairie plant populations.

Whilst the data obtained by Hartnett et al. (1994) can be interpreted with reference to mycorrhizal effects, the application of benomyl to natural populations of grasses is known to influence other fungi associated with roots. For example, in a series of experiments, benomyl has been applied to natural populations of *V. ciliata* growing in the Breckland, an area of eastern England characterized by low rainfall and sandy soils of low P fertility (<11 mg kg^{-1}; Newsham, Fitter & Watkinson, 1994). The fungicide has been shown to reduce mycorrhizal colonization of *V. ciliata* roots, but generally to have few effects on biomass and fecundity (Carey, Fitter & Watkinson, 1992; West et al., 1993). This general lack of response can be attributed to the simultaneous reduction of both beneficial mycorrhizal and deleterious pathogenic fungi in roots. In subsequent experiments, the fecundity (seed number) of *V. ciliata* was shown to be improved by benomyl application at one site where mycorrhizal colonization was very low (Newsham et al., 1994). When fecundity was compared with the abundance

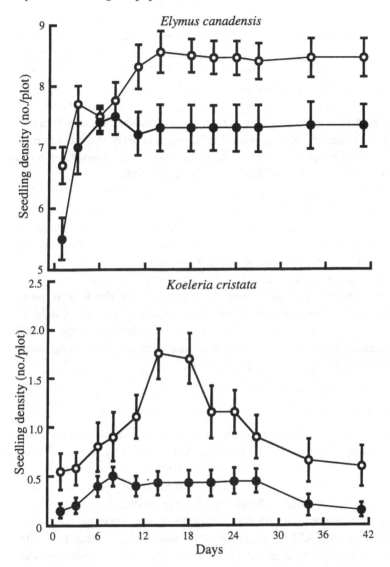

Fig. 11.4. Seedling density (mean no. per plot ± SEM) of *Elymus canadensis* and *Koeleria cristata* as a function of time. Plots were treated without (○) and with (●) the fungicide benomyl to inhibit mycorrhizal colonization of roots. Redrawn from Hartnett *et al.* (1994).

of the cosmopolitan pathogen *Fusarium oxysporum* isolated from roots using an agar plating technique, plants that were densely infected by the pathogen (but which were asymptomatic) were found to have lower fecundity than those that were less densely infected. Analysis of covariance applied to the fecundity data demonstrated that AM colonization improved the fecundity of the grass by reducing the negative impact of *F. oxysporum*. A subsequent inoculation and transplantation experiment confirmed this interpretation (Newsham *et al.*, 1995*b*; see p. 300) and it was concluded that AM colonization was vital to natural populations of *V. ciliata* by excluding the pathogen and not by improving P acquisition. The benefit of AM fungi to the plant, in terms of seed production, has been estimated to exceed 30% (Carey *et al.*, 1992), which is likely to affect the population dynamics of the grass, as the abundance of the species is directly related to fecundity (Carey, Watkinson & Gerard, 1995).

Other field-based fungicide experiments have similarly shown that AM fungi may modify population abundance (Gange, Brown & Farmer, 1990; Gange, Brown & Sinclair, 1993). For example, Gange *et al.* (1993) used the fungicide iprodione to control mycorrhizal colonization in an early successional plant community and found that the normally non-mycotrophic *Bromus sterilis* was found only in fungicide-treated areas, whilst the typically mycotrophic *Lolium perenne* was absent from the treated plots. Moreover, Newsham *et al.* (1995*d*) showed that changes in the abundance of plant species were a function of the percentage reduction in mycorrhizal colonization brought about by the application of benomyl to a community containing *Vulpia ciliata* and *Festuca ovina*.

Conclusions

It is apparent that mycorrhizal fungi have a great potential to influence grass population dynamics. The ability of AM fungi to improve host P relations, particularly under nutrient stressed conditions, is often considered to be the most important of their effects on host plants. However, under non-nutrient-limiting conditions, functions of the mutualism that are normally considered to be subsidiary to P acquisition may become important to the host. For example, AM fungi appear to be able to alter interactions with herbivores through their effects on biomass allocation and also to alter intra- and interspecific competition. In semi-arid or seasonally dry conditions, such as those encountered in many grasslands, they may also improve the conductivity of water through host tissues and thereby reduce wilting. Nitrogen relations, meanwhile, are probably most affected in highly acidic soils where N is less mobile, and the threat of invasion of roots by soil fungi and nematodes also appears to be reduced by AM colonization. However, what is most under question is whether any of these effects do actually exist in natural grass populations, since the methods used for studying the impact of AM fungi on plants are usually inadequate for predicting actual plant responses in the natural environment. For example, if plants are grown in controlled environments, unrealistic nutrient or water regimes can result in spurious responses to colonization. Yet if plants are grown in the field to achieve more natural experimental conditions, complications arise from manipulating a complex natural system. For instance, applying bio-cides to natural ecosystems is a most unsatisfactory method by which to study plant–AM fungal interactions, since other organisms which may influence plant response – particularly pathogens – may be reduced in abundance, complicating data interpretation.

Law (1988) indicated that there was little evidence to conclude that mutu-alistic fungi could have direct impacts on host population dynamics. It is now possible to conclude that grass populations can be influenced by myc-orrhizal fungi through a wide range of impacts on host physiology and interactions with other organisms. The challenge for the future will be to demonstrate clearly that these impacts do benefit natural plant populations and to determine the further benefits derived by plants from mycorrhizal fungi which almost undoubtedly await discovery.

Acknowledgements

We are grateful to Helen Peat and David Roy for their help with the extrac-tion of data from the Ecological Flora Database. Financial support pro-vided by the Institute of Terrestrial Ecology is gratefully acknowledged by KKN.

References

Allen, E. B. & Allen, M. F. (1984). Competition between plants of different successional stages: mycorrhizae as regulators. *Canadian Journal of Botany*, **62**, 2625–9.

Allen, E. B. & Allen, M. F. (1986). Water relations of xeric grasses in the field: interactions of mycorrhizas and competition. *New Phytologist*, **104**, 559–71.

Allen, E. B. & Allen, M. F. (1988). Facilitation of succession by the nonmycotrophic colonizer *Salsola kali* (Chenopodiaceae) on a harsh site: effects of mycorrhizal fungi. *American Journal of Botany*, **75**, 257–66.

Allen, E. B. & Allen, M. F. (1990). The mediation of competition by mycorrhizae in successional and patchy environments. In *Perspectives on Plant Competition*, ed. J. B. Grace and G. D. Tilman, pp. 367–89. New York: Academic Press.

Allen, M. F. (1982). Influence of vesicular-arbuscular mycorrhizae on water movement through *Bouteloua gracilis* H. B. K. Lag ex Steud. *New Phytologist*, **91**, 191–6.

Allen, M. F. (1991). *The Ecology of Mycorrhizae.* Cambridge: Cambridge University Press.

Allen, M. F. & Allen, E. B. (1992). Mycorrhizae and plant community development: mechanisms and patterns. In *The Fungal Community, its Organisation and Role in the Ecosystem,* ed. G. C. Carroll & D. T. Wicklow, pp. 455–79. New York: Marcel Dekker.

Allen, M. F., Smith, W. K., Moore, T. S., Christensen, M. (1981). Comparative water relations and photosynthesis of mycorrhizal and non-mycorrhizal *Bouteloua gracilis* H. B. K. Lag ex Steud. *New Phytologist,* **88,** 683–93.

Anderson, R. C., Hetrick, B. A. D. & Wilson, G. W. T. (1994). Mycorrhizal dependence of *Andropogon gerardii* and *Schizachyrium scoparium* in two prairie soils. *American Midland Naturalist,* **132,** 366–76.

Armstrong, R. D., Helyar, K. R. & Christie, E. K. (1992). Vesicular-arbuscular mycorrhiza in semi-arid pastures of south-west Queensland and their effect on growth responses to phosphorus fertilizers by grasses. *Australian Journal of Agricultural Research,* **43,** 1143–55.

Augé, R. M., Duan, X., Ebel, R. C. & Stodola, A. J. W. (1994). Nonhydraulic signalling of soil drying in mycorrhizal maize. *Planta,* **193,** 74–82.

Benjamin, P. K. & Allen, E. B. (1987). The influence of VA mycorrhizal fungi on competition between plants of different successional stages in sagebrush-grassland. In *Mycorrhizae in the Next Decade, Practical Applications and Research Priorities,* ed. D. M. Sylvia, L. L. Hung & J. H. Graham, p. 144. Gainsville: IFAS.

Bethlenfalvay, G. J. & Dakessian, S. (1984). Grazing effects on mycorrhizal colonization and floristic composition of the vegetation on a semiarid range in Northern Nevada. *Journal of Range Management,* **37,** 312–16.

Black, R. & Tinker, P. B. (1979). The development of endomycorrhizal root systems. II. Effect of agronomic factors and soil conditions on the development of vesicular-arbuscular mycorrhizal infection in barley and on the endophyte spore density. *New Phytologist,* **83,** 401–13.

Boerner, R. E. J. & Harris, K. K. (1991). Effects of collembola (Arthropoda) and relative germination date on competition between mycorrhizal *Panicum virgatum* (Poaceae) and non-mycorrhizal *Brassica nigra* (Brassicaceae). *Plant and Soil,* **136,** 121–9.

Burdon, J. J. (1987). *Diseases and Plant Population Biology.* Cambridge: Cambridge University Press.

Buwalda, J. G. & Goh, K. M. (1982). Host-fungus competition for carbon as a cause of growth depressions in vesicular-arbuscular mycorrhizal ryegrass. *Soil Biology and Biochemistry,* **14,** 103–6.

Carey, P. D., Fitter, A. H. & Watkinson, A. R. (1992). A field study using the fungicide benomyl to investigate the effect of mycorrhizal fungi on plant fitness. *Oecologia,* **90,** 550–5.

Carey, P. D., Watkinson, A. R. & Gerard, F. F. O. (1995). The determinants of the distribution and abundance of the winter annual grass *Vulpia ciliata* ssp. *ambigua. Journal of Ecology,* **83,** 177–87.

Cooperband, L. R., Boerner, R. E. J. & Logan, T. J. (1994). Humid tropical leguminous tree and pasture grass responsiveness to vesicular-arbuscular mycorrhizal infection. *Mycorrhiza,* **4,** 233–9.

Crawley, M. J. (1983). *Herbivory: the Dynamics of Animal–Plant Interactions.* Oxford: Blackwell Scientific Publications.

Crawley, M. J. (1986). The population biology of invaders. *Philosophical Transactions of the Royal Society of London B.,* **314,** 711–31.

Crush, J. R. (1974). Growth responses to vesicular-arbuscular mycorrhiza in herbage legumes. *New Phytologist*, **73**, 743–9.

Fitter, A. H. (1977). Influence of mycorrhizal infection on competition for phosphorus and potassium by two grasses. *New Phytologist*, **79**, 119–25.

Fitter, A. H. (1985). Functioning of vesicular-arbuscular mycorrhizas under field conditions. *New Phytologist*, **99**, 257–65.

Fitter, A. H. & Peat, H. J. (1994). The Ecological Flora Database. *Journal of Ecology*, **82**, 415–25.

Francis, R., Finlay, R. D. & Read, D. J. (1986). Vesicular-arbuscular mycorrhiza in natural vegetation systems. IV. Transfer of nutrients in inter- and intra-specific combinations of host plants. *New Phytologist*, **102**, 103–11.

Gange, A. C., Brown, V. K. & Farmer, L. M. (1990). A test of mycorrhizal benefit in an early successional plant community. *New Phytologist*, **115**, 85–91.

Gange, A. C., Brown, V. K. & Sinclair, G. S. (1993). Vesicular-arbuscular mycorrhizal fungi: a determinant of plant community structure in early succession. *Functional Ecology*, **7**, 616–22.

Garrett, S. D. (1970). *Pathogenic Root-Infecting Fungi*. Cambridge: Cambridge University Press.

Grace, J. B. & Tilman, D. (eds.) (1990). *Perspectives on Plant Competition*. San Diego: Academic Press.

Grime, J. P., Mackey, J. M. L., Hillier, S. H. & Read, D. J. (1987). Floristic diversity in a model system using experimental microcosms. *Nature*, **328**, 420–2.

Hall, I. R. (1978). Effects of endomycorrhizas on the competitive ability of white clover. *New Zealand Journal of Agricultural Research*, **21**, 509–15.

Hall, I. R., Johnstone, P. D. & Dolby, R. (1984). Interactions between endomycorrhizas and soil nitrogen and phosphorus on the growth of ryegrass. *New Phytologist*, **97**, 447–53.

Hamel, C., Furlan, V. & Smith, D. L. (1991). N_2-fixation and transfer in a field grown mycorrhizal corn and soybean intercrop. *Plant and Soil*, **133**, 177–85.

Harley, J. L. (1975). Problems of mycotrophy. In *Endomycorrhizas*, ed. F. E. Sanders, B. Mosse & P. B. Tinker, pp. 1–24. London: Academic Press.

Hartnett, D. C., Hetrick, B. A. D., Wilson, G. W. T. & Gibson, D. J. (1993). Mycorrhizal influence on intra- and interspecific neighbourhood interactions among co-occurring prairie grasses. *Journal of Ecology*, **81**, 787–95.

Hartnett, D. C., Samenus, R. J., Fischer, L. E. & Hetrick, B. A. (1994). Plant demographic responses to mycorrhizal symbiosis in tallgrass prairie. *Oecologia*, **99**, 21–6.

Hays, R., Reid, C. P. P., St John, T. V. & Coleman, D.C. (1982). Effects of nitrogen and phosphorus on Blue Grama growth and mycorrhizal infection. *Oecologia*, **54**, 260–5.

Hetrick, B. A. D., Kitt, D. G. & Wilson, G. W. T. (1988). Mycorrhizal dependence and growth habit of warm-season and cool-season tallgrass prairie plants. *Canadian Journal of Botany*, **66**, 1376–80.

Hetrick, B. A. D., Wilson, G. W. T. & Schwab, A. P. (1994). Mycorrhizal activity in warm and cool season grasses: variation in nutrient-uptake strategies. *Canadian Journal of Botany*, **72**, 1002–8.

Iqbal, S. H. & Qureshi, K. S. (1976). The influence of mixed sowing (cereals and crucifers) and crop rotation on the development of mycorrhiza and subsequent growth of crops under field conditions. *Biologic (Lahore)*, **22**, 287–98.

Jabaji-Hare, S, Deschene, A. & Kendrick, B. (1984). Lipid-content and composition of vesicles of a vesicular-arbuscular mycorrhizal fungus. *Mycologia*, **76**, 1024–30.

Janos, D. P. (1983). Tropical mycorrhizas, nutrient cycles and plant growth. In *Tropical Rainforest: Ecology and Management*, ed. S. Sutton, T. Whitmore & A. Chadwick, pp. 327–45. Oxford: Blackwell Scientific Publications.

Jensen, A. (1982). Influence of four vesicular-arbuscular mycorrhizal fungi on nutrient uptake and growth in barley (*Hordeum vulgare*). *New Phytologist*, **90**, 45–50.

Jensen, A. (1983). The effect of indigenous vesicular-arbuscular mycorrhizal fungi on nutrient uptake and growth of barley in two Danish soils. *Plant and Soil*, **70**, 155–63.

Khan, A. G. (1972). The effect of vesicular-arbuscular mycorrhizal associations on growth of cereals. I. Effects on maize growth. *New Phytologist*, **71**, 613–19.

Khan, A. G. (1975). The effect of vesicular-arbuscular mycorrhizal associations on growth of cereals. II. Effects on wheat growth. *Annals of Applied Biology*, **80**, 27–36.

Koide, R., Li, M., Lewis, J. & Irby, C. (1988). Role of mycorrhizal infection in the growth and reproduction of wild vs. cultivated oats. *Oecologia*, **77**, 537–43.

Koide, R. & Lu, X. (1992). Mycorrhizal infection of wild oats: maternal effects on offspring growth and reproduction. *Oecologia*, **90**, 218–26.

Law, R. (1988). Some ecological properties of intimate mutualisms involving plants. In *Plant Population Ecology*, ed. A. J. Davy, M. J. Hutchings & A. R. Watkinson, pp. 315–42. Oxford: Blackwell Scientific Publications.

Little, L. R. & Maun, M. A. (1996). The '*Ammophila* problem' revisited: a role for mycorrhizal fungi. *Journal of Ecology* **84**, 1–7.

Lu, X. & Koide, R. (1991). *Avena fatua* L. seed and seedling nutrient dynamics as influenced by mycorrhizal infection of the maternal generation. *Plant, Cell and Environment*, **14**, 931–9.

May, R. M. (1981). Models for two interacting populations. In *Theoretical Ecology*, 2nd edn, ed. R. M. May, pp. 78–104. Oxford: Blackwell Scientific Publications.

McGonigle, T. P. & Fitter, A. H. (1988). Ecological consequences of arthropod grazing on VA mycorrhizal fungi. *Proceedings of the Royal Society of Edinburgh*, **94B**, 25–32.

Miller, R. M., Jarstfer, A. G., Pillai, J. K. (1987). Biomass allocation in an *Agropyron smithii–Glomus* symbiosis. *American Journal of Botany*, **74**, 114–22.

Molina, R. J., Trappe, J. M. & Strickler, G. S. (1978). Mycorrhizal fungi associated with *Festuca* in the western United States and Canada. *Canadian Journal of Botany*, **56**, 1691–5.

Morton, J. B. & Benny, G. L. (1990). Revised classification of arbuscular mycorrhizal fungi (Zygomycetes): a new order, Glomales, two new suborders, Glomineae and Gigasporineae and two new families, Acaulosporaceae and Gigasporaceae, with an emendation of Glomaceae. *Mycotaxon*, **37**, 471–91.

Morton, J. B., Bentivenga, S. P. & Wheeler, W. W. (1993). Germ plasm in the international collection of arbuscular and vesicular-arbuscular mycorrhizal fungi (INVAM) and procedures for culture development, documentation and storage. *Mycotaxon*, **48**, 491–528.

Mosse, B., Hayman, D. S. & Arnold, D. J. (1973). Plant growth responses to vesicular-arbuscular mycorrhiza. V. Phosphate uptake by three plant species from P-deficient soils labelled with ^{32}P. *New Phytologist*, **72**, 809–15.

Nelson, L. L. & Allen, E. B. (1993). Restoration of *Stipa pulchra* grasslands: effects of mycorrhizae and competition from *Avena barbata*. *Restoration Ecology*, 1, 40–50.

Newsham, K. K., Fitter, A. H. & Watkinson, A. R. (1994). Root pathogenic and arbuscular mycorrhizal fungi determine fecundity of asymptomatic plants in the field. *Journal of Ecology*, 82, 805–14.

Newsham, K. K., Fitter, A. H. & Watkinson, A. R. (1995a). Multi-functionality and biodiversity in arbuscular mycorrhizas. *Trends in Ecology and Evolution*, 10, 407–11.

Newsham, K. K., Fitter, A. H. & Watkinson, A. R. (1995b). Arbuscular mycorrhiza protect an annual grass from root pathogenic fungi in the field. *Journal of Ecology*, 83, 991–1000.

Newsham, K. K., Watkinson, A. R. & Fitter, A. H. (1995c). Rhizosphere and root-infecting fungi and the design of ecological field experiments. *Oecologia*, 102, 230–7.

Newsham, K. K., Watkinson, A. R., West, H. M & Fitter, A. H. (1995d). Symbiotic fungi determine plant community structure: changes in a lichen-rich community induced by fungicide application. *Functional Ecology*, 9, 442–7.

Peat, H. J. & Fitter, A. H. (1993). The distribution of arbuscular mycorrhizas in the British flora. *New Phytologist*, 125, 845–54.

Pendleton, R. L. & Smith, B. N. (1983). Vesicular-arbuscular mycorrhizae of weedy and colonizer plant species at disturbed sites in Utah. *Oecologia*, 59, 296–301.

Powell, C. L. (1975). The need for mycorrhizas for cocksfoot growth at high altitude: a note. *New Zealand Journal of Agricultural Research*, 18, 95.

Read, D. J. (1991). Mycorrhizas in ecosystems. *Experientia*, 47, 376–91.

Read, D. J. & Haselwandter, K. (1981). Observations on the mycorrhizal status of some alpine plant communities. *New Phytologist* 88, 341–52.

Read, D. J., Koucheki, H. K. & Hodgson, J. (1976). Vesicular-arbuscular mycorrhiza in natural vegetation systems. I. The occurrence of infection. *New Phytologist* 77, 641–53.

Reece, P. E. & Bonham, C. D. (1978). Frequency of endomycorrhizal infection in grazed and ungrazed bluegrama plants. *Journal of Range Management*, 31, 149–51.

Sanders, F. E. & Tinker, P. B. (1973). Phosphate inflow into mycorrhizal roots. *Pesticide Science*, 4, 385–95.

Skipp, R. A., Christensen, M. J. & Caradus, J. R. (1982). Invasion of white clover roots by fungi and other soil microorganisms. II. Invasion of roots in grazed pastures. *New Zealand Journal of Agricultural Research*, 25, 87–95.

Smith, S. E. & Gianinazzi-Pearson, V. (1988). Physiological interactions between symbionts in vesicular-arbuscular mycorrhizal plants. *Annual Review of Plant Physiology and Plant Molecular Biology*, 39, 221–44.

Sparling, G. P. & Tinker, P. B. (1978). Mycorrhizal infection in Pennine grassland. II. Effects of mycorrhizal infection on the growth of some upland grasses on γ-irradiated soils. *Journal of Applied Ecology*, 15, 951–8.

Subramanian, K. S., Charest, C., Dwyer, L.M. & Hamilton, R. I. (1995). Arbuscular mycorrhizas and water relations in maize under drought stress at tasselling. *New Phytologist*, 129, 643–50.

Trappe, J. M. (1987). Phylogenetic and ecologic aspects of mycotrophy in the angiosperms from an evolutionary standpoint. In *Ecophysiology of VA Mycorrhizal Plants*, ed. G. R. Safir, pp. 5–25. Boca Raton: CRC Press.

Trent, J. D., Svejcar, A. J. & Bethlenfalvay, G. J. (1993). Growth and nutrition of combinations of native and introduced plants and mycorrhizal fungi in a semiarid range. *Agriculture, Ecosystems and Environment*, **45**, 13–23.

Vietti, A. J. & van Staden, J. (1990). Histology of V-A mycorrhizal development in guayule seedlings. *Mycological Research*, **94**, 831–4.

Wallace, L. L. (1981). Growth, morphology and gas exchange of mycorrhizal and non-mycorrhizal *Panicum coloratum* L., a C_4 grass species, under different clipping and fertilization regimes. *Oecologia*, **49**, 272–8.

Wallace, L. L. (1987). Mycorrhizas in grasslands: interactions of ungulates, fungi and drought. *New Phytologist*, **105**, 619–32.

Wallace, L. L., McNoughton, S. J. & Coughenour, M. B. (1982). The effects of clipping and fertilization on nitrogen nutrition and allocation by mycorrhizal and non-mycorrhizal *Panicum coloratum* L., a C_4 grass. *Oecologia*, **54**, 68–71.

Wang, G. M., Coleman, D. C., Freckman, D. W., Dyer, M. I., McNaughton, S. J., Acra, M. A. & Goeschl, J. D. (1989). Carbon partitioning patterns of mycorrhizal versus non-mycorrhizal plants: real-time measurements using $^{11}CO_2$. *New Phytologist*, **112**, 489–93.

West, H. M. (1996). Influence of arbuscular mycorrhizal infection on competition between *Holcus lanatus* and *Dactylis glomerata*. *Journal of Ecology*, **84**, 429–38.

West, H. M., Fitter, A. H. & Watkinson, A. R. (1993). The influence of three biocides on the fungal associates of the roots of *Vulpia ciliata* ssp. *ambigua* under natural conditions. *Journal of Ecology*, **81**, 345–50.

Whittingham, J. & Read, D. J. (1982). Vesicular-arbuscular mycorrhiza in natural vegetation systems. III. nutrient transfer between plants with mycorrhizal interconnections. *New Phytologist*, **90**, 277–84.

Part three

Population biology of specific groups

12

Population dynamics in the regeneration process of monocarpic dwarf bamboos, *Sasa* species

AKIFUMI MAKITA

Introduction

The life history characteristics of bamboos differ conspicuously from other grasses in their peculiar flowering habits. Many bamboo species are long-lived monocarpic plants (McClure, 1966; Janzen, 1976; Soderstrom & Calderon, 1979; Simmonds, 1980; Veblen, 1982; Dransfield & Widjaja, 1995). In temperate regions in particular, the bamboos have long flowering intervals, estimated at 100 years or more (Janzen, 1976; Campbell, 1985). They often flower simultaneously over an extensive area and then die (Taylor & Qin 1988; Makita, 1992, Makita *et al.*, 1993).

Bamboos often form exclusive dense populations with vigorous rhizomatous vegetative reproduction. After mass flowering and death, their populations recover mainly by the development of seedling cohorts (Veblen, 1982; Taylor & Qin, 1988; Makita, 1992; Makita *et al.*, 1993). In clonal plants, the phase just after germination seems to be the most vulnerable period in their life history (Langer, Ryle & Jewiss, 1964; Kays & Harper, 1974; Cook, 1979; Solbrig, 1980; Bierzychudek, 1982; Liddle, Budd & Hutchings, 1982). Bamboos similarly suffer high mortality during seedling growth (Taylor & Qin, 1988; Makita, 1992, Makita *et al.*, 1993). After die-off of bamboos, many other species may be able to invade the bamboo stands because light conditions at ground surface are improved considerably (Nakashizuka, 1988; Taylor & Qin, 1988; Makita, 1992). Especially in forests with an undergrowth of bamboos, the mass death of bamboos provides opportunities for tree regeneration (Veblen, 1982; Nakashizuka, 1987, 1988; Taylor & Qin, 1988, 1992, 1997; Taylor, Qin & Liu, 1995; Makita, 1997; Widmer, 1997). Thus the simultaneous flowering and death of bamboos is an important episodic event in the local vegetation dynamics. Severe interspecific competition will occur in the regeneration process of bamboos, which is defined as the process from mass flowering to full recovery of the bamboo population (Makita *et al.*, 1993). Study of this process is therefore essential to clarify how monocarpic bamboo populations are maintained.

Competition between clonal plants is complicated because they can translocate resources and growth substances intraclonally (Pitelka & Ashmun, 1985; Marshall, 1990; de Kroon, 1993; Eriksson, 1993; Hara, 1994). Disadvantaged ramets in a genet may be supported by other ramets of the same genet by physiological integration (Hartnett & Bazzaz, 1985*b*; Hutchings & Bradbury, 1986; Slade & Hutchings, 1987; Hara, van der Toorn & Mook, 1993; Suzuki, 1994; Kowarik, 1995). In addition, ramet production may be regulated intraclonally to avoid severe inter-ramet competition (Hutchings, 1979; White, 1980; Pitelka, 1984). The mechanisms of density regulation in ramet populations are suggested to be different between interclonal and intraclonal interferences (Briske & Butler, 1989). Therefore, to consider the dynamics in clonal plants, we should take account of two levels of population dynamics, that is at the ramet and the genet level, and discuss how ramet density is determined.

There have been many studies on the population dynamics of clonal plants in their mature stage (Bierzychudek, 1982; Cook, 1985; van Groenendael & de Kroon, 1990; de Kroon & Kwant, 1991; Callaghan *et al.*, 1992; Svensson, Floderus & Callaghan, 1994). However, there are few reports on natural populations of clonal plants beginning at the establishment stage (Hartnett & Bazzaz, 1985*a*; Makita, 1992; Makita *et al.*, 1993). In this chapter, studies on the population dynamics of monocarpic dwarf bamboos, *Sasa* species, in the regeneration process will be reviewed. In particular, population dynamics of congeneric *Sasa* species with different growth patterns and the importance of their growth habits on density regulation will be discussed.

Life history of *Sasa*

Thirty-five species of *Sasa*, belonging to five sections, are widely distributed in Japan and adjacent regions (Suzuki, 1978). They often predominate in grassland and the understorey of forests in the temperate and boreal zones. They form dense populations, 0.3–3.0 m in height, such that establishments of other species are severely hindered under the *Sasa* canopy.

One of the most distinctive characteristics of *Sasa* is their mass flowering habit. With the onset of the reproductive phase, almost all plants over

Fig. 12.1. Photograph showing mass death after flowering of *Sasa kurilensis* in a birch forest in the Hakkoda Mountains, northern Japan. This photograph was taken one year after mass flowering. The dead *Sasa* culms were 2.5–3.0 m in height.

a wide area flower simultaneously followed by mass die-off (Fig. 12.1). There are few long-lived monocarpic clonal species that show such mass flowering habits. Other than the bamboos, the acanthaceous shrub, *Strobilantas*, in India and Malaysia, is a rare example (Robinson, 1935; van Steenis, 1942; Simmonds, 1980).

There are many reports on the mass flowering of *Sasa* (Table 12.1), though most of them are fragmentary descriptions. The area of flowering varies widely in each case, even in the same species, ranging from several to thousands of hectares. It is said that *Sasa* species flower once in a hundred years or more (Janzen, 1976; Campbell, 1985). However, the longevity of their genets has not been confirmed.

After the die-off, *Sasa* populations recover by the growth of the seedling cohorts. It will take 15–25 years from seedling emergence to reaching full size (Makita, 1992). In the full-size stage, individuals vigorously extend rhizomes and produce culms from buds on the rhizomes. Culm longevity of *Sasa* differs per species and habitat conditions, ranging between one to 15 years (Hokkaido Regional Forestry Office, 1984). Until the next mass flowering, continued death and recruitment of individual culms are repeated, and the populations are capable of maintaining stable size (Oshima, 1961).

Table 12.1. *Major records of mass flowering of* Sasa *species in Japan*

Year	Species	Locality	District	Area (ha)	Seed production (1000 m^{-2})	Seedling density (m^{-2})	References
1943	S. senanensis	Mt. Norikura	Chuubu	1,000	–	–	Muroi (1966)
1964–66	S. palmata	Mt. Ishizuchi	Shikoku	–	–	1.0–30	Yamanaka (1979)
1965–67	S. kurilensis	Chuugoku Mts.	Chuugoku	4,800	–	–	Muroi (1968)
1966	S. kurilensis	Otoineppu	Hokkaido	18	–	5.0–20	Kudoh & Ujiie (1990)
1966–68	S. kurilensis	Mt. Hyonosen	Kinki	1,000	–	–	Muroi (1966)
1968	S. kurilensis	Hachimantai	Tohoku	400	–	–	Muroi (1968)
1972	S. heterotricha	Hiroshima	Chuugoku	–	5	–	Iriguchi (1975)
1974	S. palmata	Chuugoku Mts.	Chuugoku	20,000	–	20–92	Shibanuma, Hashizume & Kondo (1977)
1974	S. kurilensis	Mt. Iizuna	Chuubu	several tens	–	–	Uchimura (1974)
1975	S. kurilensis	Ishikari	Hokkaido	20,000	10.0–15.0	0–32	Igarashi (1977) Hokkaido Regional Forestry Office (1981)
1977	S. tsuboiana	Hira Mts.	Kinki	–	3.9–13.8	14–21	Makita et al. (1988, 1993)
1979	S. kurilensis	Mt. Hakkoda	Tohoku	60	–	600–800	Makita (1992)
1995	S. kurilensis	South to Lake Towada	Tohoku	1,000	0.4–4.5	7–95	Makita, Matika & Nishiwaki (1995)
1995	S. kurilensis	Mt. Hakkoda	Tohoku	20	–	400–2,500	Makita et al. (unpub.)

The plant needs a fairly long time, maybe over 100 years, to attain the reproductive phase after gaining the full-size stage.

Growth habit

The basic frame of the bamboo plant consists of a ramifying system of segmented vegetative axes (McClure, 1966). The axes are differentiated as rhizomes, culms, and culm branches. Among them, the rhizome system constitutes the structural foundation of the bamboo genet. Morphology and growth habits of the rhizome are essential factors determining the structure and expansion pattern of a bamboo population.

McClure (1966) proposed two basic rhizome systems: pachymorph and leptomorph. The pachymorph rhizome system is seen typically in clump-forming tropical bamboos, such as the genus *Bambusa* and *Dendrocalamus*. Rhizomes of this type are short and thick, bearing a culm directly from its apex. As a result, the culms are usually close together and form rather compact clumps. In contrast, most of the temperate bamboos such as *Phyllostachys* have a leptomorph rhizome system. This systen is composed of long and slender rhizomes. It has a single lateral bud at each node and some of the buds develop into culms. Bamboos with the leptomorph rhizome system extend rhizomes horizontally into various directions, and the clump exhibits a spreading habit. Rhizomes of many genets intermingle to form a bamboo forest. Between these two distinctive rhizome systems, various intermediate types are observed (McClure, 1966; Clark, 1989). The rhizome habit is a key factor determining the ecological properties of bamboos.

Two different rhizome types can be observed among *Sasa* species. For example, *Sasa tsuboiana*, like most other temperate bamboos, has only leptomorph rhizomes, while *S. kurilensis* possesses pachymorph as well as leptomorph rhizomes (Makita, personal observation). *S. kurilensis* forms clumps with pachymorph rhizomes, interconnected by leptomorph rhizomes at some distance (Fig. 12.2). Such a difference in rhizome systems reflects in the growth habits of juveniles of the two species.

Fig. 12.3 shows the growth of *Sasa* juveniles schematically. In the first stage of seedling growth, culms are produced only by tillering from the original clump. Thereafter rhizomes are extended to produce culms from buds on the rhizomes. *S. tsuboiana* vigorously develops horizontal rhizomes from the second year after seedling emergence (Makita *et al.*, 1993). Repeated tillering from the original clump is followed by production of many culms from buds on rhizomes. This gives rise to a high number of

Fig. 12.2. Life form of a *Sasa kurilensis* sampled in a beech forest on a slope of Mt. Iizuna, central Japan in 1987. *S. kurilensis* forms clumps with pachymorph rhizome systems and each clump is connected by leptomorph rhizomes. The white bar in the centre of the photograph indicates 1 m length.

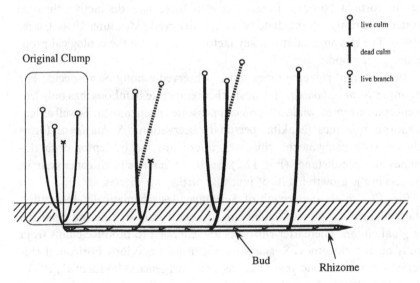

Fig. 12.3. Schematic representation of development of *Sasa* juveniles. From Makita (1996).

Table 12.2. *Correlations between genet dry weight and (A) number of culms per genet, and (B) height*

Slopes of the regression lines and correlation coefficients (r) are indicated. All the correlation coefficients were significant at the 0.1% level.

Species	n	A		B	
		Slope	r	Slope	r
S. tsuboiana[a]	45	0.661	0.921	0.107	0.566
S. kurilensis[b]	30	0.238	0.779	0.395	0.985

Notes:
[a] 4-year-old juveniles sampled in an open *Sasa* grassland and deciduous secondary forests in the Hira Mountains in April 1982.
[b] 12-year-old juveniles sampled in a beech forest on a slope of Mt. Iizuna in November 1987
Source: Data from A. Makita (unpublished).

culms per genet in this species. On the other hand, *S. kurilensis* rarely develops horizontal rhizomes during the first 10 years after germination (Kudoh & Ujiie, 1990; Makita, 1992). Juveniles produce culms only by tillering from the original clumps. The mean number of culms per genet increases little, only about 1.5 in 9-year-old juveniles of *S. kurilensis* (Makita, 1992), while 3-year-old juveniles of *S. tsuboiana* had 10–70 culms per genet (Makita *et al.*, 1993).

Differences in the growth patterns in these two species are clearly recognizable on the basis of plant weight (Table 12.2; Makita, unpublished). *S. tsuboiana* showed a much steeper slope than *S. kurilensis* for the relationship between genet weight and the number of culms per genet. In contrast, *S. kurilensis* considerably increased culm height as the genet weight increased, while the difference in height between smaller and larger juveniles in *S. tsuboiana* was more modest. That is, the growth of juveniles is represented by the number of culms for *S. tsuboiana* and by culm height for *S. kurilensis*.

Makita (1990) found that genet weight of juveniles of each species was closely correlated with leaf area, and the slopes of regression lines were similar between the two species (Makita, 1990). The way of increasing total leaf area per genet was by increasing the number of culms for *S. tsuboiana*, and by producing the taller culms with large leaf area per culm for *S. kurilensis*.

Fig. 12.4. Photographs showing the life forms of (*a*) 4-year-old *Sasa tsuboiana*
and (*b*) 12-year-old *S. kurilensis*. *S. tsuboiana* possesses many short culms with
rhizomes extending in various directions, while *S. kurilensis* is taller but with
fewer culms. The *S. kurilensis* in this photograph had not begun rhizome
extension. The heights of each species in these photographs are about 0.3 m for *S.
tsuboiana* and 1.3 m for *S. kurilensis*.

As a result of the difference in growth patterns, the growth forms of juveniles of the same size in the two species are quite different (Fig. 12.4). For
example, *S. kurilensis* juveniles of 100 g dry weight had no more than 10
culms, though *Sasa tsuboiana* bore about 100 culms at this size (A. Makita,
unpublished). Heights of juveniles of such size were 150 and 20 cm in *S.
kurilensis* and *S. tsuboiana*, respectively. Longevity of culms tends to be
longer as the culm becomes taller. Therefore, *S. kurilensis* juveniles had
larger and longer-lived culms compared with *S. tsuboiana* juveniles of the
same weight. The size at the onset of rhizome extension is also different.
Most juveniles of *S. tsuboiana* over 1 g.d.w. developed rhizomes, but many
seedlings of *S. kurilensis* over 10 g.d.w. had not begun rhizome extension.
S. tsuboiana juveniles were shorter but had proliferating culms with rhizomes running in various directions. As a result, the extent of the *S. tsuboiana* genet was much wider than that of *S. kurilensis* with fewer but taller
culms. Such differences in juvenile growth pattern have an important effect
on the population dynamics of the species.

Population dynamics at the ramet level

The aboveground vegetative unit of bamboos is a culm. Some studies were conducted to clarify changes in culm density after mass flowering of the above mentioned two *Sasa* species (Kudoh & Ujiie, 1990; Makita, 1992; Makita *et al.*, 1993). According to their results, the pattern of changes in culm density of the juvenile populations during the regeneration process is similar irrespective of species and initial density (Fig. 12.5). In the first stage, culm density increased rapidly, but after that there was a constant decline in culm density. The duration of the first stage and the rate of increase in culm density was different among the investigated stands. These attributes are probably a function of the initial seedling density and seedling growth rate (Makita, 1996).

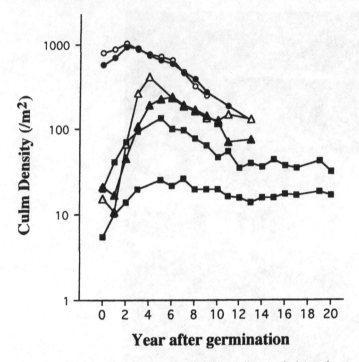

Year after germination

Fig. 12.5. Changes in culm density of the juvenile populations after mass flowering of *S. kurilensis* (○, ●, ■) and *S. tsuboiana* (△, ▲). Open symbols indicate open sites and closed ones are forest understorey. Redrawn after Kudoh & Ujiie (1990) (■); Makita (1992) (○, ●); Makita *et. al.* (1993) (△, ▲).

Makita (1996) examined the time trajectory of the relationship between culm density and mean weight per culm in juvenile populations of *S. kurilensis* and *S. tsuboiana* for ten years after flowering (Fig. 12.6). During the first stage, culm density increased with the increase in plant weight in both species. Then culm density of both species decreased with an increase in weight along the thinning slope of *c.* $-3/2$. The conclusion is that the juvenile populations of both species multiplied in culm densities until they reached the full density state and then showed a size-dependent decrease in culm density.

When annual plants or tree saplings show such a pattern as in Fig. 12.6, we conclude that the population is subject to the $-3/2$ power rule (Yoda *et al.*, 1963), namely that the reduction of density is caused by self-thinning. However, for clonal plants, such relationships in ramet level do not lead to this conclusion. If the production of ramets is regulated intraclonally, the apparent pattern of density–weight relationship for ramets may be similar

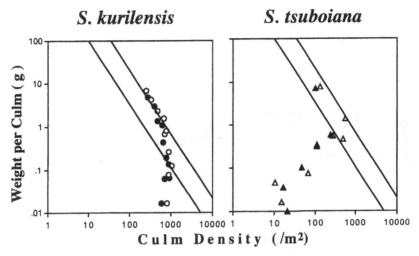

Fig. 12.6. Time trajectories of the relationship between culm density and weight for *Sasa kurilensis* and *S. tsuboiana* juvenile populations. Culm weight is estimated as the biomass of a genet (leaves, culms and rhizomes) divided by the number of culms. Dead parts were excluded from the measurement. Symbols as in Fig. 12.5. The two lines represent the upper and the lower boundaries of the −3/2 self-thinning rule

$$\log_{10} W = K - 3/2 \log_{10} \rho, \quad 3.5 \le K \le 4.3$$

where W is the mean weight (g), ρ is the density (m^{-2}) and K is a constant specific to the species and environmental condition. From Makita (1996).

to that of non-clonal populations that suffer self-thinning. Thus we should consider the regulation of ramet density with special regard to intraclonal regulation.

Population dynamics at the genet level

Seed production of *Sasa* species after mass flowering is ordinarily abundant, up to 15 000 seeds per m^2 (see Table 12.1). However, initial seedling density in the following year is relatively low except in the cases of *S. kurilensis* in the Hakkoda Mountains. Makita *et al.* (1993) observed that *S. tsuboiana* produced 4 000– 14 000 seeds m^{-2} after mass flowering in 1977. The number of seeds in the soil decreased drastically just after shedding with only 3–8% of the seeds remaining in the soil in the next spring (Fig. 12.7). Emergent seedling density in the first autumn was 14–21 m^{-2}, only 0.1–0.5% of produced seeds. *Sasa* seeds are highly nutritious (Maeda, 1977), and consumed by various animal species (Janzen, 1976). In particular, murid rodents have attracted public attention as predators of *Sasa*

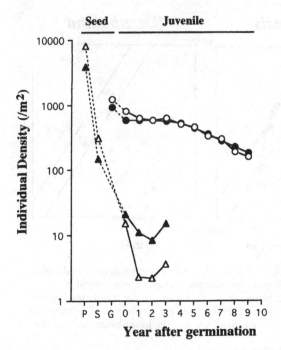

Fig. 12.7. Survivorship curve in *Sasa kurilensis* and *S. tsuboiana* populations.
Symbols are the same as in Fig. 12.5. P, S and G indicate the estimated densities
of all produced seeds, remaining seeds in litter and soil in the next spring after
flowering, and germinated seeds, respectively. Redrawn after Makita (1992) and
Makita *et al.* (1993).

seeds (Ito, 1975; Nakatsu, 1985). In the die-off stand of *S. tsuboiana*, high
densities of *Microtus montebelli* were observed (Kuwahata, 1979).
Overwinter loss of seeds by predation seems to be the main cause of low
seedling density of *Sasa* species.

It is quite difficult to clarify the changes in genet density of clonal plants
with rhizomatous vegetative reproduction because of the difficulty in
assessing the extent of a genet. There are few reports considering the
changes in genet density in natural populations. However, Makita (1992)
presented the survivorship curve of *S. kurilensis* juvenile populations for
10 years from germination; *S. kurilensis* juveniles showed little horizontal
expansion. Genet density was estimated with the mean culm density in permanent quadrats divided by mean number of culms per harvested genet.
Three phases can be recognized in the survivorship curve during the regeneration process (Fig. 12.7). The first phase is the establishment phase with
rather high genet mortality. Next, the juvenile population passes through a

low mortality period for several years (density-stable phase) and, thereafter, entered a phase of nearly constant mortality (thinning phase). The time trajectory of density–weight relationship in genets indicates that the *S. kurilensis* juvenile population follows the self-thinning rule (Makita, 1992). The established seedling populations scarcely change their genet density until the population reaches full density, then their density decreases by self-thinning. Based on growth analysis of *S. kurilensis* juveniles, Makita (1992) clarified that the mortality of juveniles can be explained from their dry matter economy: smaller genets suppressed by neighbouring ones find it difficult to produce new organs, which makes the C/F ratio – the weight ratio of non-photosynthetic and photosynthetic organs – of such genets high. Finally, the suppressed juveniles with high C/F ratio presumably die because high C/F ratio in dark conditions is fatal for the survival of juveniles (Kuroiwa *et al.*, 1964).

Genet density of *S. tsuboiana* could be estimated for only three years after seedling germination, after which time the extent of a genet could no longer be ascertained because of vigorous rhizome expansion (Makita *et al.*, 1993). Therefore it was impossible to determine whether density-dependent mortality of genets occurred during the regeneration process of *S. tsuboiana*. In *S. tsuboiana*, however, the number of culms per genet is closely related to the genet weight (see Table 12.2). Even if the smaller genets die due to self-thinning, their losses do not seem to contribute greatly to the decrease in culm density.

Density regulation

The mode of intergenet competition is an important factor that determines the mechanism of density regulation. Makita (1996) found that different growth patterns of *S. tsuboiana* and *S. kurilensis* brought about differences in the mode of intergenet competition. As a result, the mechanism regulating the density of ramets was different between species, although the apparent pattern of changes in culm (ramet) density was strikingly similar (Makita, 1996).

Rhizome extension in *S. kurilensis* was scarce. Their genets consisted of a small number of culms. Differences in genet weight were reflected in the differences in plant height. As presented schematically in Fig. 12.8, the difference in the height of neighbouring genets directly affects the light conditions for a whole genet. This situation is similar to annual plant or tree sapling populations and causes asymmetric competition. Light conditions of smaller individuals deteriorate as the population closes. Once smaller

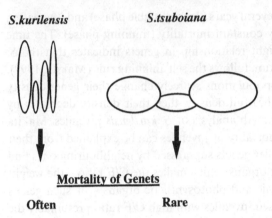

S.kurilensis   *S.tsuboiana*

Mortality of Genets

Often   Rare

Fig. 12.8. Schema of the growth forms of *Sasa kurilensis* and *S. tsuboiana*. Each circle indicates a genet of each species.

genets become shaded by other genets, they are caught in a vicious cycle that enhances the imbalance in dry matter economy and may lead to their death (Makita, 1992). In conclusion, culm density of *S. kurilensis* decreases because of the death of genets due to self-thinning after the population reaches full density conditions.

On the other hand, the mode of intergenet competition of *S. tsuboiana* differs from *S. kurilensis*. It is unlikely for a whole genet of *S. tsuboiana* to be overtopped by others (Fig. 12.8), because of the rapid rhizome expansion from an early phase of seedling growth. Even if some ramets are shaded, other ramets may grow in good conditions and be able to support the disadvantaged ramets. Hence, adversity encountered by some ramets may not be directly connected with the mortality of genets in *S. tsuboiana*, as suggested in many clonal plants.

Makita (1996) paid attention to the culm production of *S. tsuboiana*. Rhizomes of *Sasa* have a segmented structure and each rhizome node has a single bud. Observations were made of the changes in the ratio of the buds on rhizomes that grow to become culms to the total number of the rhizome nodes (the Bud Utility Ratio, BUR) during population recovery. When culm density of the *S. tsuboiana* juvenile population reached peak value in the fourth year after germination, BUR was very high, more than 50%. However, BUR decreased drastically thereafter and became around 20% in the sixth year after germination. Rhizome length per unit area was already comparable to that in the full-size population in the same area in the sixth year after mass flowering. So the change in BUR is a direct reflection of changes in culm densities. BUR in the full-size population of *S.*

tsuboiana in this area was reported to be 2–6% (Konno, 1977). These results indicate that *S. tsuboiana* decrease their culm production by reducing the BUR during the regeneration process. Bamboos seem to regulate their ramet density by controlling the ratio of sprouting as against the total number of buds on their rhizomes. *S. tsuboiana* forms a large 'bud bank' (Tappeiner *et al.*, 1991) by vigorous rhizome extension from an early stage in seedling growth. Its juveniles produce many culms with high BUR until the population reaches a full density state. Culm natality then declines by a decreasing BUR as culm size increases. The decrease in culm density during the regeneration process of *S. tsuboiana* must be caused mainly by intra-clonal regulation of culm production.

Conclusion

As shown in this chapter, growth pattern, especially horizontal expansion pattern, is an important attribute that determines the population dynamics during the regeneration process in *Sasa* species. The two *Sasa* species discussed in this chapter exhibit differential regulation processes of ramet density with quite different horizontal expansion patterns.

Clear developmental stages can be recognized in the course of *Sasa* juvenile growth: these are the tillering stage and the horizontal expansion stage with rhizome extension. Clonal plants distribute their ramets spatially to lessen the risk of genet extinction by physiological integration, and, consequently, prolong their genet longevity (Cook, 1985; Eriksson & Jerling, 1990). After attaining the horizontal expansion stage, *Sasa* species embody ecological features of clonal plants, and are unlikely to suffer high genet mortality. *S. tsuboiana* enters the horizontal expansion stage early during juvenile growth. Ramet density is regulated mainly intraclonally, as in many other clonal plants. On the other hand, *S. kurilensis* remains in the tillering stage for much longer, so that the mode of intergenet competition in this stage of *S. kurilensis* is asymmetric, as found in non-clonal plants. The difference in the mode of intergenet competition caused by the difference in developmental stages has a distinctive effect on the population dynamics of bamboos during the regeneration process.

Horizontal distribution patterns of ramets in a genet may be an impor-tant factor determining species-specific ecological features of clonal plants (Hutchings & Mogie, 1990). In bamboos, the distribution pattern of culms (ramets) is strongly regulated by the rhizome system. Rhizome habit is also an essential factor that determines the growth habit and, consequently, the mode of intergenet competition. Rhizome systems are basically different between the tropical bamboos and the temperate bamboos. Flowering habits are also different. Temperate bamboos tend to flower with longer intervals and show stronger synchrony than tropical ones. Adaptive significance of such different flowering habits must be connected with the difference in population structures caused by rhizome systems. Thus, to clarify the evolutionary trends in the life history of bamboos, further investigations must be conducted on the peculiar flowering habits of the bamboo species with various growth patterns. It is essential to analyse the life history characteristics with special reference to the mode of clonal growth.

Acknowledgement

I wish to thank Drs. Marinus J. A. Werger, Gregory P. Cheplick, and Thomas T. Lei for their critical reading of the manuscript. This study forms a part of guest scientist activities, Center for Ecological Research, Kyoto University.

References

Bierzychudek, P. (1982). Life histories and demography of shade-tolerant temperate forest herbs: a review. *New Phytologist*, **90**, 757–76

Briske, D. D. & Butler, J. L. (1989). Density-dependent regulation of ramet populations within the bunchgrass *Schizachyrium scoparium*: interclonal versus intraclonal interference. *Journal of Ecology*, **77**, 963–74.

Campbell, J. J. N. (1985). Bamboo flowering patterns: a global view with special reference to East Asia. *Journal of American Bamboo Society*, **6**, 17–35.

Callaghan, T. V., Carlsson, B. A., Jonsdottir, I. S., Svensson, B. M. & Jonasson, S. (1992). Clonal plants and environmental change. *Oikos*, **63**, 341–7.

Clark, L. G. (1989). Systematics of *Chusquea* Section *Swallenochloa*, Section *Verticillatae*, Section *Serpentes*, and Section *Longifoliae* (Poaceae–Bambusoideae). *Systematic Botany Monographs vol. 27*. Michigan: The American Society of Plant Taxonomists.

Cook, R. E. (1979). Patterns of juvenile mortality and recruitment in plants. In *Topics in Plant Population Biology*, ed. O. T. Solbrig, S. Jain, G. B. Johnson & P. H. Raven, pp. 207–31. New York: Columbia University Press.

Cook, R. E. (1985). Growth and development in clonal plant populations. In *Population Biology and Evolution of Clonal Organisms*, ed. J. B. C. Jackson, L. W. Buss & R. E. Cook, pp. 259–96. New Haven and London: Yale University Press.

de Kroon, H. (1993). Competition between shoots in stands of clonal plants. *Plant Species Biology*, **8**, 85–94.

de Kroon, H. & Kwant, R. (1991). Density-dependent growth responses in two clonal herbs: regulation of shoot density. *Oecologia*, **86**, 298–304.

Dransfield, S. & Widjaja, E. A. (1995). *Plant Resources of South-East Asia. No.7 Bamboos*. Leiden: Backhuys Publishers.

Eriksson, O. (1993). Dynamics of genets in clonal plants. *Trends in Ecology and Evolution*, **8**, 313–16.

Eriksson, O. & Jerling, L. (1990). Hierarchical selection and risk spreading in clonal plants. In *Clonal Growth in Plants, Regulation and Function*, ed. J. van Groenendael & H. de Kroon, pp. 79–94. The Hague: SPB Academic Publishing.

Hara, T. (1994). Growth and competition in clonal plants – persistence of shoot populations and species diversity. *Folia Geobotanica et Phytotaxonomica*, **29**, 181–201.

Hara, T., van der Toorn, J. & Mook, J. H. (1993). Growth dynamics and size structure of shoots of *Phragmites australis*, a clonal plant. *Journal of Ecology*, **81**, 47–60.

Hartnett, D. C. & Bazzaz, F. A. (1985*a*). The genet and ramet population dynamics of *Solidago canadensis* in an abandoned field. *Journal of Ecology*, **73**, 407–13.

Hartnett, D. C. & Bazzaz, F. A. (1985*b*). The integration of neighbourhood effects by clonal genets in *Solidago canadensis* . *Journal of Ecology*, **73**, 415–427.

Hokkaido Regional Forestry Office (1981). A research on the actual distribution of the dead *Sasa* populations after fructification. In *Distribution of Sasa Groups and their Characteristics in the Jurisdiction of Hokkaido Regional Forestry Office*, pp. 21–59. Sapporo: Hokkaido Regional Forestry Office. (in Japanese)

Hokkaido Regional Forestry Office (1984). Distribution and ecology of *Sasa*. In *Silvicultural Systems of Natural Forest in Hokkaido*, pp. 1–39. Sapporo: Hokkaido Regional Forestry Office. (in Japanese)

Hutchings, M. J. (1979). Weight–density relationships in ramet populations of clonal perennial herbs, with special reference to the $-3/2$ power law. *Journal of Ecology*, **67**, 21–33.

Hutchings, M. J. & Bradbury, I. K. (1986). Ecological perspectives on clonal perennial herbs. *BioScience*, **36**, 178–82.

Hutchings M. J. & Mogie M. (1990). The spatial structure of clonal plants: control and consequences. In *Clonal Growth in Plants, Regulation and Function*, ed. J. van Groenendael & H. de Kroon, pp. 57–76. The Hague: SPB Academic Publishers.

Igarashi, B. (1977). Flowering, fructification and dying of *Sasa austrokurilensis* in Ishikari District, Hokkaido in 1975. Trans. 88th Meeting of Japanese Forestry Society, 215–17. (in Japanese)

Iriguchi, M. (1975). Flowering and fruiting of *Sasa. Bull. Hiroshima Forestry Exp. Sta.*, **10**, 63–65. (in Japanese)

Ito, A. (1975). Outbreaks of the field vole, *Microtus montebelii*, in Kansai and Chugoku Districts. *Bull. Gov. For. Exp. Sta.*, **271**, 39–92. (in Japanese with English summary).

Janzen, D. H. (1976). Why bamboos wait so long to flower. *Annual Review Ecology and Systematics*, **7**, 347–91.

Kays, S. & Harper, J. L. (1974). The regulation of plant and tiller density in a grass sward. *Journal of Ecology*, **62**, 97–105.

Konno, Y. (1977). Ecology of the *Sasa* grassland on Mt. Hohrai. In *Reports of Natural Environmental Research of Biwako Valley* pp. 37–48. Otsu: Biwako Valley Co., Ltd. (in Japanese)

Kowarik, I. (1995). Clonal growth in *Ailanthus altissina* on a natural site in West Virginia. *Journal of Vegetation Science*, **6**, 853–6.

Kudoh, H. & Ujiie, M. (1990). Regeneration of *Sasa kurilensis* and tree invasion after mass flowering. *Bamboo Journal,* 8, 38–49.

Kuroiwa, S., Hiroi, T., Takada, K. & Monsi, M. (1964). Distribution ratio of net photosynthate to photosynthetic and non-photosynthetic systems in shaded plants. *Botanical Magazine Tokyo,* 77, 37–42.

Kuwahata, T. (1979). Independence of the 1979 outbreak of Japanese vole (*Microtus montebelli*) on the Mts. Hira, Shiga prefecture to the contemporary fruiting of Ibuki-zasa (*Sasa tsuboiana*). *Forest Pests,* 28, 42–6. (in Japanese)

Langer, H. M., Ryle, S. M. & Jewiss, O. R. (1964). The changing plant and tiller populations of timothy and meadow fescue swards I. Plant survival and the pattern of tillering. *Journal Applied Ecology,* 1, 197–208.

Liddle, M. J., Budd, C. S. J. & Hutchings, M. J. (1982). Population dynamics and neighbourhood effects in establishing swards of *Festuca rubra. Oikos,* 38, 52–9.

Maeda M. (1977). Relationship between mass flowering and death of *Sasa kurilensis* and dynamics of *Murid* rodents. *Sapporo Rinyu.,* 188, 43–54. (in Japanese)

Makita, A. (1990). Regeneration process after mass flowering of monocarpic dwarf bamboos, *Sasa tsuboiana* and *S. kurilensis. Bamboo Journal,* 8, 31–7.

Makita, A. (1992). Survivorship of a monocarpic bamboo grass, *Sasa kurilensis*, during the early regeneration process after mass flowering. *Ecological Research,* 7, 245–54.

Makita, A. (1996). Density regulation during the regeneration process in two monocarpic bamboos: self-thinning or intraclonal regulation? *Journal of Vegetation Science,* 7, 281–8.

Makita, A. (1997). The regeneration process in the monocarpic bamboo, *Sasa* species. In *The Bamboos,* ed. G. P. Chapman, pp. 135–45. London: Academic Press.

Makita, A., Konno, Y., Fujita, N., Takada, K. & Hamabata, E. (1993). Recovery of a *Sasa tsuboiana* population after mass flowering and death. *Ecological Research,* 8, 215–24.

Makita, A., Konno, Y., Fujita, N., Takada, K., Hamabata, E. & Mihara, T. (1988). Mass flowering of *Sasa tsuboiana* in Hira Mountains. *Bamboo Journal,* 6, 14–21. (in Japanese with English summary)

Makita, A., Makita, H. & Nishiwaki, A. (1995). Mass flowering of *Sasa kurilensis* to the South of Lake Towada, Northern Japan, in 1995. *Bamboo Journal,* 13, 34–41. (in Japanese with English summary)

Marshall, C. (1990). Source-sink relations of interconnected ramets. In *Clonal Growth in Plants,* ed. J. M. van Groenendael & H. de Kroon, pp. 23–41. The Hague: SPB Academic Publishing.

McClure, F. A. (1966). *The Bamboos: A Fresh Perspective.* Cambridge: Harvard University Press.

Muroi, H. (1966). Factors of bamboo flowering following the predation by voles. *Reports of Fuji Bamboo Garden,* 11, 7–38. (in Japanese)

Muroi, H. (1968). Flowering of *S. kurilensis* in Mt. Hyono-yama. *Reports of Fuji Bamboo Garden,* 13, 90–106. (in Japanese)

Nakashizuka, T. (1987). Regeneration dynamics of beech forests in Japan. *Vegetatio,* 69, 169–75.

Nakashizuka, T. (1988). Regeneration of beech (*Fagus crenata*) after the simultaneous death of undergrowing dwarf bamboo (*Sasa kurilensis*). *Ecological Research,* 3, 21–35.

Nakatsu A. (1985). Relationship between mass fruiting of *Sasa* and the outbreaks of voles and mice. *Hoppoh Ringyo,* 37, 121–4.

Oshima, Y. (1961). Ecological studies of *Sasa* communities. II. Seasonal variations of productive structure and annual net production in *Sasa* communities. *Botanical Magazine* (Tokyo), **74**, 280–90.

Pitelka, L. F. (1984). Application of the −3/2 power law to clonal herbs. *American Naturalist*, **123**, 442–9.

Pitelka, L. F. & Ashmun, J. W. (1985). Physiology and integration of ramets in clonal plants. In *Population Biology and Evolution of Clonal Organisms*, ed. J. B. C. Jackson, L. W. Buss & R. E. Cook, pp. 399–435. New Haven and London: Yale University Press.

Robinson, M. E. (1935). The flowering of *Strobilanthes* in 1934. *Journal of Bombay Natural Historical Society*, **38**, 117–22.

Shibanuma, I., Hashizume, H. & Kondo, Y. (1977). Fundamental studies on the growth and regeneration of *Sasa tectorius* Makino. *Bulletin of the Tottori University Forests*, **10**, 1–12. (in Japanese with English summary)

Simmonds, N. W. (1980). Monocarpy, calendars and flowering cycles in Angiosperms. *Kew Bulletin*, **35**, 235–45.

Slade, A. J. & Hutchings, M. J. (1987). Clonal integration and plasticity in foraging behaviour in *Glechoma hederacea*. *Journal of Ecology*, **75**, 1023–36.

Soderstrom, T. R. & Calderon, C. E. (1979). A commentary on the bamboos (Poaceae: Bambusoideae). *Biotropica*, **11**, 161–72.

Solbrig, O. T. (1980). Demography and natural selection. In *Demography and Evolution in Plant Populations*, ed. O. T. Solbrig, pp. 1–20. Oxford: Blackwell Scientific Publications.

Suzuki, J. (1994). Shoot growth dynamics and the mode of competition of two rhizomatous *Polygonum* species in the alpine meadow of Mt. Fuji. *Folia Geobotanica et Phytotaxonomica*, **29**, 203–16.

Suzuki, S. (1978). *Index of Japanese Bambusaceae*. Tokyo: Gakken.

Svensson, B. M., Floderus, B. & Callaghan, T. V. (1994). *Lycopodium annotinum* and light quality: growth responses under canopies of two *Vaccinuim* species. *Folia Geobotanica et Phytotaxonomica*, **29**, 159–66.

Tappeiner, J., Zasada, J., Ryan, P. & Newton, M. (1991). Salmonberry clonal and population structure: the basis for a persistent cover. *Ecology*, **72**, 609–18.

Taylor, A. H. & Qin, Z. (1988). Regeneration from seed of *Sinarundinaria fangiana*, a bamboo, in the Wolong giant panda reserve, Sichuan, China. *American Journal of Botany*, **75**, 1065–73.

Taylor, A. H. & Qin, Z. (1992). Tree regeneration after bamboo die-back in Chinese *Abies–Betula* forests. *Journal of Vegetation Science*, **3**, 253–60.

Taylor, A. H. & Qin, Z. (1997). The dynamics of temperate bamboo forests and panda conservation in China. In *The Bamboos*, ed. G. P. Chapman, pp. 189–203. London: Academic Press.

Taylor, A. H., Qin, Z & Liu, J. (1995). Tree regeneration in an *Abies faxoniana* forest after bamboo dieback, Wang Lang Natural Reserve, China. *Canadian Journal of Forestry Research*, **25**, 2034–9.

Uchimura, E. (1974). On the mass flowering of *Sasa kurilensis*. *Reports of Fuji Bamboo Garden*, **19**, 33–8. (in Japanese)

van Groenendael, J. M. & de Kroon, H. (1990). *Clonal Growth in Plants: Regulation and Function*. The Hague: SPB Academic Publishing.

van Steenis, C. G. G. J. (1942). Gregarious flowering of *Strobilanthes* (Acanthaceae) in Malaysia. *Ann. Roy. Bot. Gdn. Calcutta. 150th Anniv. Vol.*, 91–7.

Veblen, T. T. (1982). Growth patterns of *Chesquea* bamboos in the understory of Chilean *Nothofagus* forests and their influences in forest dynamics. *Bulletin of the Torrey Botanical Club*, **109**, 474–87.

White, J. (1980). Demographic factors in populations of plants. In *Demography and Evolution in Plant Populations,* ed. O. T. Solbrig, pp. 21–48. Oxford: Blackwell Scientific Publications.

Widmer, Y. (1997). Life history of some *Chusquea* species in old-growth oak forest in Costa Rica. In *The Bamboos,* ed. G. P. Chapman, pp. 17–31. London: Academic Press.

Yamanaka, M. (1979). Revegetation of *Sasa ishizuchiana* community after the mass death in Ishizuchi Mountains. In *Conservation Report of Mt. Ishizuchi and Omogo-Valley in Ishizuchi Quasi-National Park, Shikoku,* pp. 65–73. Tokyo: Nature Conservation Society of Japan. (in Japanese)

Yoda, K., Kira, T., Ogawa, H. & Hozumi, K. (1963). Self-thinning in overcrowded pure stands under cultivated and natural conditions (intraspecific competition among higher plants XI). *Journal Biol. Osaka City University,* **14,** 107–29.

13

Population dynamics of perennial grasses in African savanna and grassland

T. G. O'CONNOR AND T. M. EVERSON

Introduction

Perennial grasses are a predominant component of the savanna and grassland biomes of sub-Saharan Africa. Savannas constitute more than half of the land surface area of this region. True grasslands are less widespread, occurring mainly in mid-latitude or high-altitude regions or as smaller fragments depending on local edaphic or climatic conditions (O'Connor & Bredenkamp, 1997). The terms grassland and savanna disguise the variation in climate and environmental conditions associated with each. Perennial grassland (true or savanna) can be encountered in regions receiving rainfall of as little as 150 mm yr^{-1} or as much as 1200 mm yr^{-1}; they may occur in areas which experience snowfall in winter or areas of tropical climate; they may be found on leached, dystrophic, sandy soils or on extremely fertile volcanic, clay soils; they may be burned on an annual basis or have 20 or more years pass without experiencing a fire; most are associated with the remarkable adaptive radiation of African ungulates whilst others are barely subject to herbivory by large mammals; they are at the heart of a pastoral enterprise which directly supports a greater human population than any other pastoral enterprise in the world.

The variation in climate and environment is matched by a diversity of species and growth forms. In southern Africa alone there are 723 species of perennial grass (Gibbs Russell et al., 1990). A community may contain caespitose, stoloniferous or rhizomatous growth forms; swards may be a 5–cm-high grazing lawn or 3 m tall when flowering; swards may consist of scattered individuals or have a closed canopy.

The population dynamics of African perennial grasses might be expected to vary considerably in relation to the extreme variation in environmental conditions; the population response of species with different attributes is unlikely to be similar for a given set of environmental conditions. Study of

333

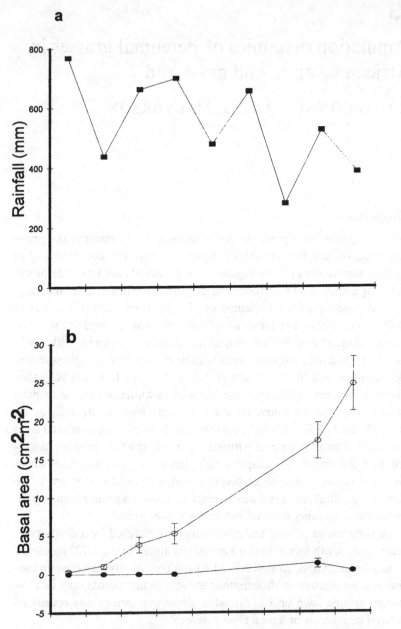

Fig. 13.1. Changes between 1986 and 1994 of (a) rainfall, and of basal area
(±S.E.; n=6 for each point) for (b) *Aristida bipartita* (open symbol, heavy grazing;
closed symbol, light grazing) and (c) *Themeda triandra* (open symbol, light
grazing; closed symbol, heavy grazing). See Table 13.4 for details of these species.
The x-axis of (c) applies to (a) and (b).

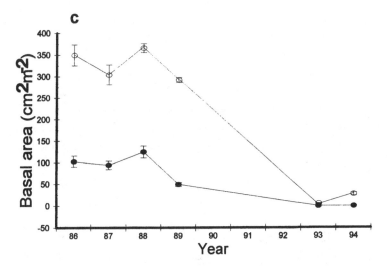

the population dynamics of African grasses is limited, fragmentary and still in its infancy; yet sufficient insight has accumulated for exploring whether there is some pattern to the relationship among population dynamics, climate, environment, extrinsic influences (grazing, fire) and species attributes of African perennial grasses.

The structure of this chapter is first to provide a description of the patterns of population change of perennial grasses in relation to environment, and then to review the current state of knowledge about population processes of African perennial grasses, highlighting the manner in which these processes are constrained by environment by contrasting a high-rainfall montane grassland with a semi-arid savanna grassland. We finally address some current ideas about plant population dynamics in the context of African perennial grasses.

Changes in abundance of perennial grasses

The semi-arid savannas and grasslands of Africa exhibit characteristically large coefficients of variation (CV) in annual rainfall (CV inversely related to mean; Tyson, 1986). Because of the conspicuous effect of rainfall variability on the population dynamics of species found in these systems (O'Connor, 1985), they are considered by some as non-equilibrium systems in which herbivory has little influence (Ellis & Swift, 1988). Changes in the abundance of individual species are driven by drought events of one or two years duration, often resulting in dramatic declines, although some species may benefit (Fig. 13.1). Population growth in years of high rainfall is,

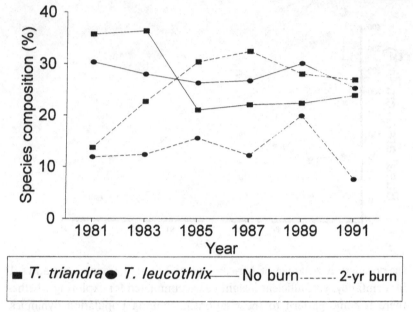

Fig. 13.2. Changes in the abundance of *Themeda triandra* and *Tristachya leucothrix* in response to frequency of defoliation by fire in a montane grassland. Data from T. M. Everson, unpublished.

however, not usually as dramatic, suggesting a hysteresis effect (Noy Meir & Walker, 1986). There is evidence to suggest that although the abundance of a species may wax or wane in response to rainfall variability, the consistent effect of grazing is small on an annual basis but cumulatively large over time (O'Connor, 1995; O'Connor & Roux, 1995), in contradiction of the non-equilibrium notion.

High-rainfall grassland or savanna is also event driven, but the event is fire. Exclusion of fire from grassland will invariably lead to the elimination of many species (O'Connor, 1985). For example, the prescribed management of biennial spring burning in montane grassland maintained the abundance of *Themeda triandra* but resulted in a decline of *Tristachya leucothrix*, whereas protection from burning resulted initially in a marked increase of *T. leucothrix* but in a decline of *T. triandra* (Fig. 13.2).

The perennial grasslands and savannas of southern Africa occur within the summer-rainfall region, but perennial grasses are also found in regions which receive a proportion of rainfall in autumn and winter. In these environments, the seasonal distribution as well as the amount of rainfall influences year-to-year changes in the abundance of perennial grasses. For the grassy, dwarf shrublands of the semi-arid Karoo, South Africa, an increased proportion of spring–summer rainfall relative to autumn–winter rainfall promotes perennial grasses at the expense of dwarf shrubs, and vice versa (Roux 1966; O'Connor & Roux, 1995).

Population processes

Tiller dynamics

The response of *T. triandra* tillers to moisture varies with climatic regime. In a semi-arid savanna, tillers of irrigated *T. triandra* plants had significantly higher mass, more elevated shoot apices and more leaves than non-irrigated plants (Danckwerts 1984; Danckwerts, Aucamp & Du Toit, 1984). (Tillers of *Sporobolus fimbriatus* responded to moisture in a similar fashion; Danckwerts, Aucamp & Du Toit, 1986.) Severe drought eliminated extant tiller populations (Danckwerts & Stuart-Hill, 1988). By contrast, soil moisture did not influence growth in a high rainfall grassland of the Drakensberg (Everson & Everson, 1987). Under the predictable conditions of montane grassland, the survivorship curve of tillers was smooth (Deevey Type II) (Fig. 13.3). Because a fixed proportion of individuals die per unit of time, mortality risk before senescence is independent of tiller age or of fluctuations in rainfall.

The developmental pattern of shoot apices influences the susceptibility of different species to defoliation at different seasons and frequencies (Booysen, Tainton & Scott, 1963; Tainton & Booysen, 1963, 1965*a,b*; Tainton, 1964; Rethman & Booysen, 1968; Rethman, 1971; Drewes & Tainton 1981). In some species (e.g. *T. triandra*) the apical buds are elevated some distance above the soil surface in mid-summer, but only elongate and develop flowers the following spring (Tainton & Booysen, 1963). These species are, therefore, adversely affected by defoliation during the nine-month period when the apical meristem is elevated. Timing of defoliation and resting periods are therefore critical for increasing flowering and herbage production (Tainton, 1964). By contrast, apical buds of *Tristachya leucothrix* remain at the soil surface until immediately prior to inflorescence maturity. The buds are, therefore, well protected for most of their developmental period and are less susceptible to defoliation during the reproductive phase. Environmental patterns such as rainfall influence the time of elevation of the shoot apex in some species (e.g. *Cymbopogon excavatus*) (Rethman & Booysen, 1967) but not in others (e.g. *Heteropogon contortus*) (Rethman & Booysen, 1968).

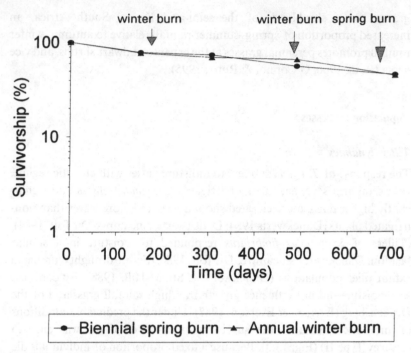

Fig. 13.3. Survivorship of a cohort of *Themeda triandra* tillers in montane
grassland when subjected to annual winter or biennial spring burns. Data from
Everson, Everson & Tainton (1985).

The population dynamics of grasses in montane vegetation depends on
interactions between the attributes of the species and environmental
factors, mainly fire. Different grass species differ in their ability to produce
tillers, a factor which has led to the recognition of distinct ecological
groups. Species which develop tillers above the soil surface (e.g. *T. triandra,
H. contortus, Trachypogon spicatus* – all decreaser species in range-science
terminology) increase with regular burning, whereas species that develop
tillers below the soil surface (e.g. *T. leucothrix, Alloteropsis semialata,
Harpochloa falx*) decrease in abundance.

Demographic studies of these species showed that their differential
responses to fire were best explained by their different rates of tiller production and mortality (Everson, Everson & Tainton, 1985). The increase of
T. triandra (maximum tiller lifespan of two years) and *H. contortus* in regularly winter- or spring-burnt grassland is a result of their high rates of
tiller initiation and limited tiller mortality because shoot apices remain
close to the surface. The annual winter-burn treatment produced more sec-

ondary tillers per primary tiller than the biennial spring-burn treatment. However, burning in summer when shoot apices are elevated has a catastrophic effect on tiller survival (<6%) because it destroys the apical meristem. The production of lateral tillers by *T. triandra* following defoliation is only prolific when defoliation occurs during the dormant season, indicating that *T. triandra* is adapted primarily to defoliation by fire and only moderately adapted to herbivory (Stuart-Hill & Mentis, 1982). By contrast *Sporobolus fimbriatus*, found mainly in semi-arid grassland, produces tillers at all times of the year and is primarily adapted to year-long herbivory (Danckwerts & Stuart-Hill, 1987).

Similarly in high-rainfall coastal grassland, tiller populations of *Cymbopogon validus* (maximum tiller longevity of two years) increased under annual or biennial burning or harvesting because an increased production of secondary tillers compensated for the increased mortality of defoliated tillers (Shackleton, 1989). Severe and frequent defoliation eliminated primary tillers, however, and prevented secondary tillers from developing (Shackleton & Mentis, 1991). Tiller mortality of both *T. triandra* and *C. validus* was marginally increased and tiller recruitment was suppressed when plants were protected from defoliation because of an accumulation of moribund material.

The inability of decreaser species to survive in fire exclusion areas in montane grassland is a result of a reduction in tiller initiation in response to reduced light beneath the dense canopy. Decreased irradiance (<30% full sunlight) eliminated *H. contortus* and *Trachypogon spicatus*, depressed tiller initiation of *T. leucothrix, A. semialata* and *T. triandra*, but had only a slight effect on *H. falx* (Everson, Everson & Tainton, 1988). Similarly, in the fertile, mesic grasslands of the Serengeti, canopy closure is the main agent responsible for reducing or inhibiting tillering (Coughenour, 1984; Coughenour, McNaughton & Wallace, 1984), by 89% for *T. triandra* (McNaughton, 1992).

Even for those species with an evolutionary history of grazing (e.g. in the Serengeti), the effect of defoliation on tiller populations has been found to interact in a complex way with many other factors. Defoliation promotes tillering of some species (McNaughton & Chapin, 1985; McNaughton, Wallace & Coughenour, 1983) but not of others (Ruess & McNaughton, 1984; Coughenour, McNaughton & Wallace, 1985*b*; McNaughton, 1985*b*, 1992), or the response depends on the frequency and height of defoliation (Edroma, 1985). The availability of nitrogen has consistently been found to stimulate tillering (Wallace, 1981; Ruess & McNaughton, 1984; Coughenour, McNaughton & Wallace, 1985*a,b*; McNaughton, 1985*b*), but increased tillering has not necessarily resulted in an increase of crown size (e.g. *T. triandra*) because there exists a negative exponential relationship between tillering and mass per tiller which, under certain conditions, has obeyed the $-3/2$ power law (Coughenour, McNaughton & Wallace, 1985 *a,b*). Competition among shoots of a plant for nitrogen would, therefore, seem a dominant process regulating growth rates of tillers. Nutrients other than nitrogen are also important – an increased availability of phosphorus stimulated tiller production of defoliated individuals of *Digitaria macroblephara* (McNaughton & Chapin, 1985). For *Panicum coloratum*, mycorrhizal infection and nitrogen, but not defoliation, increased tiller number, but only mycorrhizal infection increased crown area, thereby enhancing the tolerance of this species to grazing (Wallace, 1981). The postulation that herbivore saliva may stimulate tillering (Dyer *et al.*, 1982) has been supported for defoliated individuals of *Sporobolus ioclados* but not for those of *S. pyramidalis* (McNaughton, 1985*b*).

Although nitrogen and shading influence tillering directly they are indirectly influenced by grazing. The Serengeti supports a very high short-term grazing intensity (McNaughton, 1985*a*). This accelerates rates of nutrient cycling (McNaughton, Ruess & Seagle, 1988), thereby ensuring the maintenance of tiller populations of some species.

The tillering response of *T. triandra* to defoliation shows marked intraspecific variation depending on the history of grazing intensity (Oesterheld & McNaughton, 1988). Clones with a history of severe grazing had more, smaller tillers than clones without a history of severe grazing, illustrating these morphometric traits are in part fixed (McNaughton, 1984); clipped plants of either type produced more, smaller tillers than unclipped plants, indicating the plastic response of these traits.

How important is intertuft competition for tillering? In a dystrophic savanna, Von Maltitz (1990) used ring exclosures to show that an increase in the size available to an individual plant (four species were tested) was reflected by as much as a ten-fold increase in tiller production, number of culms with flowers and, for *Digitaria eriantha*, in the number of stolons and daughter plants. The effect of intertuft competition on tillering can be moderated by defoliation of the focal or surrounding individuals, especially where interaction between palatable and unpalatable species is involved (Edroma, 1981*a*; Brockett, 1983; Belsky, 1986*a*; Banyikwa 1988). For example, tillering of the unpalatable *Aristida junciformis* was more sensitive to defoliation than that of the palatable *Themeda triandra*, whereas yield and tiller number were reduced by competition to a greater extent for *T. triandra* than for *A. junciformis* (Morris & Tainton, 1993). The interaction between palatable and unpalatable species need not always be antagonistic. An association of *T. triandra* with less palatable plants protected it from grazing by two unselective herbivores (wildebeest and buffalo) in the Serengeti (McNaughton, 1978).

Mortality, gaps and population turnover

The mortality and turnover of tufts has not been investigated in high-rain-fall grassland.

Episodic droughts are an important mortality event for the perennial grasses of semi-arid savannas or grasslands, but the extent of mortality depends on a number of factors (Table 13.1). The percentage mortality of tufts increases with an increasing severity of drought for a single season. The pattern of tuft mortality of most species for two sequential years of drought in a semi-arid system has usually been negligible mortality during the first growing season, but almost catastrophic mortality during the second growing season and is further exacerbated by fine-textured soils or high tree densities. Extended droughts, such as have occurred in the Sahel, can eliminate the perennial grass component (Breman & Cissé, 1977; Breman *et al.*, 1980).

Drought mortality of tufts in savanna ranges between 0% and 100% among species for each community (Table 13.1). This is related to differences among species in drought tolerance resulting from differences, amongst others, in rooting patterns (Fourie & Roberts, 1977; Danckwerts & Stuart-Hill, 1988; Snyman, 1989; Snyman & Van Rensburg, 1990). Species containing aromatic compounds (*Bothriochloa radicans, Cymbopogon plurinodis*) apparently experience minimal mortality at a time when other species experience 100% mortality (Table 13.1).

Mortality of a tuft need not necessarily be complete rather a tuft may fragment into a number of tiller clusters as a result of partial dieback (O'Connor, 1994). Depending on species, 11–65% of tufts fragmented over three years, forming 2–10 daughter tufts. Small fragments (<2.5 cm diameter) rather than large individuals were more likely to die in following years. Fragments were also more likely to die than seedling recruits of the same size for some species (*Aristida bipartita, Heteropogon contortus*) but not others (*Digitaria eriantha, Themeda triandra*).

Grazing had an effect on mortality and on tuft size additional to that of drought in two savannas monitored over a number of years of a declining amount of rainfall (O'Connor, 1991*a*; Table 13.1). Severe defoliation by clipping can result in tuft mortality (Ndawula-Senyimba, 1972;

Table 13.1. *Percentage mortality of grass tufts in response to drought at various locations in South Africa* Gaps in the table indicate no data are available.

Locality	Description		Mortality (%)				Source
			Bush cleared	Uncleared			
Kalahari savanna, Northern Cape MAR[a] 425 mm		*Stipagrostis uniplumis*	0	70			Donaldson, 1967
		Schmidtia pappophoroides	26	82			
		Eragrostis lehmanniana	5	92			
Savanna, Northern Cape MAR 450 mm	Season of 246 mm, only 35 mm Nov-Feb inclusive	*Eragrostis lehmanniana*	90				Fourie & Roberts, 1977
		Themeda triandra	<5				
			Sandy/loam	Sandy clay loam	Clay loam		
Colophospermum mopane savanna, eastern Transvaal MAR 480 mm	Two successive years of drought: 1981/2 279 mm 1982/3 219 mm Variable grazing	Tree density (ha^{-1})	850	1,080	600		Scholes, 1985
		Panicum maximum	86				
		P. coloratum	84		29		
		Schmidtia pappophoroides	82				
		Digitaria eriantha		97			
		Bothriochloa radicans		74	7		
Acacia karroo savanna, eastern Cape MAR 422 mm	One season (summer-autumn) drought: 327 mm; sandy clay loam	*Cymbopogon plurinodis*	0				Danckwerts & Stuart-Hill, 1988
		Digitaria eriantha	57				
		Eragrostis chloromelas	70				
		E. obtusa	100				
		Michrochloa caffra	97				
		Panicum stapfianum	45				
		Sporobolus fimbriatus	47				
		Themeda triandra	44				

			Rainfall	Irrigation	
Grassland, Free State MAR 560 mm	Two successive years of drought contrasted with supplemented water: 1982/3 331 mm vs 581 mm 1983/4 425 mm vs 675 mm	*Digitaria eriantha* *Eragrostis chloromelas* *E. lehmanniana* *E. obtusa* *Panicum stapfianum* *Sporobolus fimbriatus* *Themeda triandra*	44 100 71 100 100 13 55	3 5 11 5	Snyman & Van Rensburg, 1990
Sandveld savanna, eastern Transvaal MAR 650 mm	Sequence of wet (821 mm), moderate (610 mm) and dry (425 mm) years Sandy loam soil Light grazing	After 3 years: *Pogonarthria squarrosa* *Digitaria eriantha* *Aristida stipitata* *Perotis patens*	65 87 25 36		O'Connor, 1991a
			Heavy grazing	Light grazing	
Savanna, eastern Transvaal MAR 650 mm	Single year of moderate drought (400 mm) Clay soils	*Aristida bipartita* *Bothriochloa insculpta* *Digitaria eriantha* *Heteropogon contortus* *Setaria incrassata* *Themeda triandra*	26 30 12 45 26 34	17 29 8 23 19 18	O'Connor, 1994

Note:
[a] MAR, mean annual rainfall.

Mufandaedza, 1976) but the defoliation regimes of these experiments may not match the frequency with which tufts are usually regrazed (Gammon & Roberts, 1978a,b, 1980) except possibly for conditions of drought and large animal biomass (Edroma, 1985).

The dieback of tufts or disturbance creates gaps which are available for colonization by seedlings or vegetative propagules. The amount of space available is a function of basal area but neither the size distribution of gaps nor the types, nature, size distributions and frequencies of small disturbances have been documented for African savannas or grasslands. Small gaps are often successfully colonized by stoloniferous or rhizomatous species (O'Connor, 1991c), whilst the centre of large gaps is usually initially colonized only by a few seedlings which benefit by an increased availability of resources and grow larger and have more tillers (Von Maltitz, 1990).

In the Serengeti, if gap creation involves the physical disturbance of the soil, seedlings are initially more important than sprouts from rootstock, stolons or rhizomes for recolonizing gaps, but the success of different strategies depends on rainfall and grazing (Belsky, 1986b,c). Vegetative propagation was most important in short grasslands dominated by clonally spreading species because of heavy grazing. Sprouting of rhizomatous grasses was abundant on shallowly disturbed ground, whereas seedlings were abundant on deeply disturbed ground. With time, recolonization by seedlings is diminished because of the denseness of the vegetation (Belsky, 1986c). For example, the response to disturbance on a midgrass site without grazing over five years involved the replacement of seed-reproducing species (annual *Chloris pycnothrix, T. triandra*) by first the stoloniferous *Digitaria macroblephara* and then the rhizomatous *Pennisetum mezianum*. The evolution of the rhizomatous habit would seem related to heavy grazing (Belsky, 1986c). Rhizomatous grasses are not, however, unaffected by defoliation: rhizomes of *Imperata cylindrica* spread 4.1 cm in a burned and 8.9 cm in an unburned grassland respectively, burning having rendered an otherwise unpalatable grass attractive to grazers (Edroma 1981b). By contrast, montane grasslands recover slowly following disturbance. Neighbouring grasses revegetated only 5% of cleared sites (1 m^2) after three years (Everson, 1994). In small patches caused by baboon foraging (0.02 m^2) seedling establishment of *T. triandra* was only 4.1 ± 2.0 seedlings m^{-2} (Everson 1994).

Mortality and recruitment lead to an appreciable degree of turnover of individuals within a relatively short period of time (Table 13.2). For seed-reproducing grasses, turnover constitutes a replacement of genets, highlighting the importance of seedling recruitment for the persistence of these

Table 13.2. *Population turnover of some savanna grass species (the fraction of the population constituted by new recruits, seedling or vegetative, after a specified period of time)*

Site	Species	%		Source
Kisokon, Kenya	*Cymbopogon popischlii*[b]	47		Bogdan, 1954
MAR[a] 873 mm	*Eragrostis superba*[b]	55		
Four years	*Themeda triandra*[b]	61		
Eastern Transvaal,		Grazing		O'Connor, 1994
South Africa		Heavy	Light	
MAR 650 mm	*Aristida bipartita*[b]	47	36	
Four years	*Bothriochloa insculpta*[c]	49	13	
	Digitaria eriantha[d]	11	5	
	Heteropogon contortus[b]	48	18	
	Themeda triandra[b]	16	7	

Notes:
[a] Mean annual rainfall.
[b] Seed reproduction.
[c] Predominantly seed reproduction, also stolons.
[d] Predominantly stolon reproduction.

species. Turnover of populations of species capable of rhizomatous or stoloniferous growth is mostly a result of recruitment of physiologically independent individuals from vegetative propagation, although some of these species also show seedling recruitment (Belsky, 1986a; Von Maltitz, 1990; O'Connor, 1991a,c).

Savanna grasses which apparently propagate mainly by clonal growth through tillering can do so in very different ways. The rhizomatous *Setaria incrassata* was found to form monodominant patches (9 m diameter) (Carter & O'Connor, 1991) which are likely to result from either propagation of a single clone or the coalescence of individual clones as they expand following recruitment from seed. Seedling recruitment seems to be a rare event for this species (0.33 seedlings $m^{-2}\,yr^{-1}$) (O'Connor, 1991c, 1994), but electrophoretic investigation revealed that extensive sexual reproduction had taken place (Carter & Robinson, 1993); clonal expansion was limited to areas of 0.25 m^2 on average. *Andropogon greenwayi* lacks stolons or rhizomes, but in the Serengeti it forms patches up to 100 m diameter through the coalescence of individual clones (<3 m diameter) (Belsky, 1986a). Individual clones expand by tillering (5 cm yr $^{-1}$ [McNaughton, 1983]) provided they are grazed, because the mats disappear if moribund material accumulates.

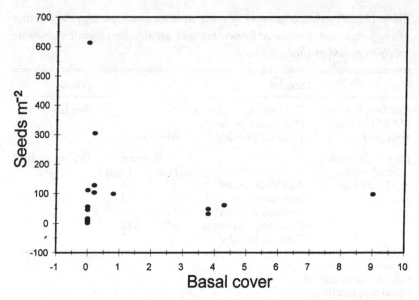

Fig. 13.4. Relationship between basal cover of a species and its seed production for a late seral stage of a highveld grassland. Data from table 2 of Jones (1968).

Seed bank dynamics

Seed production

The percentage of filled seeds can range from 0% to 86% among species in savanna (Tolsma, 1989; O'Connor & Pickett, 1992). Year-to-year variation in seed production of individual savanna species depends on the amount and seasonal distribution of rainfall (Dye & Walker, 1987; O'Connor & Pickett, 1992). Species capable of stoloniferous or rhizomatous growth do not necessarily produce fewer seeds than seed-reproducing species (O'Connor & Pickett, 1992).

Seed production is low (2–757 seeds m^{-2}) in montane grassland (Everson, 1994). In a highveld grassland, seed production was markedly lower (52-fold) for late than early seral stages (Jones, 1968). Even for late seral stages, most of the seed is produced by species which are poorly represented in the vegetation (Fig. 13.4).

Consumption of inflorescences by grazing can markedly reduce seed output (O'Connor & Pickett, 1992). The reproductive output of stoloniferous or rhizomatous species was more vulnerable to grazing than that of seed-reproducing species because they tended to invest in a few reproductive culms, each culm carrying a great number of seeds.

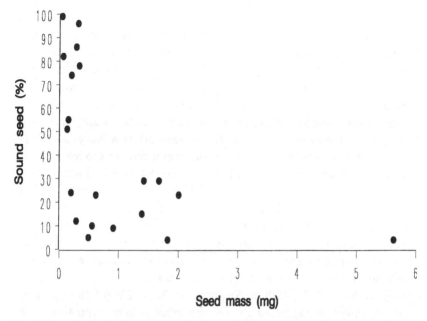

Fig. 13.5. Relationship between seed mass and the percentage of sound seed for grass species of a highveld grassland. Data from Jones (1964).

Pre- and post-dispersal predation

Pre-dispersal losses of viable seeds to insect infestation affect <10–100% of the seeds in some years although some species (possibly inversely related to caryopsis size) are unaffected (Jones, 1964, 1968; Lock & Milburn, 1970; Tolsma, 1989; O'Connor & Pickett, 1992). The number of sound seed produced by a species was negatively related to seed mass in a highveld grassland (Fig. 13.5): above a threshold size levels of sound seed were uniformly low, in part because of parasitism (Jones, 1964, 1968).

Ants and rodents are important seed predators of perennial grasses but the impact of birds is not known. In a savanna, ants and rodents consumed 64% (ants 30%) of the seeds on offer during five days (Capon & O'Connor, 1990). Similarly in montane grassland, predators (mainly ants) removed 70–98% of the seeds of *T. triandra* (Everson, 1994). Such rates of predation at a time when natural seed was abundant suggests that seed predation can influence the number of seeds available for recruitment, and also that the restricted period of seed release for *T. triandra* may achieve predator satiation. The ability of seeds to become located in cryptic or protected sites (e.g. burial by hygroscopic awn) may be important.

Dispersal

African perennial grasses are either self dispersed or wind and animals function as dispersal agents, but data are limited. In a Botswana savanna, a species with a plume-like spikelet and a light disseminule (*Enneapogon cenchroides*) had a low rate of descent and could be transported up to 13 m at a wind velocity of 10 m s^{-1}, whereas a species with a heavier disseminule and non-plumed awns (*Aristida congesta*) was transported only 2.31 m under the same conditions (Ernst, Veenendaal & Kebakile, 1992). In this savanna, transport by water was important on eroded sites. In montane grassland, most seeds of *T. triandra* (94%) fell within a 0.5-m radius of the parent plant and no seeds fell beyond 1.75 m (Everson, 1994). This suggests that large disturbances more than 2–3 m from a seed source are unlikely to be colonized by *T. triandra* seedlings.

Some species (e.g. *Tragus berteronianus*) are undoubtedly adapted for dispersal on animals, but many other species enjoy incidental epizoochory by virtue of their diaspore structures (e.g. awns of *Themeda triandra*) (Agnew & Flux, 1970; Milton, Siegfried & Dean, 1990) although these structures are best explained for serving other adaptive functions (e.g. Peart, 1979 for the awns of *T. triandra*). Substantial numbers of seed may be transported by epizoochory and they collect at the resting places of animals (Ernst *et al.*, 1992). Dispersal by large mammals would not seem important in montane grassland where the seasonal pattern of growth does not support great animal numbers.

Consumption of seed by mammalian herbivores may assist in dispersal but this facilitation seems to be restricted to awnless species (e.g. *Eragrostis* spp. by zebra; McNaughton, 1983). For example, all but 11 of 1137 seeds of monocotyledons recovered from dung of cattle feeding in a diverse grassland in seed were *Cynodon dactylon*, whose germinability is increased after passage through the ruminant gut (Jones, 1964).

Dormancy

Most studies of dormancy in African perennial grasses have been conducted for species in strongly seasonal environments characterized by hot, wet summers and dry winters, although the temperature profile of winters depends on location.

All except one (*Cynodon dactylon*, Veenendaal & Ernst, 1991) of 12 species tested from seasonal environments have shown innate dormancy, lasting up to a year, at the time of seed shed (Hacker, 1984; Hacker & Ratcliff, 1989; Ernst, Kuiters & Tolsma, 1991; Veenendaal & Ernst, 1991;

O'Connor & Pickett, 1992; Baxter, Van Staden & Granger, 1993). Innate dormancy would seem to be conferred in some species through embryonic effects or by the physical constriction of the palea and lemma (Hacker, 1984; Ernst *et al.*, 1991) or of the glumes (Veenendaal & Ernst, 1991). Induced or enforced dormancy has not, to our knowledge, been demonstrated for any African perennial grasses, although there are indications of secondary dormancy for *Cenchrus ciliaris* (Hacker & Ratcliff, 1989) and *Digitaria milanjiana* (Hacker *et al.*, 1984).

The dormancy characteristics of species or accessions of perennial grasses adapt them to their climate of origin. Innate dormancy of *T. triandra* seeds in montane grasslands protects the seeds from early germination in response to unseasonal warm periods during winter (Everson, 1994). By contrast, innate dormancy for species (including *T. triandra*) in a seasonal, semi-arid environment is considered a drought-avoidance syndrome (Angevine & Chabot, 1979) which ensures that seeds shed during summer do not germinate during winter following out-of-season rainfall at a time when subsequent growing conditions are likely to be inimical for the survival of seedlings (Veenendaal & Ernst, 1991). The influence of climate on innate dormancy is well illustrated by *D. milanjiana*: accessions from low-rainfall, seasonal environments accorded with the drought-avoidance syndrome; accessions from high rainfall, equatorial (aseasonal) environments showed no dormancy; accessions from low rainfall, equatorial environments had intermediate dormancy (Hacker, 1984; Hacker *et al.*, 1984). Similar findings were obtained for *Cenchrus ciliaris* (Hacker & Ratcliff, 1989). There is evidence for a genetic basis to the differences described for both *C. ciliaris* and *D. milanjiana*.

Because rainfall is an inappropriate cue for germination in these seasonal environments, exposure to appropriate levels or amplitudes of ambient temperature seems to be the primary cue for the breakdown of dormancy (Jones, 1964; Wright, 1973; Burger, Grunow & Rabie, 1979; Hacker, 1984; Hacker *et al.*, 1984; Hacker & Ratcliff, 1989; Baxter, Van Staden & Granger, 1993), although there is invariably a breakdown of dormancy with the passage of time (Wright, 1973; Ernst *et al.*, 1991) but it is usually accompanied by a decrease in seed viability (Hacker & Ratcliff, 1989). The match between environment and dormancy characteristics is further illustrated by the optimum temperature for the germination of seeds of *T. triandra* derived from montane grassland being less (25 °C) than for seeds derived from a warmer, low-altitude environment (30–40 °C) (Baxter *et al.*, 1993).

High-rainfall, productive environments are usually subject to fire, often on an annual basis. Plant-derived smoke can overcome the deep dormancy of eight-week-old *Themeda triandra* seeds (Baxter *et al.*, 1994), with the effects of smoke retained for up to 21 days (Baxter & Van Staden, 1994), although smoke from fuel of grass species differs in its effect of promoting germination (Baxter, Granger & Van Staden, 1995). The separate role of smoke and temperature in the enhancement of germination by fire in the field cannot be distinguished (Baxter *et al.*, 1994), and therefore the importance of smoke-induced germination awaits further experimentation. Although fire invariably stimulates germination of *T. triandra* in savannas (up to ten-fold) (Lock & Milburn, 1970; Ndawula-Senyimba, 1972), removal of vegetation and litter is sufficient to stimulate germination, whereas undisturbed vegetation inhibits recruitment (Lock & Milburn, 1970).

Seed longevity

A persistent seed bank (*sensu* Thompson & Grime, 1979) will develop only if seeds enter a state of secondary dormancy or if innate dormancy lasts for an extended period, but there is little evidence of either for African perennial grasses (see section on Dormancy above).

A study of the seed longevity of *T. triandra* using experimental seed banks (seed predation excluded) confirmed the transient nature of its seed bank, albeit delayed because of innate dormancy (O'Connor, 1997). The fate of a seed depended on the microsite it came to occupy. Less than 2% of seed shed onto the surface remained viable until the end of the growing season following the year of seed shed (<18 months) but not thereafter, whereas no buried seed remained viable. The fate of most buried seed was to germinate and die before emergence (83.5%), 12.5% emerged while the remainder decayed. By contrast, most unshaded seed on the surface emerged (61%) although 27.5% died within a few weeks; the remainder decayed. Even in the absence of seed predation, the seed bank was lost within one year of seed release.

The effect of fire on seed mortality has been little investigated. Hygroscopic awns can bury seeds of *T. triandra* at 1 cm depth, at which soil temperatures during fire are not lethal (Lock & Milburn, 1970). If fires penetrate the organic matter on the soil surface, substantial seed mortality may result: 20% of *Stipa trichotoma* seeds succumbed to a single fire in this fashion (Joubert, 1984).

Seed banks

African perennial grasses do not seem to develop persistent seed banks (Belsky, 1986*b*; O'Connor & Pickett, 1992) because of limited seed longevity and the impact of predation. As a result, the representation of grasses in the seed bank of montane grassland (7–38%) contrasts with their dominance in the aboveground vegetation (75–81%) (Everson, 1994). The size of seed banks in semi-arid grasslands is therefore related to the current year's production of seed, which depends on rainfall, the abundance of the species in the vegetation and the extent to which culms are consumed by grazing (e.g. *T. triandra* 78%, *Aristida bipartita* 15%) (O'Connor & Pickett, 1992). Late seral stages of highveld grassland have fewer seeds in the soil than early seral stages because they produce far fewer seeds (Jones, 1964). The short-lived nature of seeds results in most seeds becoming buried within 2 cm of the surface (Joubert, 1984; Veenendaal, 1991; O'Connor & Pickett, 1992), a safe depth for emergence of seedlings (Veenendaal, 1991).

Germination and seedling establishment

The influence of dormancy on germination and seedling emergence can depend quite strongly on rainfall patterns in semi-arid savannas. In Botswana, where a rainfall event of 20 mm is required for germination, several establishment opportunities for *Aristida congesta* occurred when the rains started early in September, but only one wave of germination took place when the rains arrived late in November (Veenendaal, 1991). Similarly, the timing of patch disturbance in a savanna in relation to the commencement of the rainy season determined the density of grass seedlings which established (Von Maltitz, 1990). There were almost no seedlings if the disturbance was a month after the rains. The temporal patterning of emergence of *Themeda triandra* in savanna depends on an interaction between dormancy and the microsite of the seed (O'Connor, 1997). Emergence of surface, shaded seed was spread over the season, whereas emergence of exposed or buried (2 cm depth) seeds occurred in response to one heavy rainfall event. The effect of timing of emergence on the survival of a seedling has not been investigated.

Successful recruitment of a seedling following emergence can be influenced by a host of factors. In semi-arid environments, widespread mortality of seedlings is commonplace if there is limited availability of moisture following germination (Donaldson, 1967; Ndawula-Senyimba, 1972; O'Connor, 1994, 1996). Shade (tree canopies) can therefore enhance seedling survival because of its ameliorating effect on soil moisture

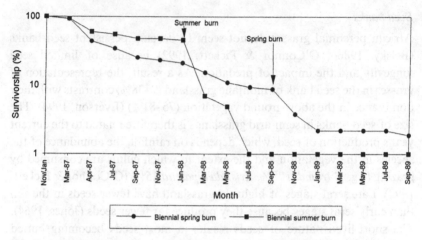

Fig. 13.6. Survivorship of a cohort of *Themeda triandra* seedlings emerging in a
montane grassland in November 1986 under a regime of biennial spring burning
(circles) and biennial summer burning (squares). Data from Everson, (1994).

(Veenendaal, 1991; O'Connor, 1996). Strong water wash during high-inten-
sity storms can uproot seedlings on eroded areas (Veenendaal, 1991).
Heavy grazing following an emergence event can preclude successful establishment (Danckwerts & Stuart-Hill, 1988). The species identity of an
established plant can also influence the number of seedlings which emerge;
in a dystrophic savanna the fewest seedlings emerged adjacent to *Digitaria
eriantha*, while the greatest number emerged adjacent to *Eragrostis rigidior*
(Von Maltitz, 1990).

The host of factors which impact upon seedling recruitment would seem
to preclude prediction of this process. However, some influences have been
found to be more important than others in a semi-arid environment
(O'Connor, 1996). Availability of seeds is paramount because without
seeds there can be no recruitment. If seeds are available, then sufficient
moisture is required for germination and growth. Only once seedlings have
emerged, can herbivory, fire and competition exert an influence.
Competition had little effect in the first season of growth, although it is
undoubtedly important in subsequent seasons. The sequential ordering of
these influences on seedling recruitment constitutes a hierarchy of effects
which facilitates prediction.

In montane grassland, the pattern of survivorship of seedlings of *T.
triandra* was a Deevey Type III which was characterized by a high mortality risk (60%) in the first winter following establishment (Fig. 13.6). Under
a regime of biennial spring burning, only 2% of a cohort survived a four-

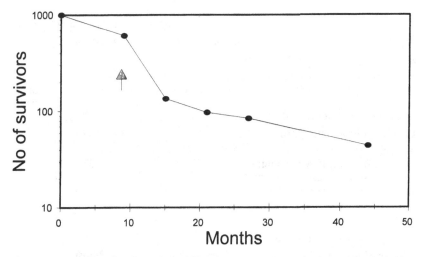

Fig. 13.7. Survivorship of a cohort of seeds and seedlings of *Themeda triandra* in a semi-arid savanna in the eastern Cape, South Africa (data from O'Connor, 1996, 1997). The arrow marks a single wave of seed germination in November 1992 at the start of the wet season. Time 0 marks seed release; seed survival is in the absence of seed predation.

year period; most seedlings succumbed in the first winter. By contrast in a summer-burn treatment, both winter and fire had a catastrophic effect on seedling survival. Low survivorship of seedlings ensures that seedlings contribute little to population dynamics. Seedlings are more vulnerable than tillers to the dry, cold conditions of winter. In contrast, the pattern of survivorship of *T. triandra* seedlings in a semi-arid environment was initially high mortality in the growing season following germination, but relatively constant mortality thereafter (Fig. 13.7).

The survival of a seedling through the dormant season depended positively on its size at the end of the first growing season in both savanna (O'Connor, 1996) and montane grassland (Everson, 1994). Caryopsis size enhances the size attained by a seedling both across species (Veenendaal, 1991; Ernst & Tolsma, 1992) and within a species (Burger, *et al.*, 1979).

Environmental context: savanna versus montane grassland

Although the tiller has been the traditional focus for the study of grass demography, we suggest that the relative importance of ramets and genets depends on the environment of the plant. To illustrate this, the population dynamics of *Themeda triandra* in montane grassland and semi-arid

Table 13.3. *The environmental characteristics of a semi-arid savanna and a montane grassland site, both in summer-rainfall regions, each supporting healthy populations of* Themeda triandra

Environmental variable	Montane grassland	Semi-arid savanna
Mean annual rainfall (mm)	1,356	650
Coefficient of variation (%)	14	29
Temperature (°C):		
January: Average daily maximum	27.9	32.3
Average daily minimum	7.3	19.6
July: Average daily maximum	21.7	25.4
Average daily minimum	1.8	5.6
Frost days per year	25	0
Length of growing season (days)	153	<85[a]
Control of growth	Temperature	Moisture

Note:
[a] Days with sufficient moisture available.

savanna are compared. Semi-arid savanna is drier and hotter than montane grassland, and has less reliable rainfall (Table 13.3).

The relative importance of tillering and seed production for the regeneration of *T. triandra* varies. In montane grassland, which is stable if burnt regularly in the dormant season, reproduction is mainly by vegetative means; seedling recruitment plays a minor role (Everson, 1994). In savanna, there is continual turnover of tufts mostly as a result of drought-induced mortality. Persistence of the population is, therefore, dependent on seedling recruitment, which highlights the role of the seedbank.

Allocation of resources for seed reproduction and vegetative reproduction (via tillers) is presumed to determine the optimal fitness of a plant. In the stable environment of montane grasslands most of the resources of *T. triandra* are allocated to asexual reproduction which perpetuates the successful genotypes in the local population. However, seed reproduction is advantageous if disturbance occurs. Seedlings which established in gaps showed increased vegetative propagation (5.2 secondary tillers per plant) when compared with seedlings establishing in dense grassland, which failed to tiller. After initial germination and growth to approximately 100 mm, one-year-old and three-year-old seedlings in dense grassland could not be distinguished. While the response to changed conditions is slow (three to five years) it appears that the investment in seed production ensures the persistence of the local population under extreme environmental conditions.

Discussion

The ability of a species to persist in the face of disturbance could depend on its vital attributes (Noble & Slatyer, 1980). In savannas, increased grazing pressure and recurrent drought could predispose some species to local extinction (O'Connor, 1991*b*) whilst aiding the population expansion of others because their attributes may result in different responses to grazing (Table 13.4). Empirical findings support this notion: drought and sustained grazing eliminated *T. triandra* (and other palatable perennial grasses) but promoted *Aristida bipartita* (Fig. 13.1). Similarly in montane grassland, *T. triandra* is prone to extinction when subjected to sustained grazing or out of season burning (Everson, 1994). *T. triandra* is replaced mainly by *Eragrostis curvula* and *E. plana* under heavy utilization by livestock; the catastrophic effect of summer burning on tiller survival facilitates the invasion of the unpalatable *Aristida junciformis*.

McNaughton (1983) has postulated that the 'proximate mechanisms regulating species abundances are many weak forces acting probabilistically, so that the cumulative effects are large, but the individual effects are minor, interactive and uncertain', which suggests that the likelihood of developing a predictive capacity for population change is limited.

We propose a conceptual model which infers that prediction of changes within savanna grasslands is an attainable goal (Table 13.5). Rainfall variation is a major determinant of population change, yet this variation is often incorrectly termed stochastic. It is, in fact, strongly patterned, with sequences of years of above- or below-average rainfall (Tyson, 1986). As a consequence, the probability of fire is also patterned over time because of the close relationship between fire and fuel load (Siegfried, 1981; Van der Walt & Le Riche, 1984), irrespective of origin of fire. By contrast, in high-rainfall grassland there is usually sufficient fuel for fire to occur every year (e.g. Berry & MacDonald, 1979). A second consequence is a pattern to the variation in grazing intensity because of a shifting herbivore:forage biomass ratio. For example, grazing intensity is heightened in a drought year following a series of wet years in which animal biomass has built up because animal mortality is usually low in the first year of drought (e.g. Scholes, 1985). Conversely, animal numbers are usually slow to recover in wet years following drought.

There is a resultant patterning of population processes over time (Table 13.5). Mortality of plants is heightened during drought thereby creating gaps for colonization by seedlings or vegetative propagules in the subsequent wet years, during which time most successful establishment takes place. Competition among individuals is expected to assume greater impor-

Table 13.4. *The population response to grazing of a species considered prone to local extinction,* Themeda triandra, *and a species with the converse characteristics,* Aristida bipartita *(both these tufted species reproduce by seed)*

Population process	*Themeda triandra*		*Aristida bipartita*		Source
	Grazing		Grazing		
	Heavy	Light	Heavy	Light	
Diaspore mass (mg) (awnless)	4.05		0.33		Capon & O'Connor, 1990
Viable seed produced m^{-2} of tuft (10^6)	0.037	0.054	0.51	0.59	O'Connor & Pickett, 1992
Seed predated in 5 days (%)	69		47		Capon & O'Connor, 1990
Seed longevity (months)	12		?		O'Connor, 1997
Seed bank (m^{-2}) (5 cm depth)	50	99	1,465	2,672	O'Connor & Picket, 1992
Maximum number of seedlings $m^{-2}yr^{-1}$	2.9	3.6	26	20	O'Connor, 1994
Seedling survival after 1 year (mean 3 years) (%)	60	59	85	71	O'Connor, 1994
Average of an individual tuft consumed by grazing	17		9		O'Connor, 1992
Tuft mortality	see Table 1				
Population turnover	see Table 2				
Population growth rate (mean of 4 seasons) (individuals $individual^{-1}\ yr^{-1}$)	0.002	0.108	0.260	0.123	O'Connor, 1993
Colonizing ability[a]	0.2		22.5		O'Connor, 1991c

Note:
[a] Ratio of percentage contribution to experimental patches after 3 years to percentage contribution to surrounding community.

Table 13.5. *Schematic of an hypothesized 20-year rainfall cycle and the associated hypothesized changes in the population dynamics of a palatable perennial grass.*

The sequence of years illustrated is not considered important but is intended to illustrate the shifting importance of certain population processes depending on prevailing and previous rainfall conditions. A regime of benign herbivory is assumed but it is not assumed to be static

Rain	4 years high	3±average	2 drought	1 average	2 dry but not drought	3±average	2 wet	1 average	2 wet
Population change	Increase to maximum	Gradual decline	Catastrophic collapse in year 2	Slight recovery	Stasis	Slow population growth	Population reaches maximum	Slight decline	Recovers rapidly to maximum
Population processes	Tuft mortality minimal	Partial die-back of tufts	Catastrophic tuft mortality	Mortality absent	Mortality absent	Tuft mortality minimal	Tuft mortality minimal	Minor partial die-back of tufts	Tuft mortality minimal
	Pronounced inter-tuft competition	Pronounced intra-tuft competition	Competition minimal	Competition minimal	Competition minimal	Inter-tuft competition increases as tufts grow	Pronounced inter-tuft competition	Intra- and inter-tuft competition	Pronounced inter-tuft competition
	Seed production high	Seed production reduced	Seed production curtailed	Seed production limited because of few plants	Seed production almost absent	Seed production increases as population increases	Seed production high	Slight reduction in seed production	Seed production high
	Seeds germinate but vegetation eliminates seedlings	Some seedlings may establish if canopy opened up	Seeds germinate but die from lack of moisture	Seed bank depleted, few seedlings recruited	Seed bank depleted, no seedling recruitment	Continued increase in seed bank and in seedling recruitment	Seed bank large and levels of seedling recruitment peak	Seedling recruitment arrested	Seeds germinate but most are eliminated by vegetation

Table 13.5. (cont.)

The sequence of years illustrated is not considered important but is intended to illustrate the shifting importance of certain population processes depending on prevailing and previous rainfall conditions. A regime of benign herbivory is assumed but it is not assumed to be static

	Seed predation marked but not important	Seed predation intensifies as seed production declines	Seed predators die back	Seed predation low	Seed predation low	Seed predators increase and seed predation intensifies	Seed predation peaks at a time when seedling recruitment peaks	Intensive seed predation	Seed predation marked but not important
Ratio of herbivore to forage biomass	Moderate	Continued increase (potentially important effects of herbivory)	Peaks in year 1 of drought, then collapses	Low	Stasis	Ratio relatively constant as both components increase	Initially low because vegetation responds more quickly	Moderate to high	Moderate

tance during a sequence of wet years when established individuals expand in size. The host of factors impacting upon a population process may show a hierarchical ordering as referred to above.

In conclusion, the writing of this paper revealed that the research of African perennial grasses has been descriptive and empirical; there is a dearth of long-term, manipulative field studies or of modelling approaches. Conceptual and mathematical models, experimental tests of key population processes, and long-term data sets are all required to advance current understanding.

References

Agnew, A. D. Q. & Flux, J. E. C. (1970). Plant dispersal by hares (*Lepus capensis* L.) in Kenya. *Ecology*, **51**, 735–7.

Angevine, M. W. & Chabot, B. F. (1979). Seed germination syndromes in higher plants. In *Topics in plant population biology*, ed. O. T. Solbrig, S. Jain, G. B. Johnson & P. H. Raven, pp. 188–206. New York: Columbia University Press.

Banyikwa, F. F. (1988). The growth response of two East African perennial grasses to defoliation, nitrogen fertilizer and competition. *Oikos*, **51**, 25–30.

Baxter, B. J. M., Granger, J. E. & Van Staden, J. (1995). Plant-derived smoke and seed germination: is all smoke good smoke? That is the burning question. *South African Journal of Botany*, **61**, 275–7.

Baxter, B. J. M. & Van Staden, J. (1994). Plant-derived smoke: an effective seed pre-treatment. *Plant Growth Regulation*, **14**, 279–82.

Baxter, B. J. M., Van Staden, J. & Granger, J. E. (1993). Seed germination response to temperature, in two altitudinally separate populations of the perennial grass *Themeda triandra*. *South African Journal of Science*, **89**, 141–4.

Baxter, B. J. M., Van Staden, J., Granger, J. E. & Brown, N. A. C. (1994). Plant-derived smoke and smoke extracts stimulate seed germination of the fire-climax grass *Themeda triandra*. *Environmental and Experimental Botany*, **34**, 217–23.

Belsky, A. J. (1986a). Population and community processes in a mosaic grassland in the Serengeti, Tanzania. *Journal of Ecology*, **74**, 841–56.

Belsky, A. J. (1986b). Revegetation of artificial disturbances in grasslands of the Serengeti National Park, Tanzania I. Colonization of grazed and ungrazed plots. *Journal of Ecology*, **74**, 419–37.

Belsky, A. J. (1986c). Revegetation of artificial disturbances in grasslands of the Serengeti National Park, Tanzania II. Five years of successional change. *Journal of Ecology*, **74**, 937–51.

Berry, A. & MacDonald, I. A. W. (1979). Fire regime characteristics in the Hluhluwe-Corridor-Umfolozi game reserve complex in Zululand. I. Area description and an analysis of causal factors and seasonal incidence of fire in the central complex with particular reference to the period 1955 to 1978. *Report to the National Programme for Environmental Sciences*, CSIR, Pretoria.

Bogdan, A. (1954). Bush-clearing and grazing trial at Kisokon, Kenya. *East African Agricultural Journal*, **19**, 253–9.

Booysen, P. de V., Tainton, N. M. & Scott, J. D. (1963). Shoot apex development in grasses and its importance in grassland management. *Herbage Abstracts*, 33, 209–13.

Breman, H. & Cissé A. M. (1977). Dynamics of Sahelian pastures in relation to drought and grazing. *Oecologia*, 28, 302–15.

Breman, H., Cissé, A. M., Djiteye, M. A. & Elberse, W. Th. (1980). Pasture dynamics and forage availability in the Sahel. *Israel Journal of Botany*, 28, 227–51.

Brockett, G. M. (1983). The effect of defoliation on the persistence of *Elionurus muticus* (Spreng.) Kunth in the highland sourveld of Natal. *Proceedings of the Grassland Society of southern Africa*, 18, 81–3.

Burger, S. J., Grunow, J. O. & Rabie, J. W. (1979). Saad- en ontkiemingsstudies op *Anthephora pubescens* Nees. [Seed and germination studies on *Anthephora pubescens* Nees.] *Agroplantae*, 11, 15–17.

Capon, M. H. & O'Connor, T. G. (1990). The predation of perennial grass seeds in Transvaal savanna grasslands. *South African Journal of Botany*, 56, 11–15.

Carter, A. J. & O'Connor, T. G. (1991). A two-phase mosaic in a savanna grassland. *Journal of Vegetation Science*, 2, 231–6.

Carter, A. J. & Robinson, E. R. (1993). Genetic structure of a population of the clonal grass *Setaria incrassata*. *Biological Journal of the Linnean Society*, 48, 55–62.

Coughenour, M. B. (1984). A mechanistic simulation analysis of water use, leaf angles, and grazing in East African graminoids. *Ecological Modelling*, 26, 203–30.

Coughenour, M. B., McNaughton, S. J. & Wallace, L. L. (1984). Simulation study of East-African perennial graminoid responses to defoliation. *Ecological Modelling*, 26, 177–201.

Coughenour, M. B., McNaughton, S. J. & Wallace, L. L. (1985a). Responses of an African graminoid (*Themeda triandra* Forsk.) to frequent defoliation, nitrogen, and water: a limit of adaptation to herbivory. *Oecologia*, 68, 105–10.

Coughenour, M. B., McNaughton, S. J. & Wallace, L. L. (1985b). Responses of an African tall-grass (*Hyparrhenia filipendula* stapf.) to defoliation and limitations of water and nitrogen. *Oecologia*, 68, 80–86.

Danckwerts, J. E. (1984). *Towards improved livestock production off sweet grassveld*. Ph.D. thesis. Pietermaritzburg: University of Natal.

Danckwerts, J. E., Aucamp, A. J. & Du Toit, P. F. (1984). Ontogeny of *Themeda triandra* tillers in the False thornveld of the eastern Cape. *Journal of the South African Grassland Society*, 1, 9–14.

Danckwerts, J. E., Aucamp, A. J. & Du Toit, P. F. (1986). Ontogeny of *Sporobolus fimbriatus* tillers in the False thornveld of the eastern Cape. *Journal of the South African Grassland Society*, 3, 96–102.

Danckwerts, J. E. & Stuart-Hill, G. C. (1987). Adaptation of a decreaser and an increaser grass species to defoliation in semi-arid grassveld. *Journal of the Grassland Society of southern Africa*, 4, 68–73.

Danckwerts, J. E. & Stuart-Hill, G. C. (1988). The effect of severe drought and management after drought on the mortality and recovery of semi-arid grassveld. *Journal of the Grassland Society of southern Africa*, 5, 218–22.

Donaldson, C. H. (1967). The immediate effects of the 1964/66 drought on the vegetation of specific study areas in the Vryburg district. *Proceedings of the Grassland Society of southern Africa*, 2, 137–41.

Drewes, R. H. & Tainton, N. M. (1981). The effect of different winter and early spring removal treatments on *Themeda triandra* in the Tall Grassveld of Natal. *Proceedings of the Grassland Society of southern Africa*, **16**, 139–43.

Dye, P. J. & Walker, B. H. (1987). Patterns of shoot growth in a semi-arid grassland in Zimbabwe. *Journal of Applied Ecology*, **24**, 633–44.

Dyer, M. I., Detling, J. K., Coleman, D. C. & Hilbert, D. W. (1982). The role of herbivores in grasslands. In *Grasses and Grasslands*, ed. J. R. Estes, R. J. Tyrl & J. N. Brunken, pp. 255–95. Norman: University of Oklahoma Press.

Edroma, E. L. (1981a). Some effects of grazing on the productivity of grassland in Rwenzori National Park, Uganda. *African Journal of Ecology*, **19**, 313–26.

Edroma, E. L. (1981b). The role of grazing in maintaining high species-composition in *Imperata* grassland in Rwenzori National Park, Uganda. *African Journal of Ecology*, **19**, 215–33.

Edroma, E. L. (1985). Effects of clipping on *Themeda triandra* Forsk and *Brachiaria platynota* (K. Schum.) Robyns in Queen Elizabeth National Park, Uganda. *African Journal of Ecology*, **23**, 45–51.

Ellis, J. E. & Swift, D. M. (1988). Stability of African pastoral ecosystems: alternate paradigms and implications for development. *Journal of Range Management*, **41**, 450–9.

Ernst, W. H. O., Kuiters, A. T. & Tolsma, D. J. (1991). Dormancy of annual and perennial grasses from a savanna of southeastern Botswana. *Acta Oecologia*, **12**, 727–39.

Ernst, W. H. O. & Tolsma, D. J. (1992). Growth of annual and perennial grasses in a savanna of Botswana under experimental conditions. *Flora*, **186**, 287–300.

Ernst, W. H. O., Veenendaal, E. M. & Kebakile, M. M. (1992). Possibilities for dispersal in annual and perennial grasses in a savanna in Botswana. *Vegetatio*, **102**, 1–11.

Everson, C. S. & Everson, T. M. (1987). Factors affecting the timing of grassland regrowth after fire in the montane grasslands of Natal. *South African Forestry Journal*, **142**, 47–52.

Everson, C. S., Everson, T. M. & Tainton N. M. (1985). The dynamics of *Themeda triandra* tillers in relation to burning in the Natal Drakensberg. *Journal of the Grassland Society of southern Africa*, **2**, 18–25.

Everson, C. S., Everson, T. M. & Tainton, N. M. (1988). Effects of intensity and height of shading on the tiller initiation of six grass species from the highland sourveld of Natal. *South African Journal of Botany*, **54**, 315–18.

Everson, T. M. (1994). *Seedling establishment of* Themeda triandra *Forssk. in the montane grasslands of Natal*. Ph.D. thesis. Pietermaritzburg: University of Natal.

Fourie, J. H. & Roberts, B. R. (1977). Seasonal dry matter production and digestibility of *Themeda triandra* and *Eragrostis lehmanniana*. *Agroplantae*, **9**, 129–33.

Gammon, D. M. & Roberts, B. R. (1978a). Patterns of defoliation during continuous and rotational grazing of the Matopos sandveld of Rhodesia 2. Severity of defoliation. *Rhodesian Journal of Agricultural Research*, **16**, 133–45.

Gammon, D. M. & Roberts, B. R. (1978b). Patterns of defoliation during continuous and rotational grazing of the Matopos sandveld of Rhodesia 3. Frequency of defoliation. *Rhodesian Journal of Agricultural Research*, **16**, 147–64.

Gammon, D. M. & Roberts, B. R. (1980). Aspects of defoliation during short duration grazing of the Matopos sandveld of Zimbabwe. *Zimbabwe Journal of Agricultural Research*, **18**, 29–38.

Gibbs Russell, G. E., Watson, L., Koekemoer, M., Smook, L., Barker, N. P., Anderson, H. M. & Dallwitz, M. J. (1990). Grasses of southern Africa. *Memoirs of the Botanical Survey of South Africa* No. 58.

Hacker, J. B. (1984). Genetic variation in seed dormancy in *Digitaria milanjiana* in relation to rainfall at the collection site. *Journal of Applied Ecology*, **21**, 947–59.

Hacker, J. B., Andrew, M. H., McIvor, J. G. & Mott, J. J. (1984). Evaluation in contrasting climates of dormancy characteristics of seed of *Digitaria milanjiana*. *Journal of Applied Ecology*, **21**, 961–9.

Hacker, J. B & Ratcliff, D. (1989). Seed dormancy and factors controlling dormancy breakdown in buffel grass accessions from contrasting provenances. *Journal of Applied Ecology*, **26**, 201–12.

Jones, R. M. (1964). *A further study of secondary succession on the highveld*. M.Sc. thesis. Johannesburg: University of the Witwatersrand.

Jones, R. M. (1968). Seed production of species in the highveld secondary succession. *Journal of Ecology*, **56**, 661–6.

Joubert, D. C. (1984). The soil seed bank under nasella tussock infestations at Boschberg. *South African Journal of Plant and Soil*, **1**, 1–3.

Lock, J. M. & Milburn, T. R. (1970). The seed biology of *Themeda triandra* Forsk. in relation to fire. In *The Scientific Management of Animal and Plant Communities for Conservation*, ed. E. Duffey & A. S. Watt, pp. 337–49. Oxford: Blackwell.

McNaughton, S. J. (1978). Serengeti ungulates: feeding selectivity influences the effectiveness of plant defense guilds. *Science*, **199**, 806–7.

McNaughton, S. J. (1983). Serengeti grassland ecology: the role of composite environmental factors and contingency in community organization. *Ecological Monographs*, **53**, 291–320.

McNaughton, S. J. (1984). Grazing lawns: animals in herds, plant form, and coevolution. *American Naturalist*, **124**, 863–86.

McNaughton, S. J. (1985a). Ecology of a grazing ecosystem: the Serengeti. *Ecological Monographs*, **55**, 259–94.

McNaughton, S. J. (1985b). Interactive regulation of grass yield and chemical properties by defoliation, a salivary chemical, and inorganic nutrition. *Oecologia*, **65**, 478–86.

McNaughton, S. J. (1992). Laboratory-simulated grazing: interactive effects of defoliation and canopy closure on Serengeti grasses. *Ecology*, **73**, 170–82.

McNaughton, S. J. & Chapin, F. S. (1985). Effects of phosphorus nutrition and defoliation on C_4 graminoids from the Serengeti plains. *Ecology*, **66**, 1617–29.

McNaughton, S. J., Ruess, R. W. & Seagle, S. W. (1988). Large mammals and process dynamics in African ecosystems. *Bioscience*, **38**, 794–800.

McNaughton, S. J., Wallace, L. L. & Coughenour, M. B. (1983). Plant adaptation in an ecosystem context: effects of defoliation, nitrogen and water on growth of an African C_4 sedge. *Ecology*, **64**, 307–18.

Milton, S. J., Siegfried, W. R. & Dean, W. R. J. (1990). The distribution of epizoochoric plant species: a clue to the prehistoric use of arid Karoo rangelands by large herbivores. *Journal of Biogeography*, **17**, 25–34.

Morris, C. D. & Tainton, N. M. (1993). The effect of defoliation and competition on the regrowth of *Themeda triandra* and *Aristida junciformis* subsp. *junciformis*. *African Journal of Range and Forage Science*, **10**, 124–28.

Mufandaedza, O. T. (1976). Effects of frequency and height of cutting on some

tropical grasses and legumes. *Rhodesian Journal of Agricultural Research*, **14**, 21–38.
Ndawula-Senyimba, M. S. (1972). Some aspects of the ecology of *Themeda triandra*. *East African Agricultural and Forestry Journal*, **38**, 83–93.
Noble, I. R. & Slatyer, R. O. (1980). The use of vital attributes to predict successional changes in plant communities subject to recurrent disturbances. *Vegetatio*, **43**, 5–21.
Noy Meir I. & Walker, B. H. (1986). Stability and resilience in rangelands. In *Rangelands: a Resource under Siege*, ed. P. J. Joss, P. W. Lynch & O. B. Williams, pp. 21–5. Canberra: Australian Academy of Sciences.
O'Connor, T. G. (1985). A synthesis of field experiments concerning the grass layer in the savanna regions of southern Africa. *South African National Scientific Programmes Report* 114. Pretoria: CSIR.
O'Connor, T. G. (1991a). Influence of rainfall and grazing on the compositional change of the herbaceous layer of a sandveld savanna. *Journal of the Grassland Society of southern Africa*, **8**, 103–9.
O'Connor, T. G. (1991b). Local extinction in perennial grasslands: a life-history approach. *American Naturalist*, **137**, 753–73.
O'Connor, T. G. (1991c). Patch colonization in a savanna grassland. *Journal of Vegetation Science*, **2**, 245–54.
O'Connor, T. G. (1992). Patterns of plant selection by grazing cattle in two savanna grasslands: a plant's eye view. *Journal of the Grassland Society of southern Africa*, **9**, 97–104.
O'Connor, T. G. (1993). The influence of rainfall and grazing on the demography of some African savanna grasses: a matrix modelling approach. *Journal of Applied Ecology*, **30**, 119–32.
O'Connor, T. G. (1994). Composition and population responses of an African savanna grassland to rainfall and grazing. *Journal of Applied Ecology*, **31**, 155–71.
O'Connor, T. G. (1995). Transformation of a savanna grassland by drought and grazing. *African Journal of Range and Forage Science*, **12**, 53–60.
O'Connor, T. G. (1996). Hierarchical control over seedling recruitment of the bunch-grass *Themeda triandra* in a semi-arid savanna. *Journal of Applied Ecology*, **33**, 1094–1106.
O'Connor, T. G. (1997). Micro-site influence on seed longevity and seedling emergence of a bunchgrass (*Themeda triandra*) in a semi-arid savanna. *African Journal of Range and Forage Science*, **14**, 7–10.
O'Connor, T. G. & Bredenkamp, G. (1997). Grassland. In *The Vegetation of Southern Africa*, ed. R. M. Cowling, D. M. Richardson & S. Pierce, pp. 215–57. Cambridge: Cambridge University Press.
O'Connor, T. G. & Pickett, G. A. (1992). The influence of grazing on seed production and seed banks of some African savanna grasslands. *Journal of Applied Ecology*, **29**, 247–60.
O'Connor, T. G. & Roux, P. W. (1995). Vegetation changes (1949–71) in a semi-arid, grassy dwarf shrubland in the Karoo, South Africa: influence of rainfall variability and grazing by sheep. *Journal of Applied Ecology*, **32**, 612–26.
Oesterheld, M. & McNaughton, S. J. (1988). Intraspecific variation in the response of *Themeda triandra* to defoliation: the effect of time of recovery and growth rates on compensatory growth. *Oecologia*, **77**, 181–6.
Peart, M. H. (1979). Experiments on the biological significance of the

morphology of seed-dispersal units in grasses. *Journal of Ecology*, **67**, 843–63.

Rethman, N. F. G. (1971). Elevation of shoot-apices of two ecotypes of *Themeda triandra* on the Transvaal highveld. *Proceedings of the Grassland Society of southern Africa*, **6**, 86–92.

Rethman, N. F. G. & Booysen, P. de V. (1967). Growth and development in *Cymbopogon excavatus* tillers. *South African Journal of Agricultural Science*, **10**, 811–22.

Rethman, N. F. G. & Booysen, P. de V. (1968). Growth and development in *Heteropogon contortus* tillers. *South African Journal of Agricultural Science*, **11**, 259–72.

Roux, P. W. (1966). Die uitwerking van seisoenreënval en beweiding op gemengde karooveld. [The influence of seasonal rainfall and grazing on mixed karooveld.] *Proceedings of the Grassland Society of southern Africa*, **1**, 103–10.

Ruess, R. W. & McNaughton, S. J. (1984). Urea as a promotive coupler of plant-herbivore interactions. *Oecologia*, **63**, 331–37.

Scholes, R. J. (1985). Drought related grass, tree and herbivore mortality in a southern African savanna. In *Ecology and Management of the World's Savannas*, ed. J. C. Tothill & J. J. Mott, pp. 350–3. Canberra: Australian Academy of Science.

Shackleton, S. E. (1989). *Autecology of* Cymbopogon validus *(Stapf) Stapf ex Burtt Davy in Mkambati Game Reserve, Transkei.* M.Sc. thesis. Johannesburg: University of the Witwatersrand.

Shackleton, S. E. & Mentis, M. T. (1991). Response of *Cymbopogon validus* tillers to three clipping frequencies. *Journal of the Grassland Society of southern Africa*, **8**, 35–6.

Siegfried, W. R. (1981). The incidence of veld-fire in the Etosha National Park, 1970–1979. *Madoqua*, **12**, 225–30.

Snyman, H. A. (1989). Evapotranspirasie en waterverbruiksdoeltreffendheid van verskillende grasspesies in die sentrale Oranje-Vrystaat. [Evapotranspiration and water use efficiency of different grass species in the central Orange Free State.] *Journal of the Grassland Society of southern Africa*, **6**, 143–51.

Snyman, H. A. & Van Rensburg, W. L. J. (1990). Korttermyn invloed van strawwe droogte op veldtoestand en waterverbruiksdoeltreffendheid van grasveld in die sentrale Oranje-Vrystaat. [Short-term effect of severe drought on veld condition and water use efficiency of grassveld in the central Orange Free State.] *Journal of the Grassland Society of southern Africa*, **7**, 249–56.

Stuart-Hill, G. C. & Mentis, M. T. (1982). Coevolution of African grasses and large herbivores. *Proceedings of the Grassland Society of southern Africa*, **17**, 122–8.

Tainton, N. M. (1964). Developmental morphology of the apical meristem of *Themeda triandra* Forsk. *South African Journal of Agricultural Science*, **7**, 93–100.

Tainton, N. M. & Booysen, P. de V. (1963). The effects of management on apical bud development and seeding in *Themeda triandra* and *Tristachya hispida*. *South African Journal of Agricultural Science*, **6**, 21–30.

Tainton, N. M. & Booysen, P. de V. (1965a). Growth and development in perennial veld grasses. I. *Themeda triandra* tillers under various systems of defoliation. *South African Journal of Agricultural Science*, **8**, 93–110.

Tainton, N. M. & Booysen, P. de V. (1965b). Growth and development in perennial veld grasses. II. *Hyparrhenia hirta* tillers under various systems of defoliation. *South African Journal of Agricultural Science,* **8**, 745–60.

Thompson, K. & Grime, J. P. (1979). Seasonal variation in the seed banks of herbaceous species in ten contrasting habitats. *Journal of Ecology,* **67**, 893–921.

Tolsma, D. J. (1989). *On the ecology of savanna ecosystems in south-eastern Botswana.* Ph.D. thesis. Amsterdam: Free University of Amsterdam.

Tyson, P. D. (1986). *Climatic Change and Variability in Southern Africa.* Cape Town: Oxford University Press.

Van der Walt, P. T. & Le Riche, E. A. N. (1984). Die invloed van veldbrand op 'n *Acacia erioloba* gemeenskap in die KGNP. [The influence of a veld fire on an *Acacia erioloba* community in the KGNP.] *Koedoe supplement 1984,* 103–6.

Veenendaal, E. M. (1991). *Adaptive strategies of grasses in a semi-arid savanna in Botswana.* Ph.D. thesis. Amsterdam: Free University of Amsterdam.

Veenendaal, E. M. & Ernst, W. H. O. (1991). Dormancy patterns in accessions of caryopses from savanna grass species in south eastern Botswana. *Acta Botanica Neerlandica,* **40**, 297–309.

Von Maltitz, G. P. (1990). *The effect of spatial scale of disturbance on patch dynamics.* M.Sc. thesis. Johannesburg: University of the Witwatersrand.

Wallace, L. L. (1981). Growth, morphology and gas exchange of mycorrhizal and nonmycorrhizal *Panicum coloratum* L., a C_4 grass species, under different clipping and fertilization regimes. *Oecologia,* **49**, 272–8.

Wright, L. N. (1973). Seed dormancy, germination environment, and seed structure of Lehman lovegrass, *Eragrostis lehmanniana* Nees. *Crop Science,* **13**, 432–5.

14

A life cycle approach to the population ecology of two tropical grasses in Queensland, Australia

D. M. ORR

Approach

In this chapter, the population ecology of two important Australian perennial grasses, *Astrebla* spp. (Mitchell grass) and *Heteropogon contortus* (black speargrass), is compared using a life cycle approach. In particular, this chapter examines the transitions between flowering, seed production, seedling recruitment and mature plant survival in relation to grazing pressure and variable summer rainfall. Information is synthesized from both published literature and from current studies. The major objective of these studies is to provide the ecological basis for the management of sustainable grazing industries in these vegetation types.

Geography

Astrebla spp. grasslands and *H. contortus* pastures occupy 29.5 and 25 million ha respectively (Weston *et al.*, 1981) and are important vegetation resources for the extensive beef cattle and sheep grazing industries of Queensland.

Most *Astrebla* spp. grasslands and *H. contortus* pastures (Fig. 14.1) lie between the 250 and 550 mm and 650 and 1000 mm annual rainfall isohyets, respectively, where summer rainfall is dominant (Orr & Holmes, 1984; Isbell, 1969). *Astrebla* spp. are restricted to uniform, alkaline cracking clay soils with high soil moisture storage when soils are wet and extensive cracking when dry. In contrast, *H. contortus* pastures occur on a wide diversity of soil types, although mainly on soils of lighter texture (Isbell, 1969).

Northern *Astrebla* grasslands are open tussock grassland; trees are sparse except for scattered *Acacia farnesiana* (mimosa) and *Eucalyptus microtheca* (coolibah) in water courses. However, in recent years, the intro-

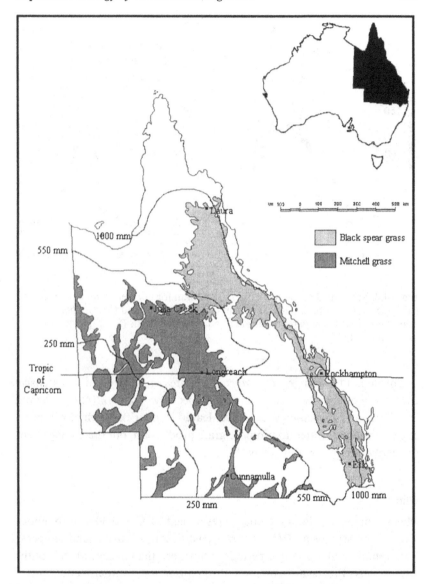

Fig. 14.1. Distribution of major areas of *Astrebla* grasslands and *H. contortus* pastures in Queensland in relation to annual rainfall isohyets.

duced *Acacia nilotica* (prickly acacia) has invaded much of these northern grasslands (Burrows *et al.*, 1990). In central and southern regions, these grasslands are open wooded tussock grasslands with scattered trees and shrubs at densities of up to 200 trees per hectare.

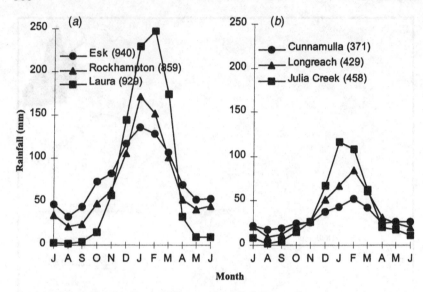

Fig. 14.2. Monthly distribution of annual rainfall for northern (squares), central (triangles) and southern (circles) sites in (a) *H. contortus* and (b) *Astrebla* grasslands. Figures in parenthesis are annual rainfall totals (See Fig. 14.1 for location of sites).

Throughout its range, *H. contortus* pastures originally had an extensive overstorey of mainly *Eucalyptus* spp., especially *E. crebra* (narrow leaved iron bark) and *E. melanophloia* (silver leaved ironbark). Much of this overstorey has been cleared to promote grass production, but the management of regrowth remains a major problem.

Climate

Mean summer maximum temperatures exceed 35° C and 30° C throughout *Astrebla* grasslands and *H. contortus* pastures, respectively, and temperatures remain high for longer periods in northern than in central and southern regions. Frosts are common throughout the southern and central regions during June and July.

High summer temperatures and evaporation rates coincide with the period of most reliable rainfall; thus plant growth is strongly regulated by the incidence of rainfall in both vegetation types. The monthly distribution of the annual rainfall total depends on geographical location, with rainfall in both northern regions generally limited to December–March (Fig. 14.2). These regions are characterized by prolonged dry winters. Further south, the annual rainfall total becomes increasingly distributed throughout the year. Winter rainfall in the southern regions can result in available moisture, but plant growth is usually limited by low temperature.

The overriding characteristic of annual rainfall throughout both vegetation types, indeed throughout much of northern Australia, is its high between-years variability: coefficients of variation are as high as 40%. Much of this variability is explained by variation in the Southern Oscillation Index (SOI), which is calculated as the normalized pressure difference between Tahiti and Darwin (Coughlan, 1988). Correlations of spring SOI values with summer rainfall are greatest for the northern *H. contortus* region and least for the southern *Astrebla* grassland region (Partridge, 1994).

Extremes in the Southern Oscillation Index and associated rainfall occurrences have been linked with population phenomena such as the death of perennial grass tussocks due to extreme drought and change from grass to legume dominance due to above-average rainfall (Taylor & Tulloch, 1985). The potential role of the Southern Oscillation Index in grazing management has been discussed by McKeon *et al.* (1990).

Pastures

Astrebla grasslands are endemic to Australia and are dominated by one or more of the four closely related species *A. lappacea* (curly Mitchell), *A. elymoides* (hoop Mitchell), *A. pectinata* (barley Mitchell) and *A. squarrosa* (bull Mitchell), whose localized dominance depends largely on soil microtopography and rainfall. *Astrebla* spp. occupy a basal area up to 4%, depending on recent summer rainfall and grazing pressure. Spaces between these perennial tussocks are occupied by a range of annual and short-lived perennial grasses and forbs (dicotyledons), whose presence depends on geographical location and current seasonal rainfall. In northern grasslands, summer annual grasses, e.g. *Iseilema* spp. (Flinders grass), can dominate, whereas further south summer and/or winter forbs can dominate. However, large variation in rainfall between years can result in major variation in pasture composition (Orr, 1986).

H. contortus is widespread throughout the sub-humid tropics and subtropics of the world. In Australia prior to European settlement, *H. contortus* was a minor component in pastures dominated by *Themeda triandra* (kangaroo grass), which was replaced by *H. contortus* following heavy grazing with sheep together with altered burning regimes (Shaw, 1957). Currently, *H. contortus* is associated with a range of perennial grasses, e.g.

Aristida spp. (wiregrasses), *Bothriochloa* spp. (bluegrasses) and *Chloris* spp., whose local distribution depends on soil, climate and grazing history.

In recent years, both vegetation types have undergone intensified use and some deleterious changes are apparent (Tothill & Gillies, 1992). For example, the proportion of *H. contortus* in these pastures has been reduced. This reduction is associated with increased grazing pressure accentuated by drought, the widespread use of cattle diet supplements and the introduction of exotic legumes and of *Bos indicus* (Brahman) cattle. In the northern region, actual cattle numbers carried during drought in the 1980s exceeded the 'safe carrying capacity' as simulated using historical rainfall data and a moderate utilization level (Pressland & McKeon, 1989). Under these conditions, the dominant *H. contortus* may be replaced by the introduced *Bothriochloa pertusa* (Indian couch grass). In the southern region, overgrazed *H. contortus* is replaced by the unpalatable native perennial *Aristida* spp.

Plant morphology

Astrebla spp. are strongly perennial C_4 tussock grasses with a short rhizome from which erect main tillers arise and grow to 75–80 cm in height, at maturity. Secondary tillers arise from the base of these main tillers and from lower leaf nodes along the main tillers. The inflorescence is a spike-like raceme and spikelets are short-pedicelled and arranged along the raceme (Tothill & Hacker, 1973). Basal diameter of individual tussocks varies to about 20–25 cm, but can reach 35 cm (D. M. Orr, unpublished). *Astrebla* spp. possess a dual root system whereby rhizomes produce roots that branch near the surface to collect moisture from light falls of rain and branch again at depth (Everist, 1964). *Astrebla* spp. achieve perenniality through rhizomes that act as a source of new main tillers, through main tillers which flower in different years and through meristematic sites on main tillers.

Heteropogon contortus is a perennial C_4 tussock grass with erect tillers to 1–1.5 m and with considerable morphological variation associated with its widespread distribution. The inflorescence is a simple raceme of spikelet pairs arranged in two rows and the spikelet possesses a hard, pointed base with the fertile lemma bearing a stout hygroscopic awn (Tothill & Hacker, 1973). Basal diameter of individual tussocks in the southern region varies up to about 10–12 cm, but some individuals can reach 25 cm (D. M. Orr, unpublished).

Seedlings of both species produce tillers in a generally circular pattern because new tillers arise mainly from the end of rhizomes that radiate from the centre of tussocks. As the tussock ages, the centre contains a high proportion of old tillers so that tussocks eventually tend to break up into separate segments (Scanlan, 1983).

Phenology and cytology

Flowering in *Astrebla* spp. is insensitive to photoperiod, and inflorescences can be produced throughout the year following rainfall effective for plant growth. Inflorescences emerge rapidly on newly formed secondary tillers, but emergence is not as rapid from newly formed main tillers. In the highly variable rainfall environment of *Astrebla* grassland, any photoperiod restriction on flowering would represent a potential limitation to seed production (Jozwik, 1970).

In contrast, *H. contortus* plants are sensitive to a short day photoperiod. North of 20° S, plants display an obligate short day requirement, with flowering restricted to late summer, which is near the end of the period of most reliable rainfall. South of 20° S, plants display either a quantitative or obligate short day requirement (Tothill, 1966; Tothill & Knox, 1968). This facultative response in subtropical regions permits more flexibility in flowering times, which is consistent with rainfall being more widely distributed throughout the summer (Tothill, 1966).

Astrebla spp. are autogamous and, although F1 hybrids occur they are sterile. Chromosome number in both *A. lappacea* and *A. pectinata* is $2n = 40$ (Jozwik, 1969). *H. contortus* is an obligate apomict. Plant morphology reflects similar variation in chromosome number which, for Australian strains, range from $2n = 40-90$. Populations of *H. contortus* from tropical regions are tetraploid while populations from subtropical regions exhibit a wide range of ploidy levels (Tothill & Hacker, 1976).

Seed production

In both species, large variation in seed production occurs in response to variation in grazing pressure and rainfall. Seed production of *Astrebla* spp. varies from 0 to 750 seeds m^{-2}, with inflorescence density (5–20 inflorescences m^{-2}) the major factor determining seed production (Orr & Evenson, 1991a; Orr, 1991). In central grasslands, increasing grazing on *Astrebla* spp. plants stimulates the production of secondary tillers which, in turn, produce inflorescences and result in a stimulation of seed production (Orr & Evenson, 1991b). No evidence of this effect was apparent in either northern or southern grasslands, and it has been attributed to a more even distribution of rainfall in central than in northern or southern grasslands promoting the production and growth of secondary tillers (Orr, 1991).

Large variation in both the number of spikelets per inflorescence (10–20) (Orr & Evenson, 1991a; Orr, 1991) and seeds per spikelet (0–6) (Myers, 1942) are characteristic of seed production throughout *Astrebla* grasslands. Some of the variation in seeds per spikelet may be due to ambient temperatures at the time of seed set (Orr, 1991), since flowering is not restricted by photoperiod.

Seed production in both *A. lappacea* and *A. pectinata* in northern grasslands is promoted by spring fire and above-average summer rainfall, resulting in a tenfold increase in inflorescence density (Scanlan, 1980). This result suggests that fire may stimulate tiller development in the same way that grazing stimulates tillering in the central grasslands. Spring burning does not promote seed production in *H. contortus* (Campbell, 1995).

Seed production of *H. contortus* in the southern region in autumn varied from 50 to 5000 seeds m^{-2} at stocking rates of 0.3 to 0.9 animals ha^{-1} over a range of summer rainfall conditions (D. M. Orr, unpublished). Similarly, seed production of up to 6300 seeds m^{-2} has been recorded in the northern region (Howden 1988).

Inflorescence density (0–600 inflorescences m^{-2}) again is the major factor determining seed production; the number of seeds per inflorescence (10–12) varies little between seasons (D. M. Orr, unpublished). In all regions, inflorescence density and consequently seed production is reduced with increased stocking rates (McIvor *et al.*, 1996; D. M. Orr, unpublished). Rainfall conditions during seed set influences seed viability. After 12 months of laboratory storage, viability of seed harvested after drought during seed set was 25% compared with 60% for seed harvested after above-average rainfall conditions during seed set (D. M. Orr, unpublished).

Seed dispersal

Seed of both *Astrebla* spp. and *H. contortus* is dispersed rapidly following seed set. For *Astrebla* spp., seed fall can be complete within 7 weeks of the rainfall that produced the flowering and seed production (Roe, 1941). For *H. contortus* in the southern region, seed fall commences in late February, with more than 80% of the total seed production falling within two months (Campbell, 1995).

Seed of *Astrebla* spp. remains within the spikelet, which usually remains on the soil surface unless incorporated into the soil by the trampling action of grazing animals. Seed that falls down larger soil cracks is probably lost, since it cannot emerge from soil depths greater than 2.5 cm (Watt & Whalley, 1982*a*).

The hygroscopic awn of *H. contortus* functions to move the seed both horizontally and vertically. This increases contact between the seed and the soil surface, which results in improved germination. Backwardly directed bristles on the base of the sharply pointed callus anchor the seed and counter the force of the germinating radicle as it pushes into the soil (Peart, 1979). Because of this seed movement, 90% of the seed is buried below the soil surface by spring (Campbell, 1995).

Soil seed banks

Although few data are available, large variation in the size of the germinable soil seed banks measured in spring seems characteristic of both *Astrebla* spp. and *H. contortus* pastures. Both species form transient seed banks (*sensu* Grime, 1979). In central *Astrebla* spp. grasslands, a maximum of 800 germinable seeds m^{-2} was recorded under moderate grazing during average summer rainfall, although levels of 50–400 seeds m^{-2} may be more common (Orr & Evenson, 1991*a*). In both northern and southern grasslands, maximum levels of 250–300 seeds m^{-2} have been recorded under a range of grazing and summer rainfall conditions (Orr, 1991; Orr, 1992).

For *H. contortus*, a maximum level of 1000 germinable seeds m^{-2} has been recorded under light grazing following average summer rainfall in the southern pastures, but levels of 100–500 seeds m^{-2} are more common (D. M. Orr, unpublished). In the central region, levels of 0–100 seeds m^{-2} have been recorded at a range of stocking rates over three years of generally below-average rainfall (Orr & Paton, 1993) and levels of 0–60 seeds m^{-2} have been recorded in the northern region (McIvor, 1987; Howden, 1988).

Comparing *H. contortus* seed production in autumn with germinable seed banks in spring at the same site between 1991 and 1995 indicates that at least 40% of the seed produced in autumn fails to appear in the germinable seed bank even after accounting for variation in seed viability (D. M. Orr, unpublished). Reasons for this loss of seed are not readily apparent. Studies of seed removal by ants during winter (Howden, 1988) indicate that ants do not selectively remove seed of *H. contortus*. Studies of the diet selected by grazing animals (Gardener, 1980; R. M. Jones, unpublished) also indicate that animals do not remove seed of *H. contortus*.

Seed dormancy

In the laboratory, fresh seed of *Astrebla* spp. and *H. contortus* are dormant at seed set and retain this dormancy for 6–8 months (Myers, 1942; Tothill, 1977; Silcock, Williams & Smith, 1990). Dormancy in *H. contortus* is controlled by both endogenous and exogenous mechanisms (Tothill, 1977; Howden, 1988). This dormancy is an adaptive characteristic since freshly ripened seed germinating during late summer rain would desiccate during the dry season (Tothill, 1977).

In the field, dormancy is overcome by exposure to increasing soil temperatures, particularly at the end of the dry season, and thus seeds are ready to germinate at the start of the wet season (Mott, 1978). Dormancy in *H. contortus* is also overcome by spring burning since both heat and smoke cause seed to become non-dormant (Campbell, 1995). Under normal summer rainfall, most seeds of both *Astrebla* spp. and *H. contortus* germinate in the wet season following seed production (Tothill, 1977; Orr, 1991). However, extreme summer drought in *Astrebla* grassland can prevent germination until the subsequent summer, and it may result in exceptional recruitment events.

In the laboratory, seeds of neither *Astrebla* spp. nor *H. contortus* retain long-term viability, and germination declines substantially after 4 years in *H. contortus* (Tothill, 1977) and after 5 and 8 years in *A. pectinata* and *A. lappacea,* respectively (Silcock *et al.*, 1990). In a glasshouse experiment to determine seed longevity in soil, soil cores taken from the field have been watered for eight weeks each summer for four successive summers. Results indicate that, of the total seeds recovered, 90% of both *Astrebla* spp. and *H. contortus* seeds germinated in the summer following seed production (D. M. Orr, unpublished). Therefore, in the field, very little seed remains in the soil despite some longevity demonstrated in the laboratory.

Seed germination

Germination of *Astrebla* spp. and *H. contortus* seed occurs over a wide range of temperatures and is independent of light. *Astrebla* spp. germinate between 20 °C and 40 °C (Jozwik, Nicholls & Perry, 1970; Watt & Whalley, 1982*b*; Orr, 1986), with the most rapid germination occurring at 35°C for *A. lappacea* from the southern extremity of *Astrebla* grassland (Watt and Whalley, 1982*b*) and at 40 °C for the same species from the southern grassland of Queensland (Orr, 1986). These differences suggest the possibility of geographic variation in temperature responses. Germination of *H. contortus* occurs between 25 °C and 35 °C, with the most rapid germination occurring at 30 °C and 35 °C (Tothill, 1977).

Seeds of *A. lappacea* can achieve 40% germination at constant soil water potentials as low as −7.5 bars (Watt & Whalley, 1982*b*). Seeds of *Astrebla* spp. possess the characteristic of 'hydropedesis', which is a partially germinated state in which viability is maintained over extended periods of desiccation (Watt, 1982).

Seedling establishment

In the laboratory, maximum seedling emergence of *A. lappacea* occurs after three consecutive wet days, but seed wet for only one day displays 'hydropedesis' and germinates when soil moisture again becomes available even after four dry days (Lambert *et al.*, 1990). Seedlings can survive on primary roots for at least three weeks before further moisture is necessary for the development of secondary roots, which are formed at the soil surface (Lambert *et al.*, 1990).

Secondary root growth in cracking clay soils can be restricted by soil surface drying although this problem in some Australian native grasses is overcome by having a well-developed primary root system. Furthermore, such species also possess a lower shoot:root ratio and roots are better able to locate moisture during periods of moisture stress (Watt, 1981).

In the field, large-scale recruitment of *Astrebla* spp. (<10 seedlings m^{-2}) in southern grasslands occurs infrequently (about once in 40 years) (Williams & Roe, 1975; Roe, 1987; Orr, 1991). Williams & Roe (1975) suggested that above-average rainfall spread over several days in November or December may be required for recruitment. In central *Astrebla* grasslands, recruitment was recorded in most years during a ten year study. However, only one recruitment event exceeded five seedlings m^{-2}, and this occurred only under grazing and not in exclosure (Orr & Evenson, 1991*a*). In northern grasslands, the summer rainfall pattern was believed to be more favourable for recruitment than that in central and southern grasslands. However, the lack of regular recruitment in these northern grasslands was attributed to competition from the annual grass *Iseilema* spp. (Williams, 1978).

The seedling recruitment recorded in central grasslands (Orr & Evenson, 1991*a*) shows it depends on an adequate germinable soil seed bank being present at the commencement of the rainfall which results in that recruitment. Under controlled conditions, *A. lappacea* is more competitive than *Iseilema* spp. However, in the field the relatively high densities of *Iseilema* spp. that build up during a series of 'wet' summers preclude establishment of *Astrebla* spp. (Orr, 1986). In this context, an exceptional recruitment event of *Astrebla* spp. recorded in northern grassland (see below) occurred

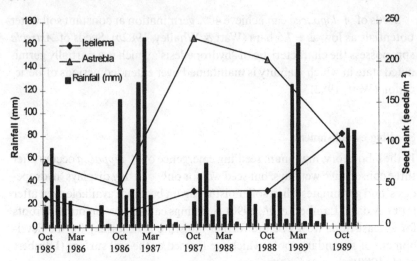

Fig. 14.3. Changes in the germinable seed banks (seeds m^{-2}) of *Astrebla* spp. and
Iseilema spp. in relation to monthly rainfall (mm) between October 1985 and
October 1989 in northern *Astrebla* grassland ('Toorak', Julia Creek) (Oct,
October; Mar, March).

when the soil seed bank of *Iseilema* spp. was 50 seeds m^{-2} compared with a
maximum of 3500 seeds m^{-2} recorded at that site (D. M. Orr, unpublished).

Exceptional recruitment events of *Astrebla* spp. of 15 and 30 seedlings
m^{-2} have been recorded in southern and northern grasslands respectively
(Orr, 1991; Orr, 1992) following severe drought in the previous summer. In
both events (and the one other event reported (Roe, 1941)), the pattern of
seed production, drought and rainfall was similar. For the 1989 event in
northern grasslands, above-average rainfall over the 1985–86 summer produced seed and severe drought over the 1987–88 summer prevented this
seed from germinating. Above-average rainfall over the 1988–89 summer
(Fig. 14.3) was associated with this exceptional recruitment event (see Fig.
14.4).

Seedling establishment of up to 300 seedlings m^{-2} for *H. contortus* has
been recorded soon after seedling emergence following regular spring
burning (Shaw, 1957). Because burning in spring causes seed to become
non-dormant (Campbell, 1995), seedlings in spring-burnt pastures germinate earlier than seedlings in unburnt pasture and are better able to establish over the summer provided rainfall is adequate. Seedling densities are
lower in unburnt pastures (Shaw, 1957; Campbell, 1995).

Most *H. contortus* seedlings emerge with the first rainfall event of
summer; emergence decreases as the summer progresses (McIvor &

Gardener, 1991; Orr & Paton 1997). However, seedling survival can be adversely affected by dry periods during summer (Orr & Paton 1997) particularly in the southern region, where the summer rainfall season is less well defined than in the northern region.

Thus, the recruitment of *Astrebla* spp. seedlings is highly episodic whereas recruitment of *H. contortus* seedlings occurs annually (Fig. 14.4). Variability in *H. contortus* recruitment between years is affected by both seasonal rainfall and stocking rates.

Seedling survival

Few quantitative data exist on seedling survival beyond the establishment summer. For *A. lappacea* in the central region, plant survival after five years varied from 2% to 40%, and much of this variation was attributed to differences in the pattern of rainfall in the establishment year (Orr & Evenson, 1991*a*).

A comparison of seedling survival under grazing in northern *Astrebla* grassland indicates that survival of seedlings from the recruitment event in 1989 (see Fig. 14.4) decreases with increasing grazing pressure (Fig. 14.5). In contrast, for *H. contortus* pastures in the central region, seedling survival was around 20% after three years, with little apparent difference between years or between stocking rates within years. Reasons are not apparent for this difference in the effect of stocking rate on seedling survival.

Growth and development

In the laboratory, temperature has a major influence on tiller production in *Astrebla* spp. with optimum tiller numbers produced at day/night temperatures of 28/23 °C. Differences in tillering potential between the four species are related to minor differences in soil microhabitat. *A. lappacea* and *A. pectinata* grow on drier slopes and produce more tillers than *A. elymoides* and *A. squarrosa,* which grow in more mesic depressions (Jozwik, 1970).

In the field the production of new tillers in *Astrebla* spp. depends on the amount of rainfall. For *A. lappacea,* rainfall events of around 40 mm result in the growth only of secondary tillers, which arise from the lower nodes on existing main tillers. Rainfall events in excess of 75 mm result in the growth of both new main tillers from the rhizome and new secondary tillers from the lower nodes on existing main tillers (Everist, 1964). Similar growth responses occur in *A. pectinata* (Jozwik *et al.,* 1970). Main tillers survive for up to three years (Scanlan, 1983).

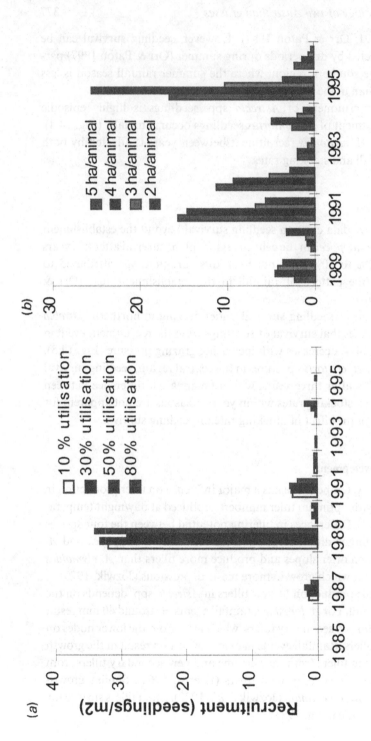

Fig. 14.4. Seedling recruitment (seedlings m^{-2}) of (a) *Astrebla* spp. between 1985 and 1996 under four levels of utilization of end of summer forage in northern *Astrebla* grassland ('Toorak', Julia Creek) and (b) *H. contortus* between 1989 and 1996 under four stocking rates in central *H. contortus* pasture ('Galloway Plains', Calliope).

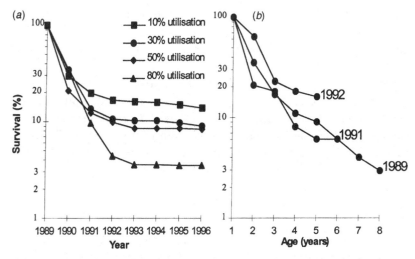

Fig. 14.5. Seedling survival (%) of (*a*) *Astrebla* spp. seedlings between 1989 and 1996 under four levels of utilization of end of summer forage in northern *Astrebla* grassland ('Toorak', Julia Creek) and (*b*) three annual cohorts of *H. contortus* in central *H. contortus* pasture ('Galloway Plains', Calliope).

Secondary tillers are the major source of dry matter in years of below average rainfall (Everist, 1964; Jozwik *et al.*, 1970). Because of this, Jozwik *et al.* (1970) advised against grazing *A. pectinata* below 15–20 cm since this would destroy the sites from which secondary tillers arise. This residual plant height corresponds to removal of only 30% of plant material (Orr, 1980). In contrast, Scanlan (1983), working during a series of summers with above-average rainfall, reported an inverse relationship between new main tiller production and the presence of old main tillers of *A. lappacea*. Consequently, the complete removal of old tillers by mowing or by burning (Scanlan, 1980) or by short-term heavy grazing (Hall & Lee, 1980) result in a greater production of new main tillers when rainfall is above average.

Much less is known about the tillering process in *H. contortus*. Nevertheless, the major period for new tiller production is at the start of the summer growing season, when 60% of the tiller buds present at the base of old tillers give rise to new tillers irrespective of moisture conditions experienced during the previous winter season (Mott *et al.*, 1992). Under the watered and fertilized conditions of this experiment, plants of *H. contortus* clipped to a height of 5 cm at intervals of seven days re-established a photosynthetic surface by the development of new tillers. This experiment (Mott *et al.*, 1992) also indicated the mechanism underlying the historical replacement of the pre-European dominant *Themeda triandra* by *H.*

contortus. Tiller production in *T. triandra* is synchronous with virtually all new tillers produced at the start of the summer growing season. Therefore, defoliation at this time disrupts the supply of new tillers, leading to plant death.

Under low nutrient conditions in the field, Mott *et al.* (1992) reported a similar burst of tillering in *H. contortus,* but with fewer tillers produced per plant. In the southern region, the basal area of *H. contortus* was reduced after spring burning for four years when this burning was followed by grazing at normal stocking rates. In contrast, no reduction in basal area occurred when spring burning for four years was followed either by grazing at a reduced stocking rate or by deferring grazing for six months. Total perennial grass basal area remained similar under these treatments (Orr & Paton, 1997).

Astrebla spp. and *H. contortus* are both physiologically well adapted to the environments in which they occur. *A. lappacea* exhibits a high degree of control over water loss at reduced leaf water potential. This enables it to survive severe desiccation; however, plants are able to use moisture when it is available (Doley & Trivett, 1974). Similarly, *H. contortus* has high water use efficiency under conditions of high soil moisture availability, and it adjusts its stomatal response to leaf water deficits under conditions of declining soil moisture availability (Wilson *et al.,* 1980).

Astrebla spp. and *H. contortus* are both tolerant of moderate defoliation (Jozwik *et al.,* 1970; Mott *et al.,* 1992; Grice & McIntyre, 1995). Furthermore, moderate defoliation of *A. lappacea* may be beneficial to seed production and leaf growth (Orr, 1980), and it may encourage new tiller development (Scanlan, 1983). Moderate defoliation may also be beneficial to *H. contortus.*

Plant longevity

Most tussocks of *Astrebla* spp. and *H. contortus* are relatively short lived. For example, over 90% of the *Astrebla* spp. tussocks recruited in an exceptional recruitment in northern grassland (Fig. 14.4) had died within five years. Similar mortality of *Astrebla* spp. has been reported in both southern (Williams & Roe, 1975) and central grasslands (Orr & Evenson, 1991*a*). For *H. contortus,* up to 60% of individual tussocks in the southern region die within five years (Mott *et al.,* 1985). Nevertheless, the maximum recorded lifespan for *Astrebla* spp. is 23 years in the southern region (Williams & Roe, 1975). Maximum lifespans of 35 and 48 years for *A. lappacea* have been projected from recordings of plant survival over ten years

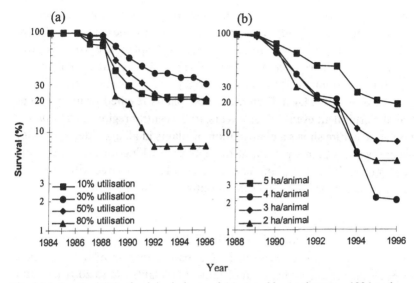

Fig. 14.6. Survival (%) of original plants of (*a*) *Astrebla* spp. between 1984 and 1996 under four levels of utilization of end of summer forage in northern *Astrebla* grassland ('Toorak', Julia Creek) and (*b*) *H. contortus* between 1988 and 1996 under four stocking rates in central *H. contortus* pasture ('Galloway Plains', Calliope).

under both ungrazed and grazed conditions in northern and central grassland respectively (Orr & Evenson, 1991*a*; Orr, 1991). These differences in lifespan are attributed to a distribution of rainfall more favourable for plant growth in central than in northern or southern grasslands (Orr & Evenson, 1991*a*). No estimate of the maximum lifespan of *H. contortus* tussocks is available.

Survival of the original (mixed age) populations of both *Astrebla* spp. and *H. contortus* at a range of grazing pressures (Fig. 14.6) indicates that survival of both species is influenced by grazing pressure. However, an examination of the irregular decline in *Astrebla* spp. survival at heavy grazing (80% utilization) in relation to seasonal rainfall, indicates that drought interacts with heavy grazing pressure to further reduce tussock survival. In contrast, the regular decline in survival of *H. contortus* suggests that survival in this species is probably influenced primarily by grazing pressure.

Plant age structure

Few data are available on plant age structure in either vegetation type. After 10 years, in central *Astrebla* grassland the age structure of *Astrebla* spp. was dominated by plants present at the start of the study, although plants recruited during the study were apparent in younger age categories (Orr & Evenson, 1991a). However, an examination of the plant age structure after 13 years in northern *Astrebla* grassland (Fig. 14.7) revealed that the age structure was dominated by plants in the 8 years of age category, and data presented in Fig. 14.4 indicate that these plants resulted from an exceptional recruitment event. After 9 years, in the central region the *H. contortus* age structure shows a contribution of plants in all age categories, with generally more plants present at the lightest stocking rate. The generally greater contribution of plants aged 5 years or less compared with those >5 years indicates the importance of recruitment in these pastures.

Plant population ecology and grazing management

From the foregoing discussion, it is apparent that *Astrebla* spp. persist through the prolonged life span of individual plants (up to 20 years), and that new individuals result from irregular recruitment. Increasing stocking rate appears to reduce seedling survival rather than seedling recruitment. In contrast, *H. contortus* persists through more regular seedling recruitment, which compensates for poor survival of individual plants. Increasing stocking rate reduces seedling recruitment through reduced seed production.

This demographic understanding provides a basis for the management of grazing pressures. In both vegetation types, grazing should ensure the survival of mature tussocks, which are the source of seed production for future seedling recruitment. For *Astrebla* spp., exceptional recruitment events need to be recognized and grazing pressure reduced to maximize survival of the resulting plants. For *H. contortus*, grazing should provide that plants produce seed most years to provide regular recruitment. However, evidence of deleterious changes in both vegetation types (Tothill & Gillies, 1992) suggests, at least in some situations, that grazing pressures have exceeded long-term carrying capacities.

A demographic understanding of the interaction between fire and grazing pressure in *H. contortus* pastures has provided grazing management principles for the restoration of *H. contortus* dominance (Orr *et al.*, 1997; Orr & Paton, 1997). Under these principles, burning in spring followed either by reduced grazing pressure or by resting for six months will promote seed production and seedling recruitment. Increased seedling recruitment increases the frequency of occurrence of *H. contortus* in the pasture, and reduced grazing pressure allows these seedlings to develop and increase the proportion of *H. contortus* in the pasture.

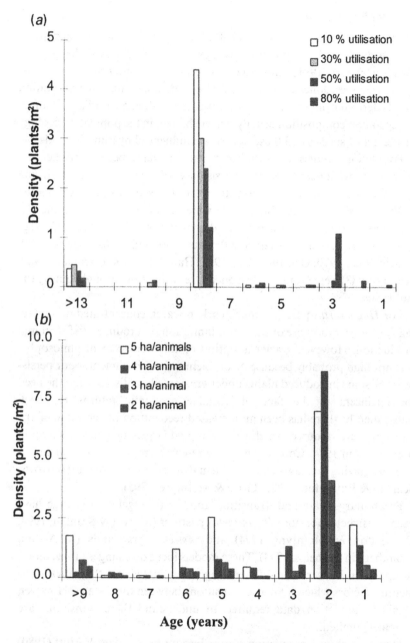

Fig. 14.7. Age structure (plants m^{-2}) of plants in 1996 of (*a*) *Astrebla* spp. in northern *Astrebla* grassland ('Toorak', Julia Creek) under four levels of utilization of end of summer forage and (*b*) *H. contortus* in central *H. contortus* pasture ('Galloway Plains', Calliope) under four stocking rates.

384   D. M. Orr

Need for long-term research

Evidence of deleterious changes (Tothill & Gillies, 1992) and limited quantitative data on the population ecology of major species under grazing (McIvor & Orr, 1991) emphasize the need for long-term research. For example, Jones, Jones & McDonald (1995) demonstrated that vegetation trends derived from short-term experiments are often misleading. Changes in vegetation composition usually reflect changes in the population ecology of the major species and these in turn are influenced by rainfall trends.

Population studies in *Astrebla* grasslands have been sporadic and prompted by concern for the persistence of these grasslands during drought. For example, vegetation studies in *Astrebla* grasslands during the 1930s (e.g. Everist, 1935; Roe, 1941) were prompted by drought and the resultant increase in grazing pressure during the early 1930s. Similarly, vegetation studies commenced during the late 1960s and 1970s (Jozwik, 1970; Jozwik *et al.*, 1970; Orr, 1980; Orr, 1986; Hall & Lee, 1980; Scanlan, 1980; Scanlan 1983) were prompted by concern resulting from drought during the late 1960s.

For *H. contortus* pastures, much early research concentrated on replacing *H. contortus* because of perceived limitations in productivity for animal production. However, grazier adoption of this 'replacement philosophy' was not high probably because of problems of establishment, poor persistence of some introduced plants under grazing, plant diseases and the need for fertilizers in the face of variable economic returns for beef. Subsequently, there has been an increased recognition of the value of the native pasture resource, which was prompted largely by a major slump in beef prices in 1974. Consequently, the research emphasis has changed to one of grazing management of the native pasture resource (Burrows, Scanlan & Rutherford, 1988; Grice & McIntyre, 1995).

Preliminary 'State and Transition' models of vegetation change have been constructed for both *H. contortus* pastures (McIvor & Scanlan, 1994; Orr, Paton & McIntyre, 1994) and *Astrebla* grasslands (McArthur, Chamberlain & Phelps, 1994). These models focus on changes in vegetation composition between the different vegetation states and they attempt to identify factors which influence transitions between these states. However, detailed population data required to understand these transitions are extremely limited.

How long should population studies be continued? Jones & Mott (1980) suggest that population studies continue until final measurements are made on plants recruited during the study rather than on the original individuals. For *H. contortus,* 15 years may be sufficient to document a range of annual seedling cohorts. However, for *Astrebla* spp., a minimum period of 50 years may be necessary, depending on episodic recruitment events and trends in seasonal rainfall. Jones *et al.* (1995) suggest that long-term grazing studies play a valuable role in both the development and validation of pasture models, particularly for modelling plant persistence and vegetation change.

Conclusions

Both *Astrebla* grasslands and *H. contortus* pastures are important resources for the grazing industries of Queensland. Despite this, limited quantitative data exist on the population ecology of the major species. Clearly, *Astrebla* spp. persist through the prolonged lifespan of individual plants because seedling recruitment is episodic. For *H. contortus*, regular recruitment is required as the lifespan of individual tussocks is relatively short. The present challenge is to develop a quantitative understanding of population ecology under grazing within the constraint of highly variable rainfall. A quantitative understanding of the processes resulting in seedling recruitment would be particularly useful given the overall importance of seedling recruitment in maintaining plant age structure. The need for quantitative population data is vital especially if grazing pressures continue to increase. An improved quantitative understanding of the population ecology of these grasses will lead to improved guidelines for grazing management.

Acknowledgements

Much of the research referenced here has benefited from funds provided by the Australian Wool Research and Promotion Organisation and the Meat Research Corporation.

References

Burrows, W. H., Carter, J. O., Scanlan, J. C. & Anderson, E. R. (1990). Management of savannas for livestock production in north east Australia: contrasts across the tree–grass continuum. *Journal of Biogeography*, **17**, 503–12.
Burrows, W. H., Scanlan, J. C. & Rutherford, M. T. (1988). Native pastures in Queensland: the resources and their management. Information Series Q187023, Queensland Department of Primary Industries, Brisbane.
Campbell, S. D. (1995). Plant mechanisms that influence the balance between *Heteropogon contortus* and *Aristida ramosa* in spring burnt pastures. Ph.D. thesis, University of Queensland.

Coughlan, M. J. (1988). Season climate outlooks. *The Changing Climate and Central Queensland Agriculture*, ed. E. R Anderson, pp. 17–26 Australian Institute of Agricultural Science, Rockhampton, Queensland.

Doley, D. and Trivett, N. B. A. (1974). Effects of low water potentials on transpiration and photosynthesis in Mitchell grass (*Astrebla lappacea*). *Australian Journal of Plant Physiology*, 1, 539–50.

Everist, S. L. (1935). Response during 1934 season of Mitchell and other grasses in western and central Queensland. *Queensland Agricultural Journal*, 43, 374–87.

Everist, S. L. (1964). The Mitchell grass country. *Queensland Naturalist*, 17, 45–50.

Gardener, C. J. (1980). Diet selection and liveweight performance of steers on *Stylosanthes hamata*–native grass pastures. *Australian Journal of Agricultural Research*, 31, 379–92.

Grice, A. C. & McIntyre, S. (1995). Speargrass (*Heteropogon contortus*) in Australia: dynamics of species and community. *Rangelands Journal*, 17, 3–25.

Grime, J. P. (1979). Plant strategies and vegetation processes. New York: John Wiley.

Hall, T. J. & Lee, G. R. (1980). Response of an *Astrebla* spp. grassland to heavy grazing by cattle and light grazing by sheep in north-west Queensland. *Australian Rangeland Journal*, 2, 83–93.

Howden, S. M. (1988). Some aspects of the ecology of four tropical grasses with special emphasis on *Bothriochloa pertusa*. Ph.D. thesis, Griffith University, Brisbane.

Isbell, R. F. (1969). The distribution of black spear grass (*Heteropogon contortus*) in tropical Queensland. *Tropical Grasslands*, 3, 35–42.

Jones, R. M., Jones, R. J. & McDonald, C. K. (1995). Some advantages of long-term grazing trials with particular reference to changes in botanical composition. *Australian Journal of Experimental Agriculture*, 35, 1029–38.

Jones, R. M. & Mott, J. J. (1980). Population dynamics in grazed pastures. *Tropical Grasslands*, 14, 218–24.

Jozwik, F. X. (1969) Some systematic aspects of Mitchell grasses (*Astrebla* F. Muell.). *Australian Journal of Botany*, 17, 359–74.

Jozwik, F. X. (1970). Response of Mitchell grasses (*Astrebla* F. Muell.) to photoperiod and temperature. *Australian Journal of Agricultural Research*, 21, 395–405.

Jozwik, F. X., Nicholls, A. O. & Perry, R. A. (1970). Studies on the Mitchell grasses (*Astrebla*). Proceedings XI International Grassland Congress, pp. 48–51. Canberra: Australian National Press.

Lambert, F. J., Bower, M., Whalley, R. D. B., Andrews, A. C. & Bellotti, W. D. (1990). The effects of soil moisture and planting depth on emergence and seedling morphology of *Astrebla lappacea* (Lindl.) Domin. *Australian Journal of Agricultural Research*, 41, 367–76.

McArthur, S. R., Chamberlain, H. J. & Phelps, D. G. (1994). State and transition models for rangelands. 12. A general state and transition model for the mitchell grass, bluegrass-browntop and Queensland bluegrass pastures zones of northern Australia. *Tropical Grasslands*, 28, 274–8.

McIvor, J. G. (1987). Changes in germinable seed levels in soil beneath pastures near Townsville, North Queensland. *Australian Journal of Experimental Agriculture*, 27, 283–9.

McIvor, J. G. & Gardener, C. J. (1991). Soil seed densities and emergence patterns in pastures in the seasonally dry tropics of north eastern Australia. *Australian Journal of Ecology*, 16, 159–69.

McIvor, J. G. & Orr, D. M. (1991). Sustaining productive pastures in the tropics. 3. Managing native grasslands. *Tropical Grasslands*, **25**, 91–7.

McIvor, J. G. & Scanlan, J. C. (1994). State and transition models for rangelands. 8. A state and transition model for the northern speargrass zone. *Tropical Grasslands*, **28**, 256–9.

McIvor, J. G., Singh, V., Corfield, J. P. & Jones, R. J. (1996). Seed production by native and naturalised grasses in north-east Queensland: effects of stocking rate and season. *Tropical Grasslands*, **30**, 262–9.

McKeon, G. M., Day, K. A., Howden, S. M., Mott, J. J., Orr, D. M. & Scattini, W. J. (1990). Northern Australia savannas: management for pastoral production. *Journal of Biogeography*, **17**, 355–72.

Mott, J. J. (1978). Dormancy and germination in five native grass species from savanna woodland communities of the Northern Territory. *Australian Journal of Botany*, **26**, 621–31.

Mott, J. J., Ludlow, M. M., Richards, J. H. & Parsons, A. D. (1992). Effects of moisture supply in the dry season and subsequent defoliation on persistence of the savanna grasses *Themeda triandra, Heteropogon contortus* and *Panicum maximum*. *Australian Journal of Agricultural Research*, **43**, 241–60.

Mott, J. J., Williams, J., Andrew, M. H. & Gillison, A. N. (1985). Australian savanna ecosystems. In *Ecology and Management of the World's Savannas*, ed J. C. Tothill & J. J. Mott, pp. 56–82. Canberra: Australian Academy of Science.

Myers, A. (1942). Germination of seed of curly Mitchell grass (*Astrebla lappacea* Domin.). *Journal of the Australian Institute of Agricultural Science*, **8**, 31–2.

Orr, D. M. (1980). Effects of sheep grazing *Astrebla* grasslands in central western Queensland II. Effects of seasonal rainfall. *Australian Journal of Agricultural Research*, **31**, 807–20.

Orr, D. M. (1986). Factors affecting the vegetation dynamics of *Astrebla* grasslands. Ph.D. thesis, University of Queensland, Brisbane.

Orr, D. M. (1991). Trends in the recruitment of *Astrebla* spp. in relation to seasonal rainfall. *Rangeland Journal*, **13**, 107–17.

Orr, D. M. (1992). *Astrebla* grasslands – a resilient ecosystem. In *Working Papers of the 7th Biennial Conference, Australian Rangeland Society, Cobar, New South Wales*, pp. 320–1. (Mimeo).

Orr, D. M. & Evenson, C. J. (1991*a*). Effects of sheep grazing *Astrebla* spp. grassland in central western Queensland III. Dynamics of *Astrebla* spp. under grazing and exclosure between 1975 and 1986. *Rangelands Journal*, **13**, 36–46.

Orr, D. M. & Evenson, C. J. (1991*b*). Effect of a single clipping on the seed production response of *Astrebla* spp. *Rangelands Journal*, **13**, 57–60.

Orr, D. M. & Holmes, W. E. (1984). Mitchell grasslands. In *Management of Australia's Rangelands*, ed. G. N. Harrington, A. D. Wilson & M. D. Young, pp. 241–54. Melbourne: CSIRO.

Orr, D. M. & Paton, C. J. (1993). Impact of grazing pressure on the plant population dynamics of *Heteropogon contortus* (black speargrass) in subtropical Queensland. In *Proceedings XVIIth International Grassland Congress, Palmerston North*, pp. 1908–10.

Orr, D. M. & Paton, C. J. (1997). Using fire to manage species composition in *Heteropogon contortus* (black speargrass) pastures 2. Enhancing the effects of fire with grazing management. *Australian Journal of Agricultural Research*, **48**(6), 795–802.

Orr, D. M., Paton, C. J. & Lisle, A. T. (1997). Using fire to manage species
    composition in *Heteropogon contortus* (black speargrass) pastures 1. Burning
    regimes. *Australian Journal of Agricultural Research*, 48(6), 803–10.
Orr, D. M., Paton, C. J. & McIntyre, S. (1994). State and transition models for
    rangelands. 10. A state and transition model for the southern black
    speargrass zone of Queensland. *Tropical Grasslands*, 28, 266–9.
Partridge, I. J. (1994). Will it rain?: the effects of the Southern Oscillation and El
    Nino on Australia. Second Edition. Information Series Q94015, Queensland
    Department of Primary Industries, Brisbane.
Peart, M. H. (1979). Experiments on the biological significance of the
    morphology of seed-dispersal units in grasses. *Journal of Ecology*, 67,
    843–63.
Pressland, A. J. & McKeon, G. M. (1989). Monitoring animal numbers and
    pasture condition for drought administration – an approach. *Proceedings 5th
    Australian Soil Conservation Conference, Perth*, pp. 17–27.
Roe, R. (1941). Studies on the Mitchell grass association in south-western
    Queensland 1. Some observations on the response of Mitchell grass pastures
    to good summer rains following the 1940 drought. *Journal of the Council of
    Scientific and Industrial Research (Australia)*, 14, 253–9.
Roe, R. (1987). Recruitment of *Astrebla* spp. in the Warrego region of south-
    western Queensland. *Tropical Grasslands*, 21, 91–2.
Scanlan, J. C. (1980). Effects of spring wildfires on *Astrebla* (Mitchell grass)
    grasslands in north-west Queensland under varying levels of growing
    seasonal rainfall. *Australian Rangeland Journal*, 2, 162–8.
Scanlan, J. C. (1983). Changes in tiller and tussock characteristics of *Astrebla
    lappacea* (curly Mitchell grass) after burning. *Australian Rangeland Journal*,
    5, 13–19.
Shaw, N. H. (1957). Bunch spear grass dominance in burnt pastures in south-
    eastern Queensland. *Australian Journal of Agricultural Research*, 8, 326–34.
Silcock, R. G., Williams, L. M. & Smith, F. T. (1990). Quality and storage
    characteristics of the seeds of important native species in south-west
    Queensland. *Australian Rangeland Journal*, 12, 14–20.
Taylor, J. A. & Tulloch, D. (1985). Rainfall in the wet tropics: extreme events at
    Darwin and similarities between years during the period 1870–1983 inclusive.
    Australian Journal of Ecology, 10, 281–95.
Tothill, J. C. (1966). Phenological variation in *Heteropogon contortus* and its
    relation to climate. *Australian Journal of Botany*, 14, 35–47.
Tothill, J. C. (1977). Seed germination studies with *Heteropogon contortus*.
    Australian Journal of Ecology, 2, 477–84.
Tothill, J. C. & Gillies, C. (1992). The pasture lands of northern Australia: their
    condition, productivity and sustainability. Tropical Grassland Society of
    Australia, Occasional Publication No. 5.
Tothill, J. C. & Hacker, J. B. (1973). The grasses of southern Queensland.
    University of Queensland Press, St. Lucia.
Tothill, J. C. & Hacker, J. B. (1976). Polyploidy, flowering phenology and climate
    adaption in *Heteropogon contortus* (Gramineae). *Australian Journal of
    Ecology*, 1, 213–22.
Tothill, J. C. & Knox, R. B. (1968). Reproduction in *Heteropogon contortus* 1.
    Photoperiodic effects on flowering and sex expression. *Australian Journal of
    Agricultural Research*, 19, 869–78.
Watt, L. A. (1981). Establishment of grasses on cracking clay soils: seedling

morphology characteristics. *Proceedings of the XVII International Grassland Congress, Kentucky*, pp. 451–3.

Watt, L. A. (1982). Germination characteristics of several grass species as affected by limiting water potentials imposed through a cracking black clay soil. *Australian Journal of Agricultural Research*, **33**, 223–31.

Watt, L. A & Whalley, R. D. B. (1982*a*). Effect of sowing depth and seedling morphology on establishment of grass seedlings on cracking black earth. *Australian Rangeland Journal*, **4**, 52–60.

Watt, L. A. & Whalley, R. D. B. (1982*b*). Establishment of small-seeded perennial grasses on black clay soils in north-western New South Wales. *Australian Journal of Botany*, **30**, 611–23.

Weston, E. J., Harbison, J. Leslie, J. K., Rosenthal, K. M. & Mayer, R. J. (1981). Assessment of the agricultural and pastoral potential of Queensland. *Technical Report No. 27, Agriculture Branch, Queensland Department of Primary Industries, Brisbane.*

Williams, O. B. (1978). Plant demography of Australian arid rangeland and implications for management, research and land policy. *Proceedings of the 1st International Rangelands Congress, Denver*, pp. 185–6.

Williams, O. B. & Roe R. (1975).Management of arid grassland for sheep: plant demography of six grasses in relation to climate and grazing. *Proceedings of the Ecological Society of Australia*, **9**, 142–56.

Wilson, J. R., Ludlow, M. M., Fisher, M. J. & Schulze, E. D. (1980). Adaption to water stress of the leaf water relations of four tropical forage grasses. *Australian Journal of Plant Physiology*, **7**, 207–20.

Index

To avoid redundancy, only scientific names for species are included. Because of their broad generality and applicability in this book, terms like *population* and *growth* are avoided.